MATHEMATICS FOR NURSES
With Clinical Applications

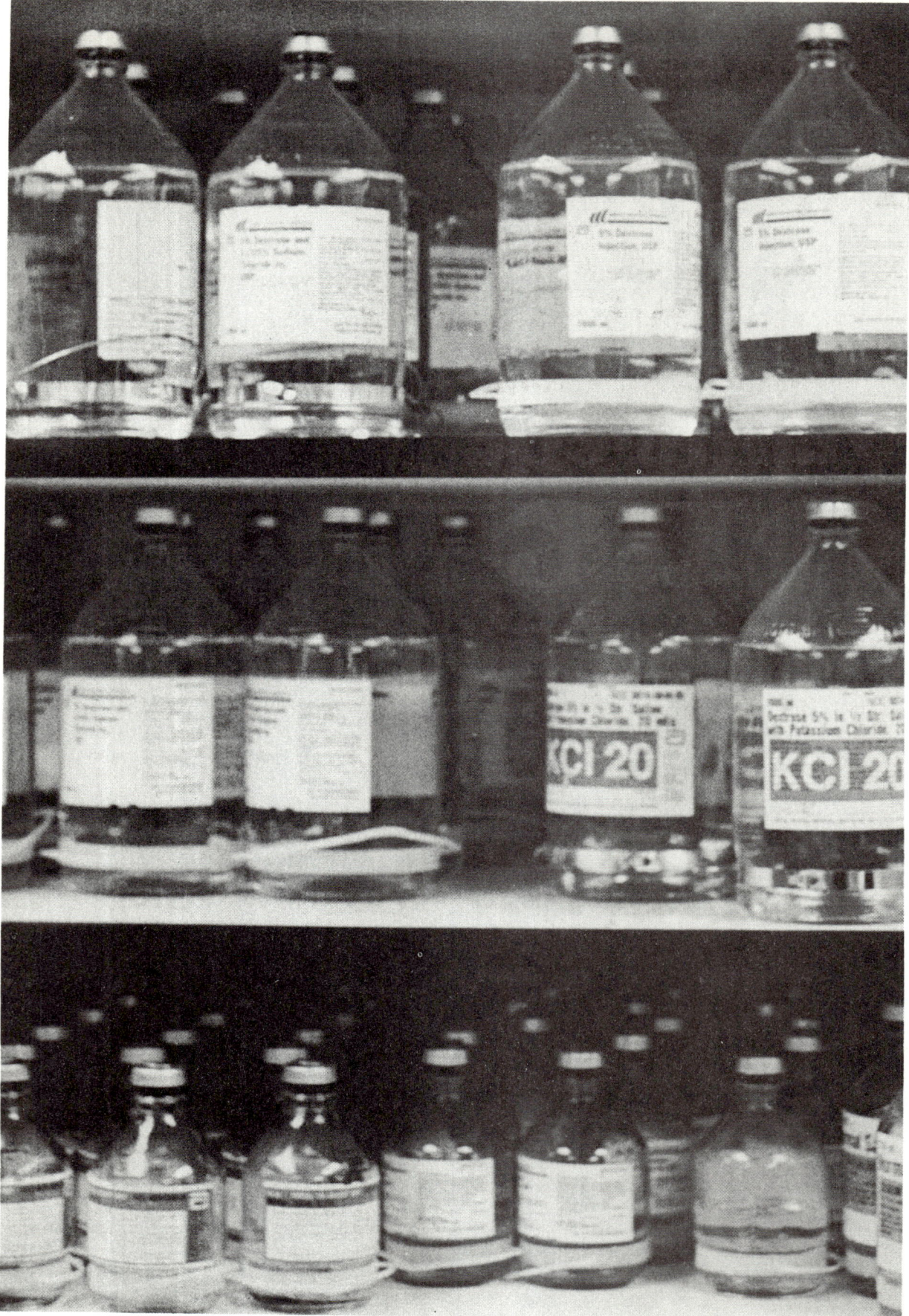

MATHEMATICS FOR NURSES
With Clinical Applications

Mary K. Miller

Brooks/Cole Publishing Company
Monterey, California

Contemporary Undergraduate Mathematics Series
Consulting Editor: Robert J. Wisner

Brooks/Cole Publishing Company
A Division of Wadsworth, Inc.

© 1981 by Wadsworth, Inc., Belmont, California 94002.
All rights reserved.
No part of this book may be reproduced,
stored in a retrieval system, or transcribed, in any form or by any means—
electronic, mechanical, photocopying, recording, or otherwise—
without the prior written permission of the publisher,
Brooks/Cole Publishing Company,
Monterey, California 93940, a division of Wadsworth, Inc.

Printed in the United States of America
10 9 8

Library of Congress Cataloging in Publication Data

Miller, Mary K   1951–
  Mathematics for nurses with clinical applications.

  Includes index.
  1. Nursing—Mathematics. 2. Pharmaceutical
arithmetic. I. Title. [DNLM: 1. Drugs—Administration and dosage—Nursing texts. 2. Mathematics—Nursing texts. QV16 M649m]
RT68.M54   513'.024613   80-26040
ISBN 0-8185-0429-3

Acquisition Editor: *Craig Barth*
Production Editor: *Stacey C. Sawyer*
Production Assistant: *Jennifer Young*
Design: *Katherine Minerva*
Cover and Frontispiece Photos: © *Stan Rice*
Illustrations: *Susan R. Spence*
Production Art Assistant: *Donna Peacock*
Typesetting: *Omegatype, Champaign, Illinois*

This book is dedicated
to the patient,
who deserves the very best
we have to offer.

Preface

Mathematics for Nurses with Clinical Applications exposes nursing students to the mathematics required for success in their profession. In addition to teaching arithmetic, the text also presents a survey of the various types of mathematical problems encountered in clinical surroundings. It is not assumed that the nursing students using this text have a background in mathematics.

This book is suitable for use in traditional lecture classes, small groups, and for self-instruction. Problems are carefully graded and paired by odds and evens—that is, types of problems are presented in pairs, so that the student has two chances to work each type of problem. Readiness-Review problems refer students to the appropriate text examples for help. Each chapter includes a list of mathematical and medical terms along with the numbers of the pages that explain the terms.

Mathematics for Nurses is divided into five parts. Part One, Arithmetic, uses a sound mathematical approach to develop basic mathematical knowledge. Students may use the chapters in Part One to learn arithmetic or to review areas of weakness.

Part Two, Pharmacy Skills, introduces the various measurement systems (Chapters 6 and 7) and teaches students how to convert from one system to another (Chapter 8). In addition, Part Two explains common Latin and medical abbreviations.

Part Three, Oral Dosages, and Part Four, Parenteral Dosages, apply the mathematical skills developed in Parts One and Two. These chapters include syringe diagrams to simulate actual dosages.

Part Five, Applications, totally integrates mathematics with nursing. The first section in Part Five, Application I: Pediatric and Geriatric Dosages, explains the traditional formulas used in calculating doses for infants and children and presents an extensive, step-by-step approach on how to use the West Nomogram. This section also covers dosage calculations for the elderly based on kidney function. The second section, Application II: Electrolyte Solutions, and the third section, Application III: Total Parenteral Nutrition, are unique to this text. Along with mathematical problems associated with the topics of electrolytes and parenteral nutrition, these sections clearly explain calculations involving milliequivalents.

The general format of each chapter comprises:
- Student Objectives
- Explanations
- Examples
- Section Readiness Reviews
- Section Review Problems
- Chapter Readiness Review
- Chapter Summary Terms
- Chapter Review Problems

The nursing student will find this book to be a useful tool in developing dosage-calculation skills. Toward the end of the nursing student's studies, this text will serve as a valuable review for licensing-board exams, and, even later, it will be an aid in performing everyday nursing duties.

Many people have assisted me in developing this book. I would like to thank the following people for their professional and critical reviews of the manuscript: Joan Gary, Parkland College; Anne Grams, University of Tennessee; Carole Bauer, Triton College; Joan Mahmud, Bergen Community College; Lee Marsh, Kalamazoo Valley Community College; and Paul Olsen, Shelby State Community College. I would also like to offer my special appreciation to James Kinney, George Corley Wallace State Community College, who reviewed the manuscript at different stages and tested it extensively in class.

In addition, I would like to express my gratitude to Robert J. Wisner of New Mexico State University for his direction and encouragement as consulting editor. The care and thoughtful guidance of those people who believed in this project, especially Jack Thornton, Craig Barth, Konrad Kerst, and the rest of the staff at Brooks/Cole, are much appreciated.

Finally, loving thanks go to Charles, who encouraged my individuality in this writing endeavor.

Mary K. Miller
Pharm. D.

Contents

One
Arithmetic

Chapter 1 Whole Numbers 2

Section 1.1 Addition of Whole Numbers 3
Section 1.2 Subtraction of Whole Numbers 5
Section 1.3 Multiplication of Whole Numbers 8
Section 1.4 Division of Whole Numbers 11
Section 1.5 Word Problems 15

Chapter 1 Readiness Review 18
Chapter 1 Summary 19
Chapter 1 Review Problems 19

Chapter 2 Fractions 21

Section 2.1 Types of Fractions 22
Section 2.2 Addition of Fractions 28
Section 2.3 Subtraction of Fractions 37
Section 2.4 Equality and Comparison of Fractions 39
Section 2.5 Addition and Subtraction of Mixed Numbers 44
Section 2.6 Multiplication of Fractions 49
Section 2.7 Division of Fractions 52

Chapter 2 Readiness Review 54
Chapter 2 Summary 58
Chapter 2 Review Problems 58

Chapter 3 Decimals 62

Section 3.1 Reading and Writing Decimals 63
Section 3.2 Changing Fractions to Decimals and Decimals to Fractions 66
Section 3.3 Addition and Subtraction of Decimals 72
Section 3.4 Rounding Off Decimals 79
Section 3.5 Multiplication and Division of Decimals 83

Chapter 3 Readiness Review 91
Chapter 3 Summary 93
Chapter 3 Review Problems 94

Chapter 4 Ratio, Proportion, and Dimensional Analysis 97

Section 4.1 Ratio 98
Section 4.2 Proportion 101
Section 4.3 Dimensional Analysis 108

Chapter 4 Readiness Review 113
Chapter 4 Summary 114
Chapter 4 Review Problems 114

Chapter 5 Percent and Preparation of Solutions 117

Section 5.1 Converting Decimals to Percents and Percents to Decimals 118
Section 5.2 Converting Fractions to Percents and Percents to Fractions 120
Section 5.3 Percent-Proportion Formula 123
Section 5.4 Preparation of Solutions—Percentage Strength 127
Section 5.5 Preparation of Solutions—Ratio Strength and Dilution of Stock Solutions 131

Chapter 5 Readiness Review 135
Chapter 5 Summary 137
Chapter 5 Review Problems 137

TWO Pharmacy Skills

Chapter 6 Metric System: Weight and Measure 140

Section 6.1 Units of Length 142
Section 6.2 Units of Volume 146
Section 6.3 Units of Weight 151
Section 6.4 Celsius Temperature 157

Chapter 6 Readiness Review 159
Chapter 6 Summary 160
Chapter 6 Review Problems 161

Chapter 7 Apothecaries' and Household Systems 163

Section 7.1 Roman Numerals 164
Section 7.2 Apothecaries' Fluid Measure 167
Section 7.3 Apothecaries' Weight 173
Section 7.4 Household Fluid Measure 179

Chapter 7 Readiness Review 189
Chapter 7 Summary 191
Chapter 7 Review Problems 191

Chapter 8 Conversions and Medical Abbreviations 195

Section 8.1 Conversion of Weight 196
Section 8.2 Conversion of Length and Volume 200
Section 8.3 Latin Abbreviations 204
Section 8.4 Common Medical Abbreviations 207

Chapter 8 Readiness Review 211
Chapter 8 Summary 212
Chapter 8 Review Problems 213

Three
Oral Dosages

Chapter 9 Tablets, Capsules, and Oral Solutions 218

Section 9.1 Tablets and Capsules 219
Section 9.2 Oral Solutions 223

Chapter 9 Readiness Review 229
Chapter 9 Summary 230
Chapter 9 Review Problems 230

Four
Parenteral Dosages

Chapter 10 Injectable Drugs in Solution 234

Section 10.1 Injectable Drug Calculations—Overview 235
Section 10.2 Hypodermic Tablets 247
Section 10.3 Injectable Drugs Requiring Reconstitution 252

Chapter 10 Readiness Review 260
Chapter 10 Summary 263
Chapter 10 Review Problems 263

Chapter 11 Intravenous Fluids and Medication 269

Section 11.1 General Concepts and Drops Per Minute 270
Section 11.2 I.V. Running Time 274
Section 11.3 Intravenous Admixtures 277

Chapter 11 Readiness Review 281
Chapter 11 Summary 282
Chapter 11 Review Problems 282

Chapter 12 Insulin and Heparin 284

Section 12.1 Insulin 285
Section 12.2 Heparin 295

Chapter 12 Readiness Review 299
Chapter 12 Summary 301
Chapter 12 Review Problems 301

Five Applications

Application I Pediatric and Geriatric Dosages 308

Section A. Pediatric Dosages 309
Section B. Geriatric Dosages 321

Application II Electrolyte Solutions 334

Section A. Electrolytes 335
Section B. Valence and Milliequivalents 336

Application II Review Problems 341

Application III Total Parenteral Nutrition 343

Overview 344

Application III Review Problems 356

Appendices 359
Answers 361
Index 390

MATHEMATICS FOR NURSES
With Clinical Applications

One
Arithmetic

1 Whole Numbers

OBJECTIVES After studying this chapter, you should be able to:

1. Recognize whole numbers.
2. Perform the operations of addition, subtraction, multiplication, and division using whole numbers.
3. Solve word problems involving whole numbers.

Section 1.1 Addition of Whole Numbers

> **Definition:** The numbers 0, 1, 2, 3, 4, 5, 6, 7, 8, 9, 10, 11, 12, 13, 14, and so on are *whole numbers*.

> **Rule: Addition of Whole Numbers**
> *Step 1:* Arrange the numbers to be added (addends) into columns so that ones are lined up with ones, tens are lined up with tens, and so on.
> *Step 2:* Add each column separately, right to left, carrying when necessary.

Example 1: Add the whole numbers 11 and 15.
Solution:
Step 1: Arrange the numbers to be added into columns so that ones are lined up with ones and tens are lined up with tens.

$$\begin{array}{r} \text{tens ones} \\ 11 \\ +15 \end{array}$$

Step 2: Add each column separately.

$$\begin{array}{r} 11 \text{ (addend)} \\ +15 \text{ (addend)} \\ \hline 26 \text{ (sum)} \end{array}$$

No carrying required. The sum is 26.

Example 2: Add the following whole numbers: 27, 49, and 71.
Solution:
Step 1: Arrange the numbers into columns.

$$\begin{array}{r} \text{tens ones} \\ 27 \\ 49 \\ +71 \end{array}$$

Step 2: Add each column separately.

$$\begin{array}{r} \overset{1}{27} \\ 49 \\ +71 \\ \hline 7 \end{array}$$

The sum of the ones column is 17. This number can also be written as 1 ten plus 7 ones. Put the 7 at the bottom of the ones column and carry the 1 ten over to the tens column. The sum of the tens column is 14 tens, or 1 hundred and 4 tens.

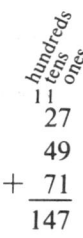

$$\begin{array}{r} \overset{1\,1}{}27 \\ 49 \\ +\ 71 \\ \hline 147 \end{array}$$

The sum is 147.

Example 3: Add the following whole numbers: 55, 125, 71, and 88.
Solution:
Step 1: Arrange the numbers into columns.

$$\begin{array}{r} 55 \\ 125 \\ 71 \\ +\ 88 \end{array}$$

Step 2: Add each column separately.

$$\begin{array}{r} \overset{1}{55} \\ 125 \\ 71 \\ +\ 88 \\ \hline 9 \end{array}$$

Add the ones column first, the tens column second, and the hundreds column last. The ones column adds up to 19. This number can also be written as 1 ten plus 9 ones. Put the 9 at the bottom of the ones column and carry the 1 ten over to the tens column. Carry as shown above.

$$\begin{array}{r} \overset{2\,1}{55} \\ 125 \\ 71 \\ +\ 88 \\ \hline 339 \end{array}$$

The tens column adds up to 23 tens. This number may be written as 20 tens plus 3 tens. 20 tens can also be thought of as 2 hundreds. Carry the 2 hundreds over to the hundreds column. The sum is 339. Do you agree?

Now work the problems in the Section Readiness Review. The Section Readiness Review will give you experience in working the types of problems that you've studied in each section. This review will prepare you for the Section Review Problems and the Chapter Readiness Review, which in turn will prepare you for the problems at the end of each chapter. Each Section Readiness Review will refer you to the appropriate example in the text. If you need help in solving a problem, go back to the corresponding example and reread it carefully. Use the blanks provided to record your answers.

Section 1.1 Readiness Review

Add the following whole numbers:

1. $17 + 22 =$ _____ (see Example 1)
2. $43 + 89 + 77 =$ _____ (see Example 2)
3. $100 + 400 + 50 =$ _____ (see Example 3)
4. $177 + 711 + 343 =$ _____ (see Example 3)
5. $1112 + 324 + 7654 =$ _____ (see Example 3)
6. $11,111 + 99 + 4444 =$ _____ (see Example 3)

Section readiness review answers (not given in order):

6. 15,654   3. 550   5. 9090
2. 209   1. 39   4. 1231

Section 1.1 Review Problems

Add the following whole numbers (answers to the odd-numbered problems are at the end of the book):

1. $8 + 21$
2. $7 + 10$
3. $5 + 37$
4. $16 + 48$
5. $13 + 8 + 24$
6. $32 + 7 + 65$
7. $18 + 21 + 12$
8. $14 + 52 + 17$
9. $18 + 486 + 34$
10. $15 + 122 + 73$
11. $742 + 19 + 573$
12. $576 + 209 + 337$
13. $924 + 821 + 146$
14. $345 + 762 + 107$
15. $116 + 181 + 863$
16. $814 + 171 + 119$
17. $600 + 800 + 3000$
18. $3879 + 428 + 2750$
19. $1345 + 511 + 8114$
20. $20,017 + 14 + 6744$

Section 1.2 Subtraction of Whole Numbers

> **Rule: Subtraction of Whole Numbers**
> *Step 1:* Arrange the numbers to be subtracted into columns so that ones are subtracted from ones, tens are subtracted from tens, and so on.
> *Step 2:* Subtract each column separately, right to left, borrowing when necessary.

6   Chapter 1 Whole Numbers

Example 1: Subtract 13 from 29.
Solution:
Step 1: Arrange the numbers to be subtracted into columns so that ones are lined up with ones, tens are lined up with tens, and so on.

$$\begin{array}{r} \text{tens ones} \\ 29 \\ -13 \end{array}$$

Step 2: Subtract each column separately.

$$\begin{array}{rl} 29 & \text{(minuend)} \\ -13 & \text{(subtrahend)} \\ \hline 16 & \text{(difference)} \end{array}$$

The difference is 16.

Example 2: Subtract 54 from 73.
Solution:
Step 1: Arrange the numbers into columns.

$$\begin{array}{r} \text{tens ones} \\ 73 \\ -54 \end{array}$$

Step 2: Immediately you will notice that 4 can't be subtracted from 3 and still give an answer that is a whole number. In order to subtract these numbers, borrow 1 ten from the tens column and move it over to the ones column, converting it to 10 ones (notice that 7 in the tens column becomes 6). The 3 is now 13.

$$\begin{array}{r} \text{tens ones} \\ 6\;1 \\ \cancel{7}\cancel{3} \\ -5\;4 \\ \hline 1\;9 \end{array}$$

Subtract the ones column—that is, $13-4=9$; then subtract the tens column, $6-5=1$. The difference is 19. Check your answer by adding the difference to the subtrahend $(19+54)$; if they add up to the minuend (73), your answer is correct.

Example 3: Find the difference between 75 and 202 $(202-75)$.
Solution:
Step 1: Arrange the numbers into columns.

$$\begin{array}{r} 202 \\ -\;\;75 \end{array}$$

Step 2: You will note that, in the ones column, 5 can't be subtracted from 2 and, in the tens column, 7 can't be subtracted from 0 and still provide answers that are whole numbers. Because it is not possible to borrow from the tens column in this example (from 0), go one column further and borrow from the hundreds. Borrow 1 hundred from the hundreds, leaving 1 hundred in the hundreds column; move the borrowed hundred over to the tens column, thus supplying 10 tens. Now borrow 1 of these tens and move it over to the ones column, leaving the tens column with one less ten and adding 10 ones to the ones column.

$$\begin{array}{r} \text{hundreds tens ones} \\ 202 \\ -\ 75 \\ \downarrow \\ {}^{1}\ {}^{10}\ \\ 2\cancel{0}\ 2 \\ -\ 75 \\ \downarrow \\ {}^{9} \\ {}^{1}\ \cancel{10}\ 1 \\ 2\cancel{0}\ 2 \\ -\ 75 \\ \hline 127 \end{array}$$

The difference is 127.

When you work subtraction problems with larger numbers, borrow from the thousands, ten thousands, and so on in the same manner as from the hundreds and tens in Example 3. The main thing to remember is to keep the columns of numbers straight so that ones are subtracted from ones, tens subtracted from tens, and so on.

Section 1.2 Readiness Review

Subtract the following whole numbers:

1. $38 - 7 =$ _____ (see Example 1)
2. $125 - 18 =$ _____ (see Example 2)
3. $3900 - 492 =$ _____ (see Example 3)
4. $5556 - 4124 =$ _____ (see Example 1)
5. $10,000 - 501 =$ _____ (see Example 3)
6. $9813 - 9666 =$ _____ (see Example 2)

Section readiness review answers (not given in order):

1. 31   6. 147   3. 3408
4. 1432   2. 107   5. 9499

8   Chapter 1 Whole Numbers

Section 1.2 Review Problems

Subtract the following whole numbers (answers to the odd-numbered problems are at the back of the book):

1. $24-13$	**2.** $49-34$	**3.** $23-14$
4. $45-39$	**5.** $92-83$	**6.** $112-47$
7. $332-66$	**8.** $500-92$	**9.** $204-56$
10. $375-215$	**11.** $214-124$	**12.** $6666-200$
13. $1668-1450$	**14.** $1136-1090$	**15.** $11,111-2222$
16. $56,780-1234$	**17.** $32,844-6255$	**18.** $12,087-5618$
19. $87,359-77,360$	**20.** $60,606-51,342$	

Section 1.3 Multiplication of Whole Numbers

The sum of three 7s is 21; that is, $7+7+7=21$. This fact may be written in addition form, as shown, or in a shorter form. The shorter form is called *multiplication*. In this example, 7 is multiplied by 3. Two symbols that indicate multiplication are $\times$ or $\cdot$ (a dot) placed between the numbers to be multiplied. Both symbols are used in this text.

$$7 \quad \cdot \quad 3 \quad = \quad 21$$
(multiplicand) (multiplier)   (product)
$$7 \quad \times \quad 3 \quad = \quad 21$$

7 (multiplicand)
$\times$ 3 (multiplier)
21 (product)

Rule: Multiplication of Whole Numbers
Step 1: Arrange the numbers to be multiplied into columns so that ones are lined up with ones, tens are lined up with tens, and so on.
Step 2: Multiply to find the product. If the multiplier has more than one digit, you will have partial products (which are explained in Example 1); be sure to place them in the correct positions and then add them to find the product.

Example 1: Multiply 26 by 12.
Solution:
Step 1: Arrange the numbers being multiplied into columns so that ones are lined up with ones, tens are lined up with tens, and so on.

$$\begin{array}{r} \text{tens ones} \\ 26 \\ \times \underline{12} \end{array}$$

Section 1.3 Multiplication of Whole Numbers 9

Step 2: Multiply through, being careful to place the partial products in the correct positions.

```
         tens
          ones
          26   (multiplicand)
       ×  12   (multiplier)
          52   (first partial product)
          26   (second partial product—moved over one place to the left)
         312   (product)
```

In this example, the multiplier, 12, has two digits: a 1 in the tens column and a 2 in the ones column. Because the multiplier has more than one digit, you must use partial products to solve this problem. First, multiply the 26 by the multiplier digit in the ones column: 26 multiplied by 2 equals 52. This is the first partial product. Second, multiply the 26 by the multiplier digit in the tens column: 26 multiplied by 1 equals 26, which is the second partial product. Put the second partial product one place to the left of the first one, because the second product actually represents the number 260; that is, the 26 on top (multiplicand) is multiplied by 1 ten of the bottom number (multiplier) to get 260, or $26 \times 10 = 260$.

```
         tens
          ones
          26
       ×  12
          52
         26∅   ← This zero is left out.
         312
```

The product is 312.

Example 2: Multiply the following numbers: 310 and 125.
Solution:
Step 1: Arrange the numbers to be multiplied into columns.

```
    310
  × 125
```

Step 2:

```
            hundreds
             tens
              ones
           310    (multiplicand)
        ×  125    (multiplier)
          1550    (first partial product)
           620    (second partial product—moved over one place to the left)
           310    (third partial product—moved over two places to the left)
        38,750    (answer or product)
```

Because the multiplier has more than one digit, you must use partial products. First, multiply 310 by the 5 in the ones column: 1550 is the first partial product. Second, multiply 310 by the 2 in the tens column: 620 is the second partial product, and it belongs one place to the left. Finally, multiply 310 by the 1 in the hundreds column: 310 is the third partial product, and it belongs two places to the left. Add all the partial products together to get a product of 38,750.

Example 3: Multiply the following numbers: 531 and 203.
Solution:
Step 1: Arrange the numbers to be multiplied into columns.

$$\begin{array}{r} 531 \\ \times\,203 \end{array}$$

Step 2:

```
         hundreds
          tens
           ones
       531    (multiplicand)
    ×  203    (multiplier)
      1593    (first partial product)
       000    (second partial product—moved over one place to the left)
      1062    (third partial product—moved over two places to the left)
    107,793   (product)
```

Multiply through, being careful to place the partial products in the correct positions. Note that the second partial product is a row of zeros. This row of zeros may be left out as shown:

```
       531
    ×  203
      1593    (first partial product)
      1062    (third partial product—remains two places to the left)
    107,793   (product)
```

Add the partial products; the product is 107,793.

Example 4: Multiply the following numbers: 157 × 1200.
Solution:
Step 1: Arrange the numbers to be multiplied into columns so that the zeros in the multiplier are to the right as shown. Draw a box around the zeros to separate them from the 12; the zeros will be used later.

$$\begin{array}{r} 157 \\ \times\ 12\,\boxed{00} \end{array}$$

Step 2: Multiply through and add the partial products. Transfer the zeros to the answer and remove the box.

```
        157
     ×  12 00
        314
        157
       1884 00
```

Step 3: The product is 188,400. The two zeros in the box can be thought of as 100. Thus, $1884 \times 100 = 188,400$.

Section 1.3 Readiness Review

Multiply the following numbers:

1. $14 \times 6 =$ _____ (see Example 1)
2. $2435 \times 105 =$ _____ (see Example 3)
3. $132 \times 18 =$ _____ (see Examples 1 & 2)
4. $479 \times 207 =$ _____ (see Example 3)
5. $1441 \times 57 =$ _____ (see Examples 1 & 2)
6. $533 \times 400 =$ _____ (see Example 4)

Section readiness review answers (not given in order):

5. 82,137	4. 99,153	1. 84
2. 255,675	3. 2376	6. 213,200

Section 1.3 Review Problems

Multiply the following numbers (answers to the odd-numbered problems are at the back of the book):

1. 18×3
2. 13×7
3. 24×14
4. 52×19
5. 35×24
6. 14×41
7. 137×82
8. 255×67
9. 759×107
10. 185×703
11. 543×62
12. 757×575
13. 161×116
14. 453×989
15. 354×524
16. 1478×600
17. $17,542 \times 1300$
18. 275×160
19. 514×400
20. 783×1000

Section 1.4 Division of Whole Numbers

The last arithmetic operation to be covered is division of whole numbers. The process of division can be thought of as a short method for repeated subtraction. For example, take the number 20 and see how many 4s are in it. Subtracting

$$\begin{array}{ccccc} 20 & 16 & 12 & 8 & 4 \\ -4 & -4 & -4 & -4 & -4 \\ \hline 16 & 12 & 8 & 4 & 0 \end{array}$$

we see that there are five 4s in the number 20. However, repeated subtraction is not the most practical way to find the answer to this problem; using division, a more practical method, we can also set this problem up in the following manner:

$$\text{(divisor)} \; 4 \overline{)20} \quad \begin{array}{l}5 \; \text{(answer or quotient)} \\ \text{(dividend)}\end{array}$$

4 divided into 20 equals 5

The number being divided is called the *dividend* (20). The number doing the dividing is called the *divisor* (4). The answer is the *quotient* (5).

There are several division symbols. You are probably familiar with this one: $\overline{)}$. Three more symbols that indicate division are (1) the ÷ sign—for example,

$$\underset{\text{(dividend)}}{20} \; \div \; \underset{\text{(divisor)}}{4} \; = \; \underset{\text{(quotient)}}{5}$$

(2) the horizontal bar with a number on top and a number on the bottom—for example,

$$\begin{array}{l}\text{(dividend)} \\ \text{(divisor)}\end{array} \; \frac{20}{4} = 5 \; \text{(quotient)}$$

and (3) the slanted bar with a number to the left and a number to the right—for example,

$$\underset{\text{(dividend)}}{20} \; / \; \underset{\text{(divisor)}}{4} \; = \; \underset{\text{(quotient)}}{5}$$

Read any of these as 20 divided by 4 equals 5.

Rule: Division of Whole Numbers
Step 1: Determine how many times the divisor divides into the dividend.
Step 2: Find the quotient.
Step 3: Express any remainder in "fractional-part" form.

Example 1: Solve the following division problem: $25\overline{)875}$
Solution:
Step 1: Determine how many times the divisor divides into the dividend. The number 25 cannot be divided into 8 and have a whole-number answer, but 25 does divide into 87 three times.

$$\begin{array}{r}3 \\ 25\overline{)875} \\ \underline{75} \\ 12\end{array}$$

Multiply 25 by 3 to obtain a product of 75. Then subtract 75 from 87 and you have a remainder of 12. (The *remainder* is the number left over after the process of subtracting is finished, and it must always be *less* than the divisor.) Now bring the 5 down from the dividend and place it to the right of the remainder, as shown.

Section 1.4 Division of Whole Numbers 13

$$\begin{array}{r} 35 \\ 25\overline{)875} \\ \underline{75} \\ 125 \\ \underline{125} \\ 0 \end{array}$$

Step 2: Determine how many times 25 divides into 125. Then multiply and subtract to find the remainder. The quotient is 35.

Step 3: There is a remainder of 0. To check the answer, multiply the divisor (25) by the quotient (35): if the product is the same as the dividend (875), the answer is correct.

Example 2: Solve the following division problem: $16\overline{)775}$
Solution:
Step 1: Determine how many times the divisor divides into the dividend. The number 16 can't be divided into 7, but it does divide into 77 four times.

$$\begin{array}{r} 4 \\ 16\overline{)775} \\ \underline{64} \\ 13 \end{array}$$

Multiply 16 by 4: the product is 64. Subtract 64 from 77 and you have a remainder of 13. Bring the 5 down and complete the division process.

$$\begin{array}{r} 48 \\ 16\overline{)775} \\ \underline{64} \\ 135 \\ \underline{128} \\ 7 \end{array}$$

Step 2: The quotient is 48.
Step 3: The remainder is 7. Express the remainder as a fractional part of the divisor—that is,

$$\frac{\text{REMAINDER}}{\text{DIVISOR}}$$

In this example, the remainder is 7/16; thus, the complete answer is 48 7/16.

Example 3: Solve the following division problem: $15\overline{)3045}$
Solution:
Step 1: Divide 15 into 30; then multiply, subtract, and bring down the 4.

$$\begin{array}{r} 2 \\ 15\overline{)3045} \\ \underline{30} \\ 4 \end{array}$$

The number 15 doesn't divide into 4, so place a zero in the quotient and bring down the 5, as shown. Now finish the division process.

$$\begin{array}{r} 203 \\ 15\overline{)3045} \\ \underline{30} \\ 45 \\ \underline{45} \\ 0 \end{array}$$

Step 2: The quotient is 203.
Step 3: There is no remainder. Thus, $3045 \div 15 = 203$.

As you know, the divisor does not always go into the dividend a whole number of times, and so you may have a remainder when you solve a division problem. The fractional-part form, however, is not the only way to express the remainder; you may also use the remainder form or the decimal-part form.

$$\begin{array}{r} 31\ \text{R}1 \\ 25\overline{)776} \\ \underline{75} \\ 26 \\ \underline{25} \\ 1\ \text{(remainder)} \end{array}$$

This is the remainder form.

$$\begin{array}{r} 31\,^1/_{25} \\ 25\overline{)776} \\ \underline{75} \\ 26 \\ \underline{25} \\ 1\ \text{(remainder)} \end{array}$$

This is the fractional-part form. Chapter 2 covers fractions in detail.

$$\begin{array}{r} 31.04 \\ 25\overline{)776.00} \\ \underline{75} \\ 26 \\ \underline{25} \\ 100 \\ \underline{100} \\ 0\ \text{(remainder)} \end{array}$$

This is the decimal-part form. Chapter 3 gives you more information on decimals.

Remember, the remainder, fractional-part, and decimal-part forms all express the same answer. You will choose different forms under different circumstances.

Section 1.4 Readiness Review

Solve the following division problems, using the fractional-part form to show any remainder:

1. $7\overline{)112} =$ _____ (see Example 1)

2. $8\overline{)152} =$ _____ (see Example 1)

3. $12\overline{)1236} =$ _____ (see Example 3)

4. $16\overline{)6416} =$ _____ (see Example 3)
5. $123\overline{)1245} =$ _____ (see Example 2)
6. $372\overline{)4007} =$ _____ (see Example 2)

Section readiness review answers (not given in order):

5. $10\frac{15}{123}$ 6. $10\frac{287}{372}$ 1. 16
2. 19   3. 103   4. 401

Section 1.4 Review Problems

Solve the following division problems, using any remainder (answers to the odd-numbered the book):

1. $5\overline{)205}$ 2. $8\overline{)648}$
4. $25\overline{)6250}$ 5. $123\overline{)10,94}$
7. $4\overline{)55}$ 8. $3\overline{)35}$
10. $11\overline{)11,011}$ 11. $12\overline{)79}$
13. $40\overline{)4873}$ 14. 38
16. $32\overline{)15,325}$ 17. 1(
19. $481\overline{)904,945}$ 20. 60

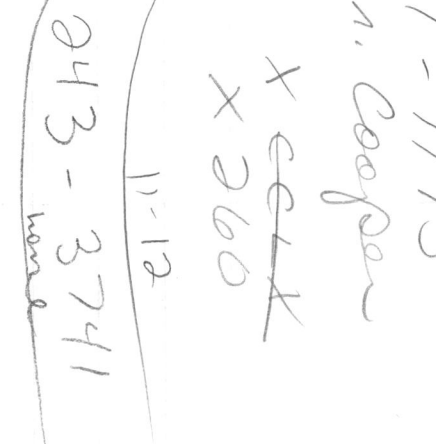

Section 1.5 Word Problems

To be successful at solving word problems, do the following:

Rule: Solving Word Problems

Step 1: Read the entire problem and try to get a general idea of what it's about.

Step 2: Read the problem again. Decide what the problem really is asking and what information you need to solve it. Keep in mind that extra information may be included in the problem.

Step 3: Read the problem a third time and estimate an answer; this will give you something to compare your computed answer to. If the estimate and the computed answer are close, then your computed answer is probably correct. You do not have to come up with the exact answer yet, just a reasonable estimate.

Step 4: Work the problem. Use textbook examples whenever possible as a guide to solving it.

Step 5: If your answer is reasonable, check your work by any method you wish; however, if your answer seems "way off," carefully reread the problem. If, after several tries, you're still having trouble, put an end to your frustration and get help from your instructor. We can all learn from mistakes.

Example 1: Dr. Margaret Jones, a neuropathologist, must study microscope slides prepared from frozen sections of tissue. One day, she read 43 slides, the next day 127, and the third day 79. How many slides did she read during the three days?

Solution:
Step 1: Read the problem through: you discover that the doctor studied microscope slides for three days.
Step 2: The main thing the problem asks you to do is determine the total number of slides the doctor has read. The fact that she read the slides during three successive days is of no importance in getting the answer.
Step 3: One estimate of the answer is 250 slides. (Approximately $40 + 130 + 80$ slides.)
Step 4: Work the problem. If no mistake has been made, your answer is 249 slides.

$$\begin{array}{r} \overset{1}{4}\overset{1}{3} \\ 127 \\ +79 \\ \hline 249 \text{ slides} \end{array}$$

Step 5: Compare the answer, 249 slides, with the estimate of 250 slides: the answers are similar. To check the answer, add the numbers backwards.

$$\begin{array}{r} \overset{1}{7}\overset{1}{9} \\ 127 \\ +43 \\ \hline 249 \text{ slides} \end{array}$$

The answers check.

Example 2: Urine is collected over a 24-hour period. The amount of fluids taken in by the patient is monitored. If the patient takes 1521 mℓ (stay with me—I'll discuss milliliters, abbreviated mℓ, in Chapter 6, Metric System) of liquid by mouth and puts out 1402 mℓ of urine, what is the difference between the liquid consumed and the liquid collected?

Solution:
Steps 1 and 2: Read this problem casually, then carefully, and you will see that the important question is finding the difference between the liquid consumed and the urine produced.
Step 3: The estimated answer is about 100 mℓ; that is, the liquid consumed is about 1500 mℓ, and the urine put out is about 1400 mℓ.
Step 4: Work the problem; the exact answer is 119 mℓ.

$$\begin{array}{r} 15\overset{11}{2}1 \\ -1402 \\ \hline 119 \end{array}$$

Step 5: This answer is close to the estimate of 100 mℓ. Check the answer by adding 119 to 1402; the answer is correct.

4. $16\overline{)6416} =$ _____ (see Example 3)
5. $123\overline{)1245} =$ _____ (see Example 2)
6. $372\overline{)4007} =$ _____ (see Example 2)

Section readiness review answers (not given in order):

5. $10\frac{15}{123}$ 6. $10\frac{287}{372}$ 1. 16
2. 19   3. 103   4. 401

Section 1.4 Review Problems

Solve the following division problems, using the fractional-part form to show any remainder (answers to the odd-numbered problems are at the back of the book):

1. $5\overline{)205}$ 2. $8\overline{)648}$ 3. $11\overline{)1111}$
4. $25\overline{)6250}$ 5. $123\overline{)10,947}$ 6. $459\overline{)37,179}$
7. $4\overline{)55}$ 8. $3\overline{)35}$ 9. $16\overline{)6583}$
10. $11\overline{)11,011}$ 11. $12\overline{)797}$ 12. $27\overline{)8336}$
13. $40\overline{)4873}$ 14. $38\overline{)569}$ 15. $25\overline{)9011}$
16. $32\overline{)15,325}$ 17. $10\overline{)13,759}$ 18. $922\overline{)17,774}$
19. $481\overline{)904,945}$ 20. $602\overline{)119,268}$

Section 1.5 Word Problems

To be successful at solving word problems, do the following:

> **Rule: Solving Word Problems**
> *Step 1:* Read the entire problem and try to get a general idea of what it's about.
> *Step 2:* Read the problem again. Decide what the problem really is asking and what information you need to solve it. Keep in mind that extra information may be included in the problem.
> *Step 3:* Read the problem a third time and estimate an answer; this will give you something to compare your computed answer to. If the estimate and the computed answer are close, then your computed answer is probably correct. You do not have to come up with the exact answer yet, just a reasonable estimate.
> *Step 4:* Work the problem. Use textbook examples whenever possible as a guide to solving it.
> *Step 5:* If your answer is reasonable, check your work by any method you wish; however, if your answer seems "way off," carefully reread the problem. If, after several tries, you're still having trouble, put an end to your frustration and get help from your instructor. We can all learn from mistakes.

Example 1: Dr. Margaret Jones, a neuropathologist, must study microscope slides prepared from frozen sections of tissue. One day, she read 43 slides, the next day 127, and the third day 79. How many slides did she read during the three days?

Solution:
Step 1: Read the problem through: you discover that the doctor studied microscope slides for three days.
Step 2: The main thing the problem asks you to do is determine the total number of slides the doctor has read. The fact that she read the slides during three successive days is of no importance in getting the answer.
Step 3: One estimate of the answer is 250 slides. (Approximately $40 + 130 + 80$ slides.)
Step 4: Work the problem. If no mistake has been made, your answer is 249 slides.

$$\begin{array}{r} \overset{1}{4}\overset{1}{3} \\ 127 \\ + 79 \\ \hline 249 \end{array} \text{ slides}$$

Step 5: Compare the answer, 249 slides, with the estimate of 250 slides: the answers are similar. To check the answer, add the numbers backwards.

$$\begin{array}{r} \overset{1}{7}\overset{1}{9} \\ 127 \\ + 43 \\ \hline 249 \end{array} \text{ slides}$$

The answers check.

Example 2: Urine is collected over a 24-hour period. The amount of fluids taken in by the patient is monitored. If the patient takes 1521 mℓ (stay with me—I'll discuss milliliters, abbreviated mℓ, in Chapter 6, Metric System) of liquid by mouth and puts out 1402 mℓ of urine, what is the difference between the liquid consumed and the liquid collected?

Solution:
Steps 1 and 2: Read this problem casually, then carefully, and you will see that the important question is finding the difference between the liquid consumed and the urine produced.
Step 3: The estimated answer is about 100 mℓ; that is, the liquid consumed is about 1500 mℓ, and the urine put out is about 1400 mℓ.
Step 4: Work the problem; the exact answer is 119 mℓ.

$$\begin{array}{r} 15\overset{11}{2}1 \\ -1402 \\ \hline 119 \end{array}$$

Step 5: This answer is close to the estimate of 100 mℓ. Check the answer by adding 119 to 1402; the answer is correct.

Section 1.5 Readiness Review

Solve the following word problems:

1. A patient admitted to the hospital gives the nurse a bag of medications from home. The nurse sends the bag to the pharmacy for identification. The pharmacist finds one vial with 85 green tablets, another vial with 47 white capsules, a small box with 11 suppositories, and a jar with 288 red tablets. What is the total number of drug items (tablets, capsules, and suppositories) brought into the hospital by this patient?

2. The Dietary Department of the hospital must make up 786 meal trays each day. If each patient receives three meals, how many patients are in the hospital?

Section readiness review answers:

1. 431 drug items   2. 262 patients

Section 1.5 Review Problems

(The answers to the odd-numbered problems are at the back of the book):

1. A laboratory technician must draw blood from 18 patients on the surgical floor, 14 patients in the intensive care unit, 43 outpatients, and 7 pediatric patients. How many blood samples will be drawn by this technician?

2. The physical therapy department maintains a Hubbard Tank (this is a stainless-steel, bathtub-like apparatus with whirlpool features). The Hubbard Tank can be used by one patient at a time. When a patient is finished using the tank, it must be cleaned and disinfected, which takes approximately 12 minutes. On one day, 21 patients used the tank. How many minutes were spent cleaning the Hubbard Tank?

3. A volunteer at the hospital has worked 2672 hours. How many more hours must he work to win the 3000-hour Hospital Volunteer Award?

4. In one month, the nuclear medicine department has performed 342 brain scans. If during the same month of last year it performed 123 brain scans, how many more brain scans have been performed this year?

5. The pharmacy must order 1500 child-resistant containers to dispense medications. If the pharmacist will use 1377 of these containers, how many will be left to count for inventory?

6. Last year the inhalation therapist used her "Bird Machine" for a total of 18,173 minutes. So far this year she has used it for 1779 minutes. How many more minutes must she use it this year before she matches last year's total minutes of use?

7. Dennis DelGreco, a nurse working the P.M. shift on a surgical ward, is in charge of giving "sleepers" (sleeping medication) to the patients. If Nurse DelGreco has 27 surgical patients and their physicians have ordered 2 capsules per patient, how many capsules will he administer?

8. Penny Van Duzer, the In-Service Coordinator for the hospital, has scheduled C.P.R. (cardiopulmonary resuscitation) instruction over a two-day period. If each class lasts an average of 17 minutes and there are 13 classes in one day, how many total minutes are spent in the C.P.R. class during the two days?

9. The Unit Dose Drug Distribution System in the hospital uses cassettes with individual drawers labeled with each patient's name and room number. There are 7 floors in the hospital and 40 drawers for each floor. How many individual drawers are there in the hospital?

10. A patient, Charles Miller, required nose surgery to improve his breathing. If Mr. Miller, as a postsurgical patient, had a total of 80 minutes of breathing treatments as indicated on his bill and each treatment lasted 5 minutes, how many treatments did he have while hospitalized?

Chapter 1 Readiness Review

Add the following whole numbers:

1. $3 + 24 =$ _____
2. $13 + 11 + 12 =$ _____
3. $22 + 19 =$ _____
4. $17 + 376 =$ _____
5. $60 + 300 + 1000 =$ _____
6. $3172 + 475 + 1118 =$ _____

Find the difference between the following numbers:

7. $25 - 11 =$ _____
8. $48 - 33 =$ _____
9. $47 - 38 =$ _____
10. $125 - 19 =$ _____
11. $2001 - 116 =$ _____
12. $4234 - 1999 =$ _____

Multiply the following numbers:

13. $17 \times 4 =$ _____
14. $23 \times 13 =$ _____
15. $138 \times 115 =$ _____
16. $256 \times 403 =$ _____
17. $1758 \times 1200 =$ _____
18. $15 \times 41 =$ _____

Solve the following division problems. Use the fractional-part form to show any remainder.

19. $15 \overline{)405} =$ _____
20. $12 \overline{)144} =$ _____
21. $12 \overline{)3660} =$ _____
22. $27 \overline{)2727} =$ _____
23. $186 \overline{)4007} =$ _____
24. $175 \overline{)1349} =$ _____

Chapter readiness review answers (not given in order):

21. 305	22. 101	23. $21\frac{101}{186}$	24. $7\frac{124}{175}$	17. 2,109,600
18. 615	19. 27	20. 12	13. 68	14. 299
15. 15,870	16. 103,168	9. 9	10. 106	11. 1885
12. 2235	5. 1360	6. 4765	7. 14	8. 15
1. 27	2. 36	3. 41	4. 393	

Chapter 1 Summary

The Chapter Summary comprises a list of key words in the order they appear in each section. Define each item in your own words, then compare your definitions with the text.

Key Words

addition   (p. 3)
whole numbers   (p. 3)
addend   (p. 3)
sum   (p. 3)
subtraction   (p. 5)
minuend   (p. 6)
subtrahend   (p. 6)
difference   (p. 6)
multiplication   (p. 8)
multiplicand   (p. 8)
multiplier   (p. 8)

partial product   (p. 8)
product   (p. 8)
division   (p. 11)
dividend   (p. 12)
divisor   (p. 12)
quotient   (p. 12)
remainder   (p. 12)
fractional part   (p. 14)
decimal part   (p. 14)
word problems   (p. 15)

Chapter 1 Review Problems

Add the following numbers (answers to all the problems are at the back of the book):

1. $4 + 13$
2. $8 + 77$
3. $12 + 6 + 35$
4. $23 + 8 + 57$
5. $12 + 25 + 10$
6. $11 + 43 + 14$
7. $18 + 536 + 74$
8. $17 + 121 + 89$
9. $750 + 20 + 500$
10. $675 + 109 + 228$
11. $324 + 221 + 148$
12. $814 + 131 + 1017$
13. $800 + 700 + 2000$
14. $1879 + 248 + 3750$
15. $1235 + 411 + 1446$
16. $30,016 + 13 + 7633$

Find the difference between the following numbers:

17. $53 - 14$
18. $65 - 29$
19. $82 - 43$
20. $111 - 36$
21. $304 - 77$
22. $365 - 205$
23. $244 - 174$
24. $666 - 300$
25. $1886 - 1750$
26. $1156 - 990$
27. $11,111 - 3333$
28. $12,340 - 5678$
29. $30,804 - 2655$
30. $12,078 - 5519$
31. $97,358 - 73,370$
32. $50,505 - 41,324$

Multiply the following numbers:

33. 13×8
34. 17×3
35. 34×24
36. 42×19
37. 25×14
38. 34×21
39. 127×62
40. 245×57
41. 649×27

42. 165×43	**43.** 243×32	**44.** 627×515
45. 181×108	**46.** 351×898	**47.** 234×514
48. 500×800		

Solve the following division problems. Use fractional parts to show any remainder.

49. $5\overline{)305}$	**50.** $8\overline{)848}$	**51.** $10\overline{)1010}$
52. $25\overline{)4250}$	**53.** $120\overline{)10,440}$	**54.** $325\overline{)2925}$
55. $5\overline{)44}$	**56.** $2\overline{)65}$	**57.** $13\overline{)749}$
58. $24\overline{)8226}$	**59.** $30\overline{)5748}$	**60.** $31\overline{)15,121}$
61. $10\overline{)13,659}$	**62.** $911\overline{)16,665}$	**63.** $421\overline{)909,025}$
64. $302\overline{)118,256}$		

Solve the following problems:

65. $342 + 55 + 107$	**66.** $44 - 36$	**67.** $3\overline{)3612}$
68. 15×5	**69.** 99×21	**70.** $132 + 74 + 2115$
71. $57 - 28$	**72.** $1928 - 1789$	**73.** $3 + 85$
74. $89,754 - 25,349$	**75.** $9\overline{)7218}$	**76.** $15 + 32 + 21$
77. 19×6	**78.** $22,222 + 66 + 2222$	**79.** $542 + 111 + 327$
80. $13\overline{)1014}$	**81.** $7\overline{)78}$	**82.** $7 + 21$
83. $1951 - 933$	**84.** $77 - 68$	**85.** $958\overline{)25,783}$
86. 72×12	**87.** 376×39	**88.** $54 + 72 + 89$
89. $9\overline{)125}$	**90.** $7820 - 4312$	**91.** 431×58
92. $43 + 10 + 91$	**93.** $258\overline{)59,399}$	**94.** 828×35

95. Mr. Rice, a glaucoma patient, requires eye drops for his condition. If he puts 3 drops in each eye 4 times a day, what is the total number of drops he uses in 2 days?

96. The cardiologist (heart specialist) orders a routine treadmill stress test for patients 35 years of age and older. If each test lasts an average of 5 minutes, and the doctor has 109 patients 35 years of age or older, how many minutes will the treadmill be in operation?

97. A bottle of Regular U-100 Insulin contains 1000 units of insulin. If a diabetic has used 154 units, how many units of insulin are left?

98. The pharmacist filled 100 gelatin capsules with a prescription she compounded. Then she put equal numbers of capsules into each of 4 vials. How many capsules were there in each vial?

99. The surgery schedule for the next day shows a total number of 35 operations. If there are 7 anesthesiologists to share the cases equally, how many surgeries will each one handle?

100. A baby is to receive an oral dose of a medicine. The nurse calculates the dose of the medicine based on the infant's age. If the baby requires 44 mℓ and the medicine comes in a 60-mℓ bottle, how many milliliters of medicine will be left in the bottle?

2 Fractions

OBJECTIVES After studying this chapter, you should be able to:

1. Identify the types of fractions.
2. Change mixed numbers into improper fractions and change improper fractions into mixed numbers.
3. Reduce fractions to lowest terms.
4. Determine if fractions are equal or unequal.
5. Perform the operations of addition, subtraction, multiplication, and division using fractions.

The ability to perform calculations involving fractions is an important skill for nurses to acquire. A knowledge of fractions will provide you with the foundation needed to solve many drug-dosage problems.

Section 2.1 Types of Fractions

> **Definition:** A *fraction* is a number that counts equal parts of a whole.

A fraction is a part of a whole and is written as A/B. B is the denominator, which tells the number of equal parts the whole is divided into. A is the numerator, which tells how many of the equal parts are being used.

$$\frac{A}{B} \begin{array}{l} \leftarrow \text{numerator} \\ \leftarrow \text{denominator} \end{array}$$

There are different types of fractions. Let's begin with proper fractions.

> **Definition:** A *proper fraction* has a numerator that is *less* than the denominator.

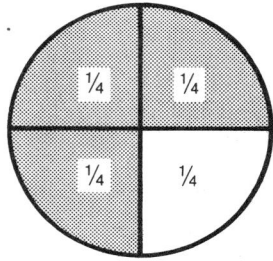

Figure 2-1.

In Figure 2-1, you see a circle divided into 4 equal parts. You will notice that 3 of the 4 parts of this circle are shaded; that is, ¾ of the whole circle is shaded. The fraction ¾ is a proper fraction.

Proper fraction: $\frac{3}{4}$ $\frac{\text{(numerator is less than the denominator)}}{\text{(denominator is greater than the numerator)}}$

Other examples of proper fractions are

$$\frac{1}{2}, \frac{1}{4}, \frac{1}{6}, \frac{2}{3}, \frac{5}{8}, \text{ and so on.}$$

> **Definition:** An *improper fraction* has a numerator that is *greater* than the denominator.

Section 2.1 Types of Fractions   23

Improper fraction: $\dfrac{8}{5}$ $\dfrac{\text{(numerator is greater than the denominator)}}{\text{(denominator is less than the numerator)}}$

Other examples of improper fractions are

$$\frac{4}{3}, \frac{12}{11}, \frac{25}{13}, \frac{11}{5}, \frac{17}{9}, \text{ and so on.}$$

Definition: A *complex fraction* is one in which either the numerator or the denominator, or both, are fractions themselves.

Complex fraction: $\dfrac{\frac{3}{5}}{6}$ $\dfrac{\text{(numerator is a fraction)}}{\text{(denominator is a whole number)}}$

Complex fraction: $\dfrac{2}{\frac{1}{3}}$ $\dfrac{\text{(numerator is a whole number)}}{\text{(denominator is a fraction)}}$

Complex fraction: $\dfrac{\frac{2}{3}}{\frac{3}{4}}$ $\dfrac{\text{(numerator is a fraction)}}{\text{(denominator is a fraction)}}$

Other examples of complex fractions are

$$\frac{3}{\frac{7}{10}}, \frac{\frac{7}{10}}{3}, \frac{\frac{7}{10}}{\frac{1}{3}}, \frac{\frac{1}{3}}{\frac{7}{10}}, \text{ and so on.}$$

The last type of fraction not only involves a fraction but includes a whole number also. This number is called a *mixed number*.

Definition: A *mixed number* consists of a whole-number portion and a fractional portion.

Mixed number: $4\dfrac{3}{4}$ (4 is the whole number, $\dfrac{3}{4}$ is the fraction)

It is important to notice that $4\,{}^3/_4$ means $4 + {}^3/_4$ and *not* $4 \times {}^3/_4$. Other examples of mixed numbers are

$$5\frac{1}{5}, 7\frac{7}{8}, 2\frac{1}{2}, 6\frac{2}{3}, \text{ and so on.}$$

Chapter 2 Fractions

Mixed numbers are sometimes changed into improper fractions. To do this, apply the following rule:

> **Rule: Changing a Mixed Number into an Improper Fraction**
> *Step 1:* Multiply the denominator of the fraction by the whole number.
> *Step 2:* Add the product found in Step 1 to the numerator of the fraction.
> *Step 3:* Write the sum found in Step 2 over the denominator of the fraction.

Example 1: Change the following mixed numbers into improper fractions:

a. $5\frac{3}{8}$ b. $6\frac{4}{5}$

c. $3\frac{1}{3}$ d. $11\frac{1}{2}$

Solution: Using the rule given for changing a mixed number into an improper fraction, proceed step by step for each fraction.

a. $5\frac{3}{8}$

According to Step 1, first multiply the denominator of the fraction, 8, by the whole number, 5. This turns out to be $8 \times 5 = 40$. The product, 40, is then added to the numerator of the fraction, 3, to give $40 + 3 = 43$, fulfilling Step 2. In Step 3, the sum found in Step 2, 43, is written over the denominator of the fraction, 8. Therefore, the mixed number $5\frac{3}{8}$ equals the improper fraction having a numerator 43 and a denominator 8.

$$5\frac{3}{8} = \frac{43}{8}$$

b. $6\frac{4}{5}$

Step 1: $5 \times 6 = 30$
Step 2: $30 + 4 = 34$
Step 3: $\frac{34}{5}$
Therefore, $6\frac{4}{5} = \frac{34}{5}$

c. $3\frac{1}{3}$

Step 1: $3 \times 3 = 9$
Step 2: $9 + 1 = 10$
Step 3: $\frac{10}{3}$
Therefore, $3\frac{1}{3} = \frac{10}{3}$

d. $11\frac{1}{2}$

Step 1: $2 \times 11 = 22$
Step 2: $22 + 1 = 23$
Step 3: $\frac{23}{2}$
Therefore, $11\frac{1}{2} = \frac{23}{2}$

Now, to change an improper fraction back to a mixed number, division is used. A fraction is a way of dividing; that is, the top number is divided by the bottom number.

Rule: Changing an Improper Fraction into a Mixed Number
Step 1: Divide the numerator of the fraction by its denominator.
Step 2: Write the quotient and express any remainder using the fractional-part method.

Example 2: Change the following improper fractions into mixed numbers

a. $\frac{43}{8}$ b. $\frac{34}{5}$

c. $\frac{10}{3}$ d. $\frac{23}{2}$

Solution: To change an improper fraction into a mixed number, divide the numerator of the fraction by the denominator, as stated in Step 1. Now write the quotient expressing any remainder in the fractional-part method, as required in Step 2. Remember, the fractional-part method means to express the answer as

$$\text{QUOTIENT } \frac{\text{REMAINDER}}{\text{DENOMINATOR}}$$

a. $\frac{43}{8}$

The numerator of this fraction, 43, is divided by the denominator, 8 (Step 1).

$$\begin{array}{r} 5 \\ 8\overline{)43} \\ \underline{40} \\ 3 \end{array}$$

Now, write the quotient, which is 5, and express the remainder in the fractional-part form (Step 2).

$$\text{QUOTIENT } \frac{\text{REMAINDER}}{\text{DENOMINATOR}}$$

$$\text{QUOTIENT} \rightarrow 5\frac{3}{8} \begin{array}{l} \leftarrow \text{REMAINDER} \\ \leftarrow \text{DENOMINATOR} \end{array}$$

Therefore, $43/8 = 5\,3/8$

b. $\frac{34}{5}$

Step 1:
$$\begin{array}{r} 6 \\ 5\overline{)34} \\ \underline{30} \\ 4 \end{array}$$

Step 2: $6\,4/5$

Therefore, $34/5 = 6\,4/5$

c. $\frac{10}{3}$

Chapter 2 Fractions

Step 1:

$$3\overline{)10}$$
$$\phantom{3\overline{)}}\underline{9}$$
$$\phantom{3\overline{)}}1$$

Step 2: $3\frac{1}{3}$

Therefore, $\frac{10}{3} = 3\frac{1}{3}$

d. $\frac{23}{2}$

Step 1:

$$2\overline{)23}$$
$$\phantom{2\overline{)}}\underline{22}$$
$$\phantom{2\overline{)2}}1$$

Step 2: $11\frac{1}{2}$

Therefore, $\frac{23}{2} = 11\frac{1}{2}$

One aspect of fractions not yet covered is how to reduce fractions to lowest terms. To learn how to reduce fractions to lowest terms, study the following rule.

Rule: Reducing a Fraction to Lowest Terms
Divide the numerator and the denominator by the largest number (greater than 1) that can divide them both.

Example 3: Reduce the following fractions to lowest terms.

a. $\frac{6}{8}$ **b.** $\frac{3}{15}$

c. $\frac{14}{35}$ **d.** $\frac{9}{10}$

Solution:

a. $\frac{6}{8}$

The largest number that is a divisor of both 6 and 8 is 2.

$$\frac{6 \div 2}{8 \div 2} = \frac{3}{4}$$

Therefore, $\frac{3}{4}$ equals the lowest terms of the fraction $\frac{6}{8}$.

b. $\frac{3}{15}$

The largest number that is a divisor of both 3 and 15 is 3.

$$\frac{3 \div 3}{15 \div 3} = \frac{1}{5}$$

c. $\frac{14}{35}$ **d.** $\frac{9}{10}$

$$\frac{14 \div 7}{35 \div 7} = \frac{2}{5} \qquad\qquad \frac{9 \div 1}{10 \div 1} = \frac{9}{10}$$

STOP HERE! This fraction is already reduced to lowest terms, because the largest number that will divide into 9 and 10 is 1.

Section 2.1 Readiness Review

Identify the following types of fractions according to the first five definitions in Section 2.1:

1. $\frac{2}{3}$ is a _____ fraction.

2. $\frac{3}{\frac{2}{3}}$ is a _____ fraction.

3. $\frac{3}{2}$ is a _____ fraction.

4. $1\frac{2}{3}$ is a _____ _____ .

Convert the following mixed numbers to improper fractions (see Example 1):

5. $1\frac{5}{8} =$ _____   6. $7\frac{2}{9} =$ _____

7. $5\frac{3}{5} =$ _____   8. $12\frac{1}{2} =$ _____

Convert the following improper fractions to mixed numbers (see Example 2):

9. $\frac{23}{5} =$ _____   10. $\frac{17}{2} =$ _____

11. $\frac{16}{11} =$ _____   12. $\frac{82}{3} =$ _____

Reduce the following fractions to lowest terms (see Example 3):

13. $\frac{4}{12} =$ _____   14. $\frac{81}{90} =$ _____

15. $\frac{13}{39} =$ _____   16. $\frac{20}{65} =$ _____

Section readiness review answers (not given in order):

4. mixed number
7. $\frac{28}{5}$
1. proper
6. $\frac{65}{9}$
12. $27\frac{1}{3}$
15. $\frac{1}{3}$
5. $\frac{13}{8}$
2. complex
11. $1\frac{5}{11}$
16. $\frac{4}{13}$
8. $\frac{25}{2}$
13. $\frac{1}{3}$
3. improper
10. $8\frac{1}{2}$
14. $\frac{9}{10}$
9. $4\frac{3}{5}$

28   Chapter 2 Fractions

Section 2.1 Review Problems

Identify the following types of fractions (answers to the odd-numbered problems are at the back of the book):

1. $\dfrac{7}{20}$ 2. $\dfrac{17}{7}$ 3. $5\dfrac{4}{9}$ 4. $\dfrac{\frac{3}{4}}{7}$ 5. $\dfrac{\frac{5}{6}}{\frac{3}{8}}$

6. $\dfrac{65}{32}$ 7. $\dfrac{9}{8}$ 8. $\dfrac{4}{5}$ 9. $\dfrac{15}{17}$ 10. $17\dfrac{4}{87}$

Convert the following mixed numbers to improper fractions:

11. $1\dfrac{1}{2}$ 12. $2\dfrac{2}{3}$ 13. $11\dfrac{11}{12}$ 14. $10\dfrac{9}{10}$ 15. $51\dfrac{2}{13}$

Convert the following improper fractions to mixed numbers:

16. $\dfrac{5}{4}$ 17. $\dfrac{3}{2}$ 18. $\dfrac{6}{5}$ 19. $\dfrac{76}{25}$ 20. $\dfrac{14}{9}$

Reduce the following fractions to lowest terms:

21. $\dfrac{10}{15}$ 22. $\dfrac{18}{36}$ 23. $\dfrac{9}{45}$ 24. $\dfrac{1}{39}$ 25. $\dfrac{10}{90}$

Section 2.2 Addition of Fractions

> **Definition:** Fractions that have the same denominator are called *like fractions*.

The fractions ⅔ and ⅓ are like fractions, but ⅔ and ½ are not.

> **Rule: Adding Like Fractions**
> *Step 1:* Add the numerators.
> *Step 2:* Place the sum found in Step 1 over the denominator.
> *Step 3:* Reduce the answer to lowest terms.
> *Step 4:* Convert improper fractions to mixed numbers.

Example 1:   Find the sum:
$$\dfrac{3}{4} + \dfrac{3}{4}$$

Solution: The first step in solving this problem is to add the numerators.
$$\dfrac{3}{4} + \dfrac{3}{4} = \dfrac{3+3}{4} = \dfrac{6}{4}$$

The sum found in Step 1, which is 6, is placed over the denominator, 4 (Step 2). The answer, 6/4, is then reduced to lowest terms (Step 3).

$$\frac{6 \div 2}{4 \div 2} = \frac{3}{2}$$

The improper fraction 3/2 is converted to a mixed number (Step 4).

$$2\overline{)3} \begin{array}{c} 1 = 1\frac{1}{2} \\ \underline{2} \\ 1 \end{array}$$

The final answer is 1½.

Example 2: Add the following fractions:

$$\frac{1}{15} + \frac{2}{15} + \frac{3}{15}$$

Solution:

Step 1 and *Step 2:*
$$\frac{1}{15} + \frac{2}{15} + \frac{3}{15} = \frac{1+2+3}{15} = \frac{6}{15}$$

Step 3:
$$\frac{6 \div 3}{15 \div 3} = \frac{2}{5}$$

Step 4: Not required. The final answer is 2/5.

Example 3: Add the following fractions:

$$\frac{3}{8} + \frac{7}{8} + \frac{10}{8}$$

Solution: The first step in solving this problem is to add the numerators.

$$\frac{3}{8} + \frac{7}{8} + \frac{10}{8} = \frac{3+7+10}{8} = \frac{20}{8}$$

The sum found in Step 1, which is 20, is placed over the denominator, 8 (Step 2). The answer 20/8 is then reduced to lowest terms (Step 3).

$$\frac{20 \div 4}{8 \div 4} = \frac{5}{2}$$

The improper fraction 5/2 is converted to a mixed number (Step 4).

$$2\overline{)5} \begin{array}{c} 2 = 2\frac{1}{2} \\ \underline{4} \\ 1 \end{array}$$

The final answer is 2½.

30   Chapter 2 Fractions

> **Definition:** *Unlike fractions* are fractions that have denominators that are *not* the same.

The fractions $\frac{1}{2}$ and $\frac{1}{3}$ are unlike fractions because their denominators are not the same. The key to adding unlike fractions is to change them into like fractions. To do this, you must first learn how to raise a fraction to higher terms.

> **Rule: Raising a Fraction to Higher Terms**
> Multiply the numerator and the denominator of the fraction by the same number. This number must be greater than 1.

Example 4: Raise $\frac{5}{6}$ to a fraction with denominator 24.
Solution: This problem may be written as

$$\frac{5}{6} = \frac{n}{24}$$

Let *n* equal the new numerator. Divide 6 into 24, and you see it goes 4 times. So, 6 must have been multiplied by 4 in order to get 24. Multiply the numerator, 5, by 4, and the new numerator *n* becomes 20.

$$\frac{5}{6} = \frac{5 \times 4}{6 \times 4} = \frac{20}{24}$$

$\frac{20}{24}$ is the answer.

Example 5: Raise $\frac{11}{12}$ to a fraction with denominator 60.
Solution: This problem may be written as

$$\frac{11}{12} = \frac{n}{60}$$

Divide 12 into 60, and it goes 5 times. Therefore, 12 must have been multiplied by 5 to get 60. Now multiply the numerator, 11, by 5, and the new numerator *n* becomes 55.

$$\frac{11}{12} = \frac{11 \times 5}{12 \times 5} = \frac{55}{60}$$

$\frac{55}{60}$ is the answer.

Example 6: Raise $\frac{3}{8}$ to a fraction with denominator 48.
Solution: This problem may be written as

$$\frac{3}{8} = \frac{n}{48}$$

Divide 8 into 48, and you see that it goes 6 times. So, 8 must have been multiplied by 6 in order to get 48. Multiply the numerator, 3, by 6, and the new numerator *n* becomes 18.

$$\frac{3}{8} = \frac{3 \times 6}{8 \times 6} = \frac{18}{48}$$

18/48 is the answer.

Example 7: Raise 7/12 to a fraction with numerator 77.

Solution:
$$\frac{7}{12} = \frac{77}{d}$$

First, divide 7 into 77, and you see that it goes 11 times. Now multiply the denominator, 12, by 11, and the new denominator becomes 132.

$$\frac{7}{12} = \frac{7 \times 11}{12 \times 11} = \frac{77}{132}$$

The fraction 7/12 is raised to 77/132.

To help you develop skill in solving fraction problems, we must cover one more concept. This is the concept of the least common denominator.

Definition: The *least common denominator* is the smallest whole number that the denominators of two or more fractions divide into evenly.

The main purpose in studying *least common denominators* here is that it provides another easy way to add unlike fractions. To determine the least common denominator, the prime-number method is used.

Definition: A *prime number* is a whole number greater than 1 that can be divided only by itself and 1.

The numbers 2, 3, 5, 7, 11, 13, 17, and 19 are examples of prime numbers. All other whole numbers greater than 1 that are not prime numbers are called *composite numbers*. A *composite number* can be divided evenly by whole numbers other than just itself and 1. The numbers 4, 6, 8, 9, 10, 12, 14, and 15 are examples of composite numbers.

Rule: The Prime-Number Method for Finding the Least Common Denominator

Step 1: Write down the denominators in a line across the page.

Step 2: Write the first prime number, 2, to the left of the row of denominators, then divide it into as many of the denominators as possible. Write each quotient under the corresponding denominator. If the prime number does not divide evenly into one of the denominators, write that denominator down again.

Step 3: Continue dividing, using the same or next prime number and writing it down to the left, until the bottom row of quotients contains all 1s.

Step 4: Multiply all the prime numbers listed on the left side of the rows of quotients to arrive at the least common denominator.

Example 8: Find the least common denominator of the following fractions, using the prime-number method:

$$\frac{7}{16}, \frac{5}{22}, \frac{13}{30}$$

Solution: First of all, write down the three denominators (Step 1).

16 22 30

Then divide as many of these numbers as possible by the first prime number, which is 2. Be sure to write the quotients under the correct number and write the prime numbers to the left (Step 2).

$$\begin{array}{r|ccc} & 16 & 22 & 30 \\ \hline 2 & 8 & 11 & 15 \end{array}$$

Notice that some of the quotients can still be divided by the prime number 2. So divide through by 2 again. Of course, you can't divide 11 or 15 by 2 evenly, so just bring the 11 and the 15 down.

$$\begin{array}{r|ccc} & 16 & 22 & 30 \\ \hline 2 & 8 & 11 & 15 \\ 2 & 4 & 11 & 15 \end{array}$$

Divide by 2 again, since 4 can be divided evenly by 2.

$$\begin{array}{r|ccc} & 16 & 22 & 30 \\ \hline 2 & 8 & 11 & 15 \\ 2 & 4 & 11 & 15 \\ 2 & 2 & 11 & 15 \end{array}$$

Divide by 2 just one more time (honest), because 2 can be divided evenly by 2.

$$\begin{array}{r|ccc} & 16 & 22 & 30 \\ \hline 2 & 8 & 11 & 15 \\ 2 & 4 & 11 & 15 \\ 2 & 2 & 11 & 15 \\ 2 & 1 & 11 & 15 \end{array}$$

Now 2 can no longer be divided evenly into any of the quotients in the bottom row, so try the next prime number, 3. The number 15 can be divided by 3.

$$\begin{array}{r|ccc} & 16 & 22 & 30 \\ \hline 2 & 8 & 11 & 15 \\ 2 & 4 & 11 & 15 \\ 2 & 2 & 11 & 15 \\ 2 & 1 & 11 & 15 \\ 3 & 1 & 11 & 5 \end{array}$$

The next prime number to try is 5.

```
     16  22  30
  2│  8  11  15
  2│  4  11  15
  2│  2  11  15
  2│  1  11  15
  3│  1  11   5
  5│  1  11   1
```

Note that the prime number 7 can't be used, because it doesn't divide evenly into 11. The next prime number to try is 11.

```
      16  22  30
   2│  8  11  15
   2│  4  11  15
   2│  2  11  15
   2│  1  11  15
   3│  1  11   5
   5│  1  11   1
  11│  1   1   1
```

The bottom row now contains all 1s (Step 3). The least common denominator can be found by multiplying all the numbers on the left side (Step 4).

$$2 \times 2 \times 2 \times 2 \times 3 \times 5 \times 11 = 2640$$

The least common denominator is 2640.

Example 9: Find the least common denominator of the fractions $4/21$, $2/9$, $11/24$; then add the fractions.

Solution:

Step 1: 21 9 24

Step 2:
```
     21  9  24
  2│  21  9  12
  2│  21  9   6
  2│  21  9   3
```

Step 3:
```
     21  9  24
  2│  21  9  12
  2│  21  9   6
  2│  21  9   3
  3│   7  3   1
  3│   7  1   1
  7│   1  1   1
```

Step 4: The least common denominator is $2 \times 2 \times 2 \times 3 \times 3 \times 7 = 504$.

The least common denominator has been found. Now it's time to add the fractions. As you know, to add unlike fractions, you must first change them to like fractions. We already know what the denominator will be in this case, because the least common denominator, 504, has already been determined. So, to add the fractions $4/21 + 2/9 + 11/24$, change these unlike fractions into like fractions by raising them to higher terms, using the least common denominator.

$$\frac{4}{21} = \frac{4 \times 24}{21 \times 24} = \frac{96}{504}$$

and

$$\frac{2}{9} = \frac{2 \times 56}{9 \times 56} = \frac{112}{504}$$

and, finally,

$$\frac{11}{24} = \frac{11 \times 21}{24 \times 21} = \frac{231}{504}$$

To add these like fractions,

$$\frac{4}{21} + \frac{2}{9} + \frac{11}{24} = \frac{96}{504} + \frac{112}{504} + \frac{231}{504} = \frac{96 + 112 + 231}{504} = \frac{439}{504}$$

The answer is $439/504$. The unlike fractions were changed to like fractions using the least common denominator.

Example 10: Find the least common denominator of the fractions $5/12$, $7/18$, $4/7$; then add the fractions.

Solution:

Step 1: 12 18 7

Step 2:
$$\begin{array}{r|ccc} & 12 & 18 & 7 \\ \hline 2 & 6 & 9 & 7 \\ 2 & 3 & 9 & 7 \end{array}$$

Step 3:
$$\begin{array}{r|ccc} & 12 & 18 & 7 \\ \hline 2 & 6 & 9 & 7 \\ 2 & 3 & 9 & 7 \\ 3 & 1 & 3 & 7 \\ 3 & 1 & 1 & 7 \\ 7 & 1 & 1 & 1 \end{array}$$

Step 4: The least common denominator is $2 \times 2 \times 3 \times 3 \times 7 = 252$.

Using the least common denominator, 252, raise the fractions $5/12$, $7/18$, and $4/7$ to higher terms.

$$\frac{5}{12} = \frac{5 \times 21}{12 \times 21} = \frac{105}{252}$$

and

$$\frac{7}{18} = \frac{7 \times 14}{18 \times 14} = \frac{98}{252}$$

and, finally,

$$\frac{4}{7} = \frac{4 \times 36}{7 \times 36} = \frac{144}{252}$$

To add these like fractions,

$$\frac{5}{12} + \frac{7}{18} + \frac{4}{7} = \frac{105}{252} + \frac{98}{252} + \frac{144}{252} = \frac{105 + 98 + 144}{252} = \frac{347}{252} = 1\frac{95}{252}$$

The answer is $1^{95}/_{252}$, a mixed number. The unlike fractions were changed to like fractions using the least common denominator.

Section 2.2 Readiness Review

Add the following fractions (reduce answers to lowest terms and express as mixed numbers) (see Examples 1, 2, & 3):

1. $\frac{7}{8} + \frac{3}{8} + \frac{5}{8} =$ _____
2. $\frac{3}{20} + \frac{23}{20} + \frac{19}{20} =$ _____
3. $\frac{49}{65} + \frac{26}{65} =$ _____
4. $\frac{61}{98} + \frac{43}{98} =$ _____

Raise the following fractions to higher terms, using the denominator or numerator given (see Examples 4, 5, 6, & 7):

5. $\frac{3}{4} = \frac{36}{d} =$ _____
6. $\frac{4}{25} = \frac{n}{125} =$ _____
7. $\frac{29}{32} = \frac{145}{d} =$ _____
8. $\frac{21}{144} = \frac{n}{432} =$ _____

Find the least common denominator of the following fractions (see Example 8):

9. $\frac{1}{3}, \frac{1}{4}, \frac{1}{5} =$ _____
10. $\frac{8}{9}, \frac{9}{10}, \frac{10}{11} =$ _____
11. $\frac{1}{2}, \frac{5}{7}, \frac{5}{6} =$ _____
12. $\frac{5}{12}, \frac{2}{5}, \frac{1}{3} =$ _____

Add the following fractions (see Examples 9 & 10):

13. $\frac{2}{3} + \frac{1}{4} + \frac{5}{6} =$ _____
14. $\frac{27}{52} + \frac{7}{13} + \frac{14}{26} =$ _____
15. $\frac{1}{12} + \frac{1}{2} + \frac{1}{18} =$ _____
16. $\frac{9}{10} + \frac{99}{100} + \frac{6}{7} =$ _____

Section readiness review answers (not given in order):

14. $1\frac{31}{52}$
15. $\frac{23}{36}$
8. $\frac{63}{432}$
3. $1\frac{2}{13}$

1. $1\frac{7}{8}$
2. $2\frac{1}{4}$
4. $1\frac{3}{49}$
9. 60

6. $\frac{20}{125}$
13. $1\frac{3}{4}$
16. $2\frac{523}{700}$
5. $\frac{36}{48}$

10. 990   **12.** 60   **7.** $\frac{145}{160}$

11. 42

Section 2.2 Review Problems

Add the following fractions (reduce answers to lowest terms and express as a mixed number or a proper fraction) (answers to the odd-numbered problems are at the back of the book):

1. $\frac{5}{27} + \frac{16}{27}$
2. $\frac{24}{25} + \frac{11}{25}$
3. $\frac{7}{32} + \frac{7}{32}$
4. $\frac{73}{81} + \frac{5}{81}$
5. $\frac{3}{4} + \frac{1}{4} + \frac{5}{4}$
6. $\frac{2}{6} + \frac{3}{6} + \frac{4}{6}$
7. $\frac{10}{12} + \frac{2}{12} + \frac{9}{12}$
8. $\frac{3}{16} + \frac{15}{16} + \frac{4}{16}$

Raise the following fractions to higher terms, using the denominator or numerator given:

9. $\frac{12}{17} = \frac{n}{34}$
10. $\frac{13}{20} = \frac{n}{80}$
11. $\frac{8}{35} = \frac{56}{n}$
12. $\frac{12}{13} = \frac{144}{d}$
13. $\frac{9}{10} = \frac{n}{1000}$

Find the least common denominator of the following fractions:

14. $\frac{3}{7}, \frac{2}{5}, \frac{1}{8}$
15. $\frac{4}{5}, \frac{5}{6}, \frac{6}{7}$
16. $\frac{11}{12}, \frac{12}{13}, \frac{13}{14}$
17. $\frac{1}{3}, \frac{7}{8}, \frac{7}{11}$
18. $\frac{9}{10}, \frac{14}{15}, \frac{3}{17}$

Add the following fractions:

19. $\frac{9}{25} + \frac{5}{12} + \frac{9}{14}$
20. $\frac{13}{27} + \frac{7}{13} + \frac{7}{30}$
21. $\frac{3}{5} + \frac{2}{7} + \frac{1}{8}$
22. $\frac{4}{7} + \frac{5}{6} + \frac{2}{3}$
23. $\frac{11}{12} + \frac{12}{13} + \frac{2}{5}$
24. $\frac{1}{4} + \frac{3}{8} + \frac{5}{12}$
25. $\frac{9}{10} + \frac{7}{10} + \frac{15}{22}$
26. $\frac{1}{6} + \frac{3}{5} + \frac{1}{8}$
27. $\frac{2}{3} + \frac{5}{6} + \frac{9}{10}$
28. $\frac{7}{10} + \frac{3}{5} + \frac{1}{3}$
29. $\frac{3}{4} + \frac{5}{12} + \frac{7}{8}$
30. $\frac{5}{6} + \frac{5}{12} + \frac{3}{8}$
31. $\frac{2}{3} + \frac{13}{14} + \frac{1}{2}$
32. $\frac{5}{9} + \frac{1}{12} + \frac{9}{10}$
33. $\frac{5}{12} + \frac{3}{8} + \frac{9}{32}$
34. $\frac{1}{5} + \frac{3}{4} + \frac{2}{11}$
35. $\frac{3}{4} + \frac{1}{7} + \frac{1}{2}$
36. $\frac{1}{12} + \frac{3}{16} + \frac{2}{5}$
37. $\frac{5}{6} + \frac{3}{8} + \frac{1}{24}$
38. $\frac{3}{7} + \frac{5}{16} + \frac{3}{8}$
39. $\frac{1}{14} + \frac{1}{8} + \frac{1}{7}$

40. $\frac{1}{4} + \frac{1}{3} + \frac{1}{5}$ 41. $\frac{21}{72} + \frac{17}{56} + \frac{16}{63}$ 42. $\frac{3}{12} + \frac{4}{16} + \frac{6}{32}$

43. $\frac{4}{21} + \frac{1}{3} + \frac{3}{14}$ 44. $\frac{5}{6} + \frac{3}{5} + \frac{3}{4}$ 45. $\frac{10}{21} + \frac{7}{10} + \frac{1}{8}$

46. $\frac{13}{14} + \frac{35}{36} + \frac{3}{4}$

Solve the following word problems:

47. Dr. Dugan has 3 patients in the hospital. On one day he spent ½ of an hour with one patient, ⅚ of an hour with another, and ⅘ of an hour with the last one. How many total hours did Dr. Dugan spend with his patients in the hospital that day?

48. Ms. Barnhart is to receive a reduced amount of medication because the physician knows that she is sensitive to drugs. The physician orders the following medication: ½ white tablet, ¼ green tablet, and ½ red tablet daily. What is the total number of tablets that Ms. Barnhart is to take daily? (Combine white, green, and red tablets.)

49. A patient arrives at the pharmacy with 3 partially filled jars of skin cream. The pharmacist estimates the first jar to be ½ full, the second to be ⅓ full, and the last to be ⅛ full. The pharmacist wants to put the contents of the partially filled jars into one jar of similar size. What is the sum of the contents of the partially filled jars? (Express your answer in terms of jars.)

50. The irrigating solution for the total hip replacement surgery is only ⅙ of a bottle. The surgeon orders more, and 2 other partially filled bottles are sent. The first of these contains ⅗ of a bottle, and the next, ⅔ of a bottle. If the surgeon used all 3 partially filled bottles, how much irrigating solution did he use? (Give your answer in terms of bottles.)

Section 2.3 Subtraction of Fractions

The method used to subtract fractions is similar to that used in adding fractions.

> **Rule: Subtracting Fractions**
> *Step 1:* If the fractions do not have the same denominator, find the least common denominator and convert them to like fractions. (If the fractions already have the same denominator, go to Step 2.)
> *Step 2:* Subtract the numerators of the like fractions.
> *Step 3:* Place this difference over the common denominator. Reduce to lowest terms.
> *Step 4:* Change any improper fraction found in Step 3 to a mixed number.

Example 1: Subtract the following fractions:

$\frac{5}{8} - \frac{1}{8}$

38 Chapter 2 Fractions

Solution:
$$\frac{5}{8} - \frac{1}{8} = \frac{5-1}{8} = \frac{4}{8} = \frac{1}{2}$$
$$\uparrow\uparrow\uparrow$$
(Step 2) (Step 3) (Step 3)

Example 2: Subtract the following fractions:

$$\frac{11}{12} - \frac{3}{12}$$

Solution:
$$\frac{11}{12} - \frac{3}{12} = \frac{11-3}{12} = \frac{8}{12} = \frac{2}{3}$$
$$\uparrow\uparrow\uparrow$$
(Step 2) (Step 3) (Step 3)

Example 3: Subtract the following fractions:

$$\frac{8}{9} - \frac{1}{3}$$

Solution: The least common denominator of $8/9$ and $1/3$ is 9: $1/3 = 3/9$ (Step 1).

$$\frac{8}{9} - \frac{1}{3} = \frac{8}{9} - \frac{3}{9} = \frac{8-3}{9} = \frac{5}{9}$$
$$\uparrow\uparrow\uparrow$$
(Step 1) (Step 2) (Step 3)

Example 4: Subtract the following fractions:

$$\frac{36}{7} - \frac{2}{3}$$

Solution: The least common denominator of $36/7$ and $2/3$ is 21; therefore,

$$\frac{36}{7} - \frac{2}{3} = \frac{108}{21} - \frac{14}{21} = \frac{108-14}{21} = \frac{94}{21} = 4\frac{10}{21}$$
$$\uparrow\uparrow\uparrow\uparrow$$
(Step 1)   (Step 2) (Step 3) (Step 4)

Section 2.3 Readiness Review

Subtract the following fractions:

1. $\dfrac{5}{8} - \dfrac{3}{8} =$ _____ (see Examples 1 & 2)

2. $\dfrac{7}{13} - \dfrac{4}{13} =$ _____ (see Examples 1 & 2)

3. $\dfrac{19}{24} - \dfrac{3}{16} =$ _____ (see Example 3)

4. $\dfrac{10}{11} - \dfrac{2}{3} =$ _____ (see Example 3)

5. $\dfrac{12}{2} - \dfrac{4}{5} =$ _____ (see Example 4)

6. $\dfrac{100}{10} - \dfrac{14}{6} =$ _____ (see Example 4)

Section readiness review answers (not given in order):

1. $\dfrac{1}{4}$ 4. $\dfrac{8}{33}$ 3. $\dfrac{29}{48}$

6. $7\dfrac{2}{3}$ 2. $\dfrac{3}{13}$ 5. $5\dfrac{1}{5}$

Section 2.3 Review Problems

Subtract the following fractions (answers to the odd-numbered problems are at the back of the book):

1. $\dfrac{3}{4} - \dfrac{1}{4}$ 2. $\dfrac{9}{10} - \dfrac{3}{10}$ 3. $\dfrac{11}{12} - \dfrac{5}{12}$

4. $\dfrac{6}{8} - \dfrac{3}{8}$ 5. $\dfrac{20}{21} - \dfrac{9}{21}$ 6. $\dfrac{14}{15} - \dfrac{5}{15}$

7. $\dfrac{5}{8} - \dfrac{1}{2}$ 8. $\dfrac{5}{8} - \dfrac{5}{11}$ 9. $\dfrac{13}{16} - \dfrac{1}{4}$

10. $\dfrac{26}{30} - \dfrac{4}{10}$ 11. $\dfrac{15}{18} - \dfrac{5}{24}$ 12. $\dfrac{55}{60} - \dfrac{1}{2}$

13. $\dfrac{22}{8} - \dfrac{1}{4}$ 14. $\dfrac{29}{10} - \dfrac{2}{3}$ 15. $\dfrac{14}{2} - \dfrac{5}{6}$

16. $\dfrac{100}{25} - \dfrac{3}{4}$ 17. $\dfrac{79}{63} - \dfrac{7}{9}$ 18. $\dfrac{40}{12} - \dfrac{5}{6}$

19. $\dfrac{20}{13} - \dfrac{4}{5}$ 20. $\dfrac{21}{7} - \dfrac{2}{3}$

Section 2.4 Equality and Comparison of Fractions

> **Definition:** Fractions are said to be equal if (and only if) their cross products are equal; for example, the fractions a/b and c/d are equal if the cross product ad ($a \times d$) equals the cross product bc ($b \times c$).
> $$\dfrac{a}{b} = \dfrac{c}{d}$$

> **Rule: Determining Equal Fractions Using Cross Products**
> *Step 1:* Multiply the numerator on the left by the denominator on the right.
> *Step 2:* Multiply the denominator on the left by the numerator on the right.
> *Step 3:* If both products are the same, the fractions are said to be equal.

Chapter 2 Fractions

Example 1: Decide whether the following fractions are equal or not. If they are, write Yes; if not, write No.

    a. Is $\dfrac{1}{2} = \dfrac{3}{6}$?   **b.** Is $\dfrac{2}{3} = \dfrac{302}{453}$?

    c. Is $\dfrac{2}{5} = \dfrac{3}{8}$?   **d.** Is $\dfrac{3}{4} = \dfrac{56}{75}$?

Solution: **a.** Is $\dfrac{1}{2} = \dfrac{3}{6}$?

Step 1: Multiply the numerator on the left by the denominator on the right.

$$\dfrac{1}{2} = \dfrac{3}{6}$$
$$1 \times 6 = 6$$

Step 2: Multiply the denominator on the left by the numerator on the right.

$$\dfrac{1}{2} = \dfrac{3}{6}$$
$$2 \times 3 = 6$$

Step 3: If the cross products are the same, the fractions are said to be equal.

     Product (Step 1)   Product (Step 2)
          6   =   6

The fractions are equal. Indicate this with a Yes.

b. Is $\dfrac{2}{3} = \dfrac{302}{453}$?

Step 1:
$$\dfrac{2}{3} = \dfrac{302}{453}$$
$$2 \times 453 = 906$$

Step 2:
$$\dfrac{2}{3} = \dfrac{302}{453}$$
$$3 \times 302 = 906$$

Step 3: $906 = 906$; Yes.

c. Is $\dfrac{2}{5} = \dfrac{3}{8}$?

Step 1:
$$\dfrac{2}{5} = \dfrac{3}{8}$$
$$2 \times 8 = 16$$

Step 2:
$$\dfrac{2}{5} = \dfrac{3}{8}$$
$$5 \times 3 = 15$$

Step 3: The cross products are not the same. 16 is not equal to 15; therefore, $\dfrac{2}{5}$ is not equal to $\dfrac{3}{8}$. No.

d. Is $\dfrac{3}{4} = \dfrac{56}{75}$?

Section 2.4 Equality and Comparison of Fractions   41

Step 1:
$$\frac{3}{4} = \frac{56}{75}$$
$$3 \times 75 = 225$$

Step 2:
$$\frac{3}{4} = \frac{56}{75}$$
$$4 \times 56 = 224$$

Step 3: The cross products are not the same. 225 is not equal to 224; therefore, 3/4 is not equal to 56/75.

To determine if one fraction is larger than another, their sizes must be compared.

> **Rule: Comparison of Fractions**
> *Step 1:* If the fractions are like fractions, go to Step 3. If the fractions are unlike fractions, find their least common denominator. (If the numbers are simple, you may be able to find the least common denominator in your head.)
> *Step 2:* Convert the fractions being compared to like fractions, using the least common denominator from Step 1.
> *Step 3:* The fraction having the greater numerator is the greater in value (larger).

Example 2: Arrange the following fractions in order of size, least to greatest:

$$\frac{1}{2}, \frac{3}{5}, \frac{5}{6}$$

Solution:

Step 1: Find the least common denominator. (These numbers are rather simple. Can you determine the least common denominator mentally? If you said 30, you're right. Let's see why.)

$$
\begin{array}{c|ccc}
 & 2 & 5 & 6 \\
\hline
2 & 1 & 5 & 3 \\
3 & 1 & 5 & 1 \\
5 & 1 & 1 & 1 \\
\end{array}
$$

$2 \times 3 \times 5 = 30$ is the least common denominator. (You might want to review Section 2.2, Example 8, on how to find least common denominators.)

Step 2: The fractions 1/2, 3/5, and 5/6 are converted to like fractions using the least common denominator from Step 1, 30.

$$\frac{1}{2} = \frac{15}{30}$$

$$\frac{3}{5} = \frac{18}{30}$$

$$\frac{5}{6} = \frac{25}{30}$$

Step 3: The like fractions, ranked from the least to the greatest numerator, are

$$\frac{15}{30} \quad \frac{18}{30} \quad \frac{25}{30}$$
$\underrightarrow{\text{least to greatest}}$

The original fractions in order of size, therefore, are

$$\frac{1}{2} \quad \frac{3}{5} \quad \frac{5}{6}$$
$\underrightarrow{\text{least to greatest}}$

Example 3: Arrange the following fractions in order of size, greatest to least:

$$\frac{7}{10}, \frac{5}{8}, \frac{4}{5}$$

Solution:

Step 1: Again, first try finding the least common denominator in your head.

$$\begin{array}{r|ccc} & 10 & 8 & 5 \\ \hline 2 & 5 & 4 & 5 \\ 2 & 5 & 2 & 5 \\ 2 & 5 & 1 & 5 \\ 5 & 1 & 1 & 1 \end{array}$$

$2 \times 2 \times 2 \times 5 = 40$

The least common denominator is 40.

Step 2:
$$\frac{7}{10} = \frac{28}{40}$$

$$\frac{5}{8} = \frac{25}{40}$$

$$\frac{4}{5} = \frac{32}{40}$$

Step 3: The like fractions, ordered greatest to least, are

$$\frac{32}{40} \quad \frac{28}{40} \quad \frac{25}{40}$$
$\underrightarrow{\text{greatest to least}}$

The original fractions, ordered largest to smallest, are

$$\frac{4}{5} \quad \frac{7}{10} \quad \frac{5}{8}$$
$\underrightarrow{\text{greatest to least}}$

Section 2.4 Readiness Review

Decide whether the following fractions are equal. If they are, write Yes; if not, write No (see Example 1):

1. Is $\frac{2}{3} = \frac{5}{8}$? _____
2. Is $\frac{3}{4} = \frac{15}{16}$? _____
3. Is $\frac{1}{2} = \frac{56}{112}$? _____
4. Is $\frac{4}{5} = \frac{17}{20}$? _____

Arrange the following fractions in order of size, least to greatest (see Example 2):

5. $\frac{3}{8}, \frac{4}{5}, \frac{1}{2}$ _____
6. $\frac{5}{6}, \frac{1}{3}, \frac{1}{4}$ _____

Arrange the following fractions in order of size, greatest to least (see Example 3):

7. $\frac{7}{12}, \frac{2}{3}, \frac{3}{4}$ _____
8. $\frac{2}{15}, \frac{5}{6}, \frac{1}{10}$ _____

Section readiness review answers (not given in order):

1. no
8. $\frac{5}{6}, \frac{2}{15}, \frac{1}{10}$
7. $\frac{3}{4}, \frac{2}{3}, \frac{7}{12}$

3. yes
5. $\frac{3}{8}, \frac{1}{2}, \frac{4}{5}$
4. no

2. no
6. $\frac{1}{4}, \frac{1}{3}, \frac{5}{6}$

Section 2.4 Review Problems

Decide whether the following fractions are equal. If they are equal, write Yes; if not, write No (answers to the odd-numbered problems are at the back of the book):

1. Is $\frac{1}{4} = \frac{2}{8}$?
2. Is $\frac{3}{5} - \frac{8}{15}$?
3. Is $\frac{4}{9} = \frac{9}{20}$?
4. Is $\frac{4}{7} = \frac{16}{28}$?
5. Is $\frac{1}{2} = \frac{13}{26}$?
6. Is $\frac{6}{13} = \frac{13}{26}$?
7. Is $\frac{3}{5} = \frac{9}{15}$?
8. Is $\frac{8}{18} = \frac{12}{28}$?
9. Is $\frac{22}{3} = \frac{44}{6}$?
10. Is $\frac{15}{18} = \frac{32}{42}$?
11. Is $\frac{22}{7} = \frac{355}{113}$?
12. Is $\frac{37}{111} = \frac{1}{3}$?

Arrange the following fractions in order of size, least to greatest:

13. $\frac{1}{2}, \frac{1}{4}, \frac{1}{3}$
14. $\frac{2}{3}, \frac{3}{5}, \frac{9}{10}$
15. $\frac{1}{2}, \frac{11}{12}, \frac{5}{6}$

16. $\dfrac{14}{20}, \dfrac{4}{5}, \dfrac{3}{4}$ 17. $\dfrac{7}{12}, \dfrac{1}{3}, \dfrac{1}{4}$ 18. $\dfrac{7}{8}, \dfrac{3}{4}, \dfrac{15}{16}$

19. $\dfrac{2}{3}, \dfrac{1}{2}, \dfrac{5}{12}$ 20. $\dfrac{2}{3}, \dfrac{1}{5}, \dfrac{8}{15}$

Arrange the following fractions in order of size, greatest to least:

21. $\dfrac{3}{4}, \dfrac{5}{6}, \dfrac{11}{12}$ 22. $\dfrac{3}{5}, \dfrac{1}{2}, \dfrac{3}{10}$ 23. $\dfrac{1}{2}, \dfrac{3}{10}, \dfrac{3}{4}$

24. $\dfrac{5}{6}, \dfrac{3}{8}, \dfrac{1}{2}$ 25. $\dfrac{4}{5}, \dfrac{3}{10}, \dfrac{3}{4}$

Section 2.5 Addition and Subtraction of Mixed Numbers

As you know, a mixed number consists of a whole-number part and a fractional part.

> **Rule: Addition of Mixed Numbers**
> *Step 1:* Add the whole-number portions.
> *Step 2:* Make sure that all the fractional parts have the same denominator and then add them. If the sum of the fractional portions is an improper fraction, convert it to a mixed number and add its whole-number portion to the other whole number.
> *Step 3:* Reduce the remaining fractional portion to lowest terms.

Example 1: Add the following mixed numbers:

$$14\dfrac{1}{6} + 2\dfrac{1}{6}$$

Solution:

$$\begin{array}{r} 14\dfrac{1}{6} \\ + \; 2\dfrac{1}{6} \\ \hline 16\dfrac{2}{6} \end{array}$$

Sum of whole numbers (Step 1)   Sum of fractions (Step 2)

The fractional part of the sum should be reduced to lowest terms, $2/6 = 1/3$. The final answer is

$$14\dfrac{1}{6} + 2\dfrac{1}{6} = 16\dfrac{2}{6} = 16\dfrac{1}{3}$$

Example 2: Add the following mixed numbers:

$$23\dfrac{5}{8} + 19\dfrac{7}{8}$$

Solution:

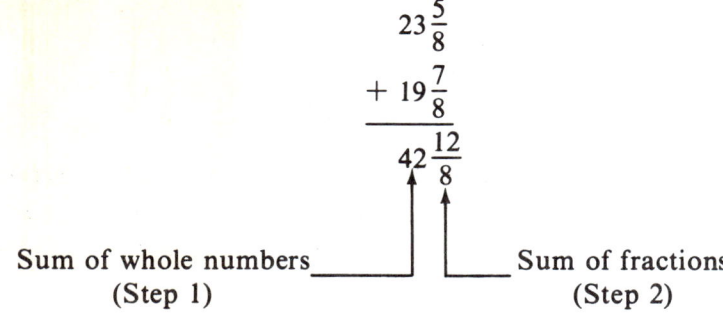

Sum of whole numbers (Step 1)   Sum of fractions (Step 2)

Change the improper fraction $^{12}/_8$ to the mixed number $1^4/_8$ (Step 2). Reduced to lowest terms, $1^4/_8$ becomes $1\frac{1}{2}$. The mixed number $1\frac{1}{2}$ is added to the whole-number portion of the sum:

$$42 + 1\frac{1}{2} = 43\frac{1}{2} \text{ (Step 2)}$$

Example 3: Add the following mixed numbers:

$$14\frac{3}{8} + 6\frac{1}{3} + 5\frac{5}{6}$$

Solution:

$$\begin{aligned} 14\frac{3}{8} &= 14\frac{9}{24} \quad \text{(Step 2)} \\ 6\frac{1}{3} &= 6\frac{8}{24} \\ +\ 5\frac{5}{6} &= 5\frac{20}{24} \\ \hline &\ 25\frac{37}{24} \end{aligned}$$

Sum of whole numbers (Step 1)   Sum of fractions (Step 2)

Change the improper fraction $^{37}/_{24}$ to the mixed number $1^{13}/_{24}$ (Step 2). Then add the mixed number $1^{13}/_{24}$ to the whole-number portion:

$$25 + 1\frac{13}{24} = 26\frac{13}{24} \text{ (Step 2)}$$

Rule: Subtraction of Mixed Numbers
Step 1: Make sure that all fractional parts have the same denominator.
Step 2: Subtract the fractional parts, borrowing from the whole-number part when necessary.
Step 3: Subtract the whole-number parts.

Example 4: Subtract the following mixed numbers:

$$20\frac{5}{6} - 5\frac{4}{6}$$

46   Chapter 2   Fractions

Solution:

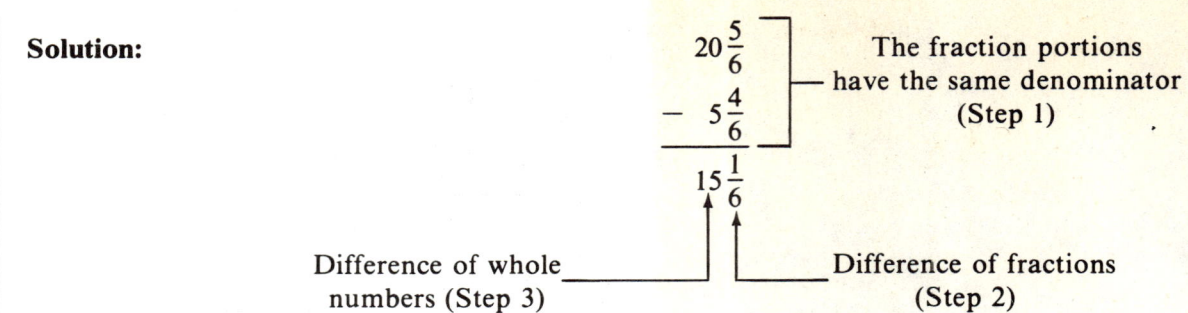

The answer is 15 1/6.

Example 5:   Subtract the following mixed numbers:

$$14\frac{8}{9} - 8\frac{3}{4}$$

Solution:

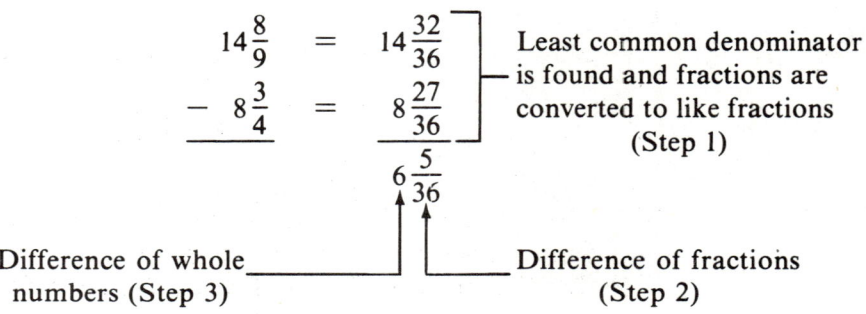

The answer is 6 5/36.

Example 6:   Subtract the following mixed numbers:

$$12\frac{1}{4} - 5\frac{3}{5}$$

Solution:

$$\begin{array}{rcl} 12\frac{1}{4} & = & 12\frac{5}{20} \\ -\ 5\frac{3}{5} & = & 5\frac{12}{20} \end{array}\right] \text{Least common denominator is found (Step 1)}$$

It is not possible to subtract 12/20 from 5/20. First, you must make 5/20 larger than 12/20. To do this, borrow from the whole-number portion, 12. Borrow 1 from the whole number and express it as a fraction having the least common denominator of all the fractional parts in the problem. Then combine the borrowed 1, in fractional form, with the original fraction, 5/20

$$12\frac{5}{20} = 11 + 1 + \frac{5}{20} = 11 + \frac{20}{20} + \frac{5}{20} = 11\frac{25}{20} \quad \text{(Step 2)}$$

Section 2.5 Addition and Subtraction of Mixed Numbers   47

Finish solving the problem.

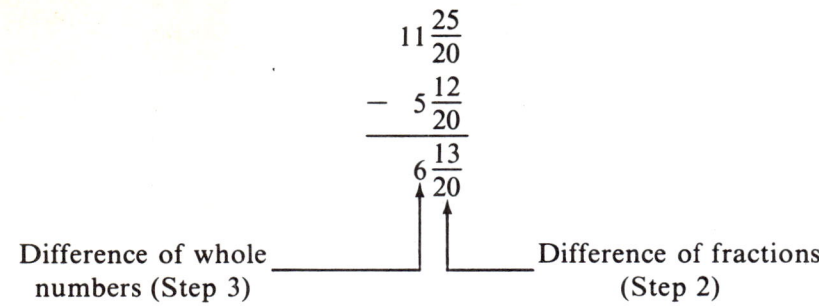

$$11\frac{25}{20}$$
$$-5\frac{12}{20}$$
$$6\frac{13}{20}$$

Difference of whole numbers (Step 3)   Difference of fractions (Step 2)

The answer is $6\frac{13}{20}$.

Example 7: Subtract the following mixed numbers:

$$74\frac{1}{10} - 18\frac{3}{5}$$

Solution:

$$74\frac{1}{10} = 74\frac{1}{10} \quad \text{(Step 1)}$$
$$-18\frac{3}{5} = 18\frac{6}{10}$$

In order to make the fractional part of the minuend large enough to subtract from, borrow 1 from the whole number and express it as a fraction having the least common denominator as its denominator (Step 2).

$$74\frac{1}{10} = 73 + 1 + \frac{1}{10} = 73 + \frac{10}{10} + \frac{1}{10} = 73\frac{11}{10}$$

Complete the problem.

$$73\frac{11}{10}$$
$$-18\frac{6}{10}$$
$$55\frac{5}{10}$$

Difference of whole numbers (Step 3)   Difference of fractions (Step 2)

The answer $55\frac{5}{10}$ reduced to lowest terms is $55\frac{1}{2}$.

Example 8: Subtract the following:

$$65 - 25\frac{3}{7}$$

48   Chapter 2 Fractions

Solution:

$$65$$
$$-25\frac{3}{7}$$

Change the whole number 65 to a mixed number with the fraction expressed as 7/7.

$$65 = 64 + 1 = 64\frac{7}{7} \text{ (Step 2)}$$

$$64\frac{7}{7}$$
$$-25\frac{3}{7}$$
$$39\frac{4}{7}$$

Difference of whole numbers (Step 3) ──── Difference of fractions (Step 2)

The answer is 39 4/7.

Section 2.5 Readiness Review

Add the following mixed numbers:

1. $3\frac{1}{3} + 4\frac{1}{3} =$ _____ (see Example 1)

2. $8\frac{7}{8} + 12\frac{5}{8} =$ _____ (see Example 2)

3. $23\frac{1}{32} + 17\frac{9}{10} =$ _____ (see Example 3)

4. $5\frac{3}{4} + 4\frac{1}{12} + 4\frac{5}{6} =$ _____ (see Example 3)

Subtract the following:

5. $44\frac{3}{4} - 20\frac{1}{4} =$ _____ (see Example 4)

6. $25\frac{4}{5} - 15\frac{2}{3} =$ _____ (see Example 5)

7. $21\frac{1}{2} - 7\frac{7}{8} =$ _____ (see Examples 6 & 7)

8. $15 - 14\frac{4}{7} =$ _____ (see Example 8)

Section readiness review answers (not given in order):

7. $13\frac{5}{8}$ 3. $40\frac{149}{160}$ 2. $21\frac{1}{2}$

1. $7\frac{2}{3}$ 5. $24\frac{1}{2}$ 4. $14\frac{2}{3}$

6. $10\frac{2}{15}$ 8. $\frac{3}{7}$

Section 2.5 Review Problems

Add the following mixed numbers (answers to the odd-numbered problems are at the back of the book):

1. $5\frac{1}{4} + 2\frac{1}{4}$
2. $7\frac{1}{3} + 8\frac{1}{3}$
3. $23\frac{5}{9} + 13\frac{1}{9}$
4. $14\frac{1}{10} + 18\frac{1}{10}$
5. $7\frac{2}{9} + 3\frac{1}{9}$
6. $15\frac{4}{13} + 12\frac{7}{13}$
7. $9\frac{5}{6} + 5\frac{1}{7}$
8. $8\frac{1}{8} + 3\frac{1}{4}$
9. $7\frac{5}{6} + 8\frac{7}{12}$
10. $15\frac{1}{10} + 20\frac{2}{3}$
11. $11\frac{2}{11} + 7\frac{1}{3}$
12. $12\frac{7}{9} + 10\frac{3}{4}$
13. $37\frac{5}{8} + 28\frac{1}{4} + 46\frac{1}{2}$
14. $481\frac{3}{5} + 10\frac{9}{10} + 41\frac{11}{20}$

Subtract the following:

15. $53\frac{9}{10} - 15\frac{3}{10}$
16. $19\frac{2}{3} - 7\frac{1}{3}$
17. $42\frac{13}{15} - 20\frac{8}{15}$
18. $30\frac{6}{7} - 8\frac{1}{49}$
19. $63\frac{1}{12} - 19\frac{5}{6}$
20. $100\frac{1}{10} - 10\frac{3}{10}$
21. $23 - 17\frac{3}{4}$
22. $62 - 51\frac{7}{8}$

Solve the following word problems:

23. It's hayfever season again, and Dr. Roy Gerard has prescribed a liquid antihistamine and decongestant combination for Patty Thornton. Her bottle of medicine contains 36 teaspoonsful. She takes 4½ teaspoonsful the first day, 4½ teaspoonsful the second day, and 3½ teaspoonsful the third day. How many teaspoonsful are left in the bottle?

24. An elderly patient is to receive a liquid vitamin supplement. If the patient gets a total of 7½ teaspoonsful one week and 8½ teaspoonsful the next, how many total teaspoonsful were given over the two weeks?

25. Mr. Richardson, a heart patient, has been put on an exercise program by his physician. The first week he rode his bicycle 2⅔ miles, the next week 3⅓ miles, the third week 1½ miles, and the last week in that month, 4 miles. How many total miles did Mr. Richardson ride his bicycle during the month?

Section 2.6 Multiplication of Fractions

> **Rule: Multiplication of Fractions**
> *Step 1:* Multiply the numerators.
> *Step 2:* Multiply the denominators.
> *Step 3:* Reduce product to lowest terms. Express improper fractions as mixed numbers.

Example 1: Multiply the following fractions:

$$\frac{7}{8} \times \frac{2}{5}$$

Solution: $\frac{7}{8} \times \frac{2}{5} = \frac{7 \times 2}{8 \times 5} = \frac{14}{40}$ (Step 1) (Step 2)

The product $^{14}/_{40}$ reduced to lowest terms is $^{7}/_{20}$ (Step 3).

In some problems involving multiplication of fractions, cancellation can be used. Cancellation requires that a number be divided into the numerator of one fraction as well as the denominator of it or another fraction. Using cancellation to solve Example 1 would involve the following:

$$\frac{7}{\underset{4}{\cancel{8}}} \times \frac{\cancel{2}^{1}}{5} =$$

The number 2 is divided into the numerator of the fraction $^{2}/_{5}$ and the denominator of the fraction $^{7}/_{8}$. The final answer is $^{7}/_{20}$.

$$\frac{7}{\underset{4}{\cancel{8}}} \times \frac{\cancel{2}^{1}}{5} = \frac{7 \times 1}{4 \times 5} = \frac{7}{20}$$

Rule: Multiplying Mixed Numbers
Step 1: Change the mixed numbers to improper fractions.
Step 2: Multiply the numerators.
Step 3: Multiply the denominators.
Step 4: Reduce the product to lowest terms. Express improper fractions as mixed numbers.

Example 2: Multiply the following:

$$4\frac{2}{3} \times 9\frac{1}{2}$$

Solution: Change the mixed numbers to improper fractions (Step 1).

$$4\frac{2}{3} \times 9\frac{1}{2} = \frac{14}{3} \times \frac{19}{2} =$$

$$\frac{\cancel{14}^{7}}{3} \times \frac{19}{\cancel{2}_{1}} = \frac{7 \times 19}{3 \times 1} = \quad \text{(cancellation used)}$$

$$\frac{7 \times 19}{3 \times 1} \quad \begin{array}{l}\text{(Step 2)}\\ \text{(Step 3)}\end{array} = \frac{133}{3} = 44\frac{1}{3} \quad \text{(Step 4)}$$

Example 3: Multiply the following:

$$6 \times 8\frac{9}{10}$$

Section 2.6 Multiplication of Fractions

Solution: The whole number 6 can be rewritten as the fraction $6/1$.

$$6 \times 8\frac{9}{10} = \frac{6}{1} \times \frac{89}{10} = \quad \text{(Step 1)}$$

$$\frac{\overset{3}{\cancel{6}}}{1} \times \frac{89}{\underset{5}{\cancel{10}}} = \frac{3 \times 89}{1 \times 5} = \quad \text{(cancellation used)}$$

$$\frac{3 \times 89}{1 \times 5} \begin{array}{l}\text{(Step 2)}\\ \text{(Step 3)}\end{array} = \frac{267}{5} = 53\frac{2}{5} \quad \text{(Step 4)}$$

Section 2.6 Readiness Review

Solve the following problems using multiplication:

1. $\frac{1}{3} \times \frac{2}{5} =$ _____ (see Example 1)

2. $\frac{9}{10} \times \frac{2}{3} =$ _____ (see Example 1)

3. $5\frac{1}{6} \times 4\frac{1}{4} =$ _____ (see Example 2)

4. $2\frac{2}{3} \times 1\frac{7}{10} =$ _____ (see Example 2)

5. $4 \times 6\frac{3}{4} =$ _____ (see Example 3)

6. $5\frac{2}{7} \times 20 =$ _____ (see Example 3)

Section readiness review answers (not given in order):

4. $4\frac{8}{15}$ 5. 27   2. $\frac{3}{5}$

1. $\frac{2}{15}$ 3. $21\frac{23}{24}$ 6. $105\frac{5}{7}$

Section 2.6 Review Problems

Solve the following problems using multiplication (answers to the odd-numbered problems are at the back of the book):

1. $\frac{3}{5} \times \frac{6}{7}$ 2. $\frac{4}{5} \times \frac{1}{8}$ 3. $\frac{11}{12} \times \frac{3}{4}$

4. $\frac{2}{3} \times \frac{1}{6}$ 5. $\frac{10}{11} \times \frac{1}{2}$ 6. $\frac{3}{4} \times \frac{2}{3}$

7. $\frac{7}{10} \times \frac{13}{16}$ 8. $\frac{22}{44} \times \frac{20}{75}$ 9. $2 \times 5\frac{1}{2}$

52   Chapter 2 Fractions

10. $6 \times 3\frac{3}{4}$　　11. $4 \times 6\frac{2}{3}$　　12. $7\frac{2}{3} \times 4$

13. $8 \times 2\frac{3}{8}$　　14. $11 \times 3\frac{1}{7}$　　15. $15\frac{3}{5} \times 4$

16. $1\frac{1}{8} \times 1\frac{5}{6}$　　17. $3\frac{3}{4} \times 2\frac{1}{2}$　　18. $2\frac{2}{3} \times 2\frac{1}{4}$

19. $4\frac{4}{5} \times 3\frac{2}{3}$　　20. $1\frac{3}{7} \times 1\frac{1}{4}$　　21. $6\frac{1}{5} \times 3\frac{1}{4}$

22. $1\frac{7}{9} \times 2\frac{2}{3}$　　23. $10\frac{1}{2} \times 12\frac{15}{16}$　　24. $1\frac{2}{3} \times 2\frac{1}{2}$

25. $3\frac{1}{3} \times 1\frac{7}{15}$

Section 2.7 Division of Fractions

> **Rule: Dividing Fractions**
> *Step 1:* Invert (flip over) the divisor and change the division sign to multiplication.
> *Step 2:* Multiply the numerators.
> *Step 3:* Multiply the denominators.
> *Step 4:* When necessary, reduce the answer to lowest terms and change improper fractions into mixed numbers.

Example 1:   Divide the following fractions:

$$\frac{5}{9} \div \frac{15}{16}$$

Solution: $\frac{5}{9} \times \frac{16}{15} =$

　　　　Divisor is inverted (flipped over) and multiplication sign replaces division sign (Step 1)

$\frac{\cancel{5}^1}{9} \times \frac{16}{\cancel{15}_3} = \frac{1 \times 16}{9 \times 3}$ (Step 2) $= \frac{16}{27}$ (cancellation used)
　　　　　　　　　　　　　(Step 3)

Answer is already expressed in lowest terms (Step 4).

　　Dividing mixed numbers is similar to multiplying mixed numbers. First, change all mixed numbers to improper fractions, then proceed to work the problem using the rule for dividing fractions.

Example 2: Divide the following mixed numbers:

$$4\frac{3}{4} \div 2\frac{1}{8}$$

Solution:

$$4\frac{3}{4} \div 2\frac{1}{8} = \frac{19}{4} \div \frac{17}{8} =$$ The mixed numbers are changed into improper fractions

$$\frac{19}{\cancel{4}_1} \times \frac{\cancel{8}^2}{17} =$$ Divisor is inverted (Step 1) and cancellation is used

$$\frac{19 \times 2}{1 \times 17} \begin{array}{l}\text{(Step 2)}\\\text{(Step 3)}\end{array} = \frac{38}{17} = 2\frac{4}{17} \quad \text{(Step 4)}$$

Example 3: Divide the following:

$$12 \div 7\frac{1}{3}$$

Solution: The whole number 12 can be rewritten as the fraction $^{12}/_1$.

$$12 \div 7\frac{1}{3} = \frac{12}{1} \div 7\frac{1}{3} = \frac{12}{1} \div \frac{22}{3} =$$ The mixed numbers are changed into improper fractions

$$\frac{\cancel{12}^6}{1} \times \frac{3}{\cancel{22}_{11}} =$$ Divisor is inverted and cancellation is used

$$\frac{6 \times 3}{1 \times 11} \begin{array}{l}\text{(Step 2)}\\\text{(Step 3)}\end{array} = \frac{18}{11} = 1\frac{7}{11} \quad \text{(Step 4)}$$

Section 2.7 Readiness Review

Solve the following problems using division:

1. $\frac{4}{8} \div \frac{15}{16} =$ _____ (see Example 1)

2. $\frac{1}{2} \div \frac{5}{6} =$ _____ (see Example 1)

3. $\frac{1}{12} \div \frac{1}{38} =$ _____ (see Example 1)

4. $5\frac{1}{4} \div \frac{1}{2} =$ _____ (see Example 2)

5. $4\frac{3}{5} \div 1\frac{1}{2} =$ _____ (see Example 2)

6. $2\frac{4}{5} \div 3 =$ _____ (see Example 3)

Section readiness review answers (not given in order):

6. $\dfrac{14}{15}$ 3. $3\dfrac{1}{6}$ 2. $\dfrac{3}{5}$

5. $3\dfrac{1}{15}$ 1. $\dfrac{8}{15}$ 4. $10\dfrac{1}{2}$

Section 2.7 Review Problems

Solve the following problems using division (answers to the odd-numbered problems are at the back of the book):

1. $\dfrac{1}{3} \div \dfrac{7}{9}$
2. $\dfrac{3}{2} \div \dfrac{1}{2}$
3. $\dfrac{2}{7} \div \dfrac{23}{49}$
4. $\dfrac{4}{3} \div \dfrac{8}{6}$
5. $\dfrac{3}{5} \div \dfrac{24}{25}$
6. $\dfrac{1}{2} \div \dfrac{2}{3}$
7. $\dfrac{5}{7} \div \dfrac{3}{49}$
8. $\dfrac{1}{3} \div \dfrac{5}{19}$
9. $\dfrac{1}{32} \div \dfrac{1}{65}$
10. $3\dfrac{1}{4} \div 2$
11. $5\dfrac{1}{2} \div 4$
12. $13\dfrac{1}{7} \div 8$
13. $2\dfrac{3}{7} \div \dfrac{3}{14}$
14. $1\dfrac{1}{6} \div 4\dfrac{2}{3}$
15. $2\dfrac{1}{5} \div 1\dfrac{2}{15}$
16. $1\dfrac{1}{12} \div 8\dfrac{2}{3}$
17. $2\dfrac{1}{8} \div 3\dfrac{3}{8}$
18. $2\dfrac{4}{5} \div 2\dfrac{6}{7}$
19. $4\dfrac{2}{3} \div 3\dfrac{1}{2}$
20. $5\dfrac{4}{5} \div 3\dfrac{7}{10}$
21. $3\dfrac{2}{3} \div 6\dfrac{3}{5}$
22. $4\dfrac{5}{16} \div \dfrac{5}{8}$
23. $10\dfrac{9}{10} \div 3\dfrac{1}{5}$
24. $12\dfrac{1}{12} \div 8\dfrac{1}{2}$
25. $25\dfrac{3}{8} \div 10\dfrac{5}{6}$

Chapter 2 Readiness Review

Identify the following types of fractions:

1. $\dfrac{1}{4}$ is a _____ fraction.

2. $\dfrac{5}{3}$ is a _____ fraction.

3. $1\dfrac{4}{9}$ is a _____ _____ .

4. $\dfrac{7}{\dfrac{1}{2}}$ is a _____ fraction.

Convert the following mixed numbers to improper fractions:

5. $2\frac{1}{3} =$ _____

6. $6\frac{5}{6} =$ _____

7. $4\frac{1}{2} =$ _____

8. $13\frac{4}{11} =$ _____

Convert the following improper fractions to mixed numbers:

9. $\frac{10}{3} =$ _____

10. $\frac{14}{5} =$ _____

11. $\frac{101}{20} =$ _____

12. $\frac{15}{2} =$ _____

Reduce the following fractions to lowest terms:

13. $\frac{3}{12} =$ _____

14. $\frac{16}{56} =$ _____

15. $\frac{13}{143} =$ _____

16. $\frac{15}{65} =$ _____

Decide whether the following fractions are equal. If they are equal, write Yes; if not, write No.

17. Is $\frac{2}{4} = \frac{13}{26}$? _____

18. Is $\frac{1}{3} = \frac{3}{15}$? _____

19. Is $\frac{4}{20} = \frac{2}{5}$? _____

20. Is $\frac{20}{24} = \frac{3}{8}$? _____

Arrange the following fractions in order of size, least to greatest:

21. $\frac{1}{3}, \frac{2}{5}, \frac{7}{8}$ _____

22. $\frac{4}{10}, \frac{5}{6}, \frac{3}{5}$ _____

Arrange the following fractions in order of size, greatest to least:

23. $\frac{12}{13}, \frac{25}{26}, \frac{1}{2}$ _____

24. $\frac{3}{10}, \frac{4}{5}, \frac{19}{20}$ _____

Add the following fractions (reduce to lowest terms and express as mixed numbers):

25. $\frac{2}{9} + \frac{4}{9} + \frac{7}{9} =$ _____

26. $\frac{1}{5} + \frac{3}{5} + \frac{4}{5} =$ _____

27. $\frac{5}{51} + \frac{3}{51} + \frac{40}{51} =$ _____

28. $\frac{4}{25} + \frac{22}{25} =$ _____

Raise the following fractions to higher terms, using the denominator or numerator given:

29. $\frac{5}{6} = \frac{40}{d} =$ _____

30. $\frac{1}{2} = \frac{n}{126} =$ _____

31. $\frac{14}{56} = \frac{12}{d} =$ _____

32. $\frac{1}{4} = \frac{n}{16} =$ _____

Chapter 2 Fractions

Find the least common denominator of the following fractions:

33. $\dfrac{1}{2}, \dfrac{1}{3}, \dfrac{1}{5}$ _____

34. $\dfrac{7}{9}, \dfrac{7}{10}, \dfrac{7}{11}$ _____

35. $\dfrac{1}{4}, \dfrac{3}{7}, \dfrac{5}{6}$ _____

36. $\dfrac{9}{12}, \dfrac{2}{5}, \dfrac{2}{3}$ _____

Add the following fractions (when possible, reduce answer to lowest terms and express as a mixed number):

37. $\dfrac{11}{12} + \dfrac{1}{4} + \dfrac{3}{5} =$ _____

38. $\dfrac{2}{9} + \dfrac{2}{3} + \dfrac{1}{2} =$ _____

39. $\dfrac{1}{5} + \dfrac{3}{20} + \dfrac{9}{10} =$ _____

40. $\dfrac{3}{10} + \dfrac{5}{100} + \dfrac{7}{25} =$ _____

Add the following mixed numbers (when possible, reduce answer to lowest terms and express as a mixed number):

41. $5\dfrac{1}{2} + 3\dfrac{1}{2} =$ _____

42. $4\dfrac{11}{12} + 5\dfrac{9}{12} =$ _____

43. $2\dfrac{1}{3} + 7\dfrac{8}{9} =$ _____

44. $4\dfrac{1}{3} + 3\dfrac{1}{10} + 5\dfrac{2}{5} =$ _____

Subtract the following:

45. $50\dfrac{2}{5} - 15\dfrac{1}{5} =$ _____

46. $13\dfrac{7}{8} - 8\dfrac{3}{5} =$ _____

47. $24\dfrac{1}{4} - 16\dfrac{5}{6} =$ _____

48. $17 - 10\dfrac{11}{20} =$ _____

Subtract the following fractions:

49. $\dfrac{10}{12} - \dfrac{9}{12} =$ _____

50. $\dfrac{14}{21} - \dfrac{8}{21} =$ _____

51. $\dfrac{14}{15} - \dfrac{3}{4} =$ _____

52. $\dfrac{9}{10} - \dfrac{2}{3} =$ _____

53. $\dfrac{10}{2} - \dfrac{1}{4} =$ _____

54. $\dfrac{30}{9} - \dfrac{11}{5} =$ _____

Multiply the following problems:

55. $\dfrac{2}{3} \times \dfrac{1}{5} =$ _____

56. $\dfrac{7}{10} \times \dfrac{3}{4} =$ _____

57. $6\dfrac{2}{5} \times 3\dfrac{3}{4} =$ _____

58. $14\dfrac{2}{3} \times 4\dfrac{3}{10} =$ _____

59. $7 \times 5\dfrac{2}{3} =$ _____

60. $5\dfrac{3}{8} \times 10 =$ _____

Divide the following problems:

61. $\dfrac{4}{9} \div \dfrac{13}{18} =$ _____

62. $\dfrac{1}{2} \div \dfrac{5}{8} =$ _____

63. $\dfrac{1}{12} \div \dfrac{3}{26} =$ _____

64. $6\dfrac{1}{6} \div \dfrac{1}{3} =$ _____

65. $7\dfrac{3}{5} \div 2\dfrac{1}{2} =$ _____

66. $3\dfrac{5}{6} \div 2 =$ _____

Chapter readiness review answers (not given in order):

65. $3\dfrac{1}{25}$ **66.** $1\dfrac{11}{12}$ **61.** $\dfrac{8}{13}$ **62.** $\dfrac{4}{5}$

63. $\dfrac{13}{18}$ **64.** $18\dfrac{1}{2}$ **57.** 24   **58.** $63\dfrac{1}{15}$

59. $39\dfrac{2}{3}$ **60.** $53\dfrac{3}{4}$ **53.** $4\dfrac{3}{4}$ **54.** $1\dfrac{2}{15}$

55. $\dfrac{2}{15}$ **56.** $\dfrac{21}{40}$ **49.** $\dfrac{1}{12}$ **50.** $\dfrac{2}{7}$

51. $\dfrac{11}{60}$ **52.** $\dfrac{7}{30}$ **45.** $35\dfrac{1}{5}$ **46.** $5\dfrac{11}{40}$

47. $7\dfrac{5}{12}$ **48.** $6\dfrac{9}{20}$ **41.** 9   **42.** $10\dfrac{2}{3}$

43. $10\dfrac{2}{9}$ **44.** $12\dfrac{5}{6}$ **37.** $1\dfrac{23}{30}$ **38.** $1\dfrac{7}{18}$

39. $1\dfrac{1}{4}$ **40.** $\dfrac{63}{100}$ **33.** 30   **34.** 990

35. 84   **36.** 60   **29.** $\dfrac{40}{48}$ **30.** $\dfrac{63}{126}$

31. $\dfrac{12}{48}$ **32.** $\dfrac{4}{16}$ **25.** $1\dfrac{4}{9}$ **26.** $1\dfrac{3}{5}$

27. $\dfrac{16}{17}$ **28.** $1\dfrac{1}{25}$ **21.** $\dfrac{1}{3}, \dfrac{2}{5}, \dfrac{7}{8}$ **22.** $\dfrac{4}{10}, \dfrac{3}{5}, \dfrac{5}{6}$

23. $\dfrac{25}{26}, \dfrac{12}{13}, \dfrac{1}{2}$ **24.** $\dfrac{19}{20}, \dfrac{4}{5}, \dfrac{3}{10}$ **17.** yes   **18.** no

19. no   **20.** no   **13.** $\dfrac{1}{4}$ **14.** $\dfrac{2}{7}$

15. $\dfrac{1}{11}$ **16.** $\dfrac{3}{13}$ **9.** $3\dfrac{1}{3}$ **10.** $2\dfrac{4}{5}$

11. $5\dfrac{1}{20}$ **12.** $7\dfrac{1}{2}$ **5.** $\dfrac{7}{3}$ **6.** $\dfrac{41}{6}$

7. $\dfrac{9}{2}$ **8.** $\dfrac{147}{11}$ **1.** proper   **2.** improper

3. mixed number   **4.** complex

Chapter 2 Summary

Define each item in your own words, then compare your definitions with the text.

Key Words

fraction (p. 22)
proper fraction (p. 22)
improper fraction (p. 22)
complex fraction (p. 23)
mixed number (p. 23)
reducing fractions to lowest terms (p. 26)
like fractions (p. 28)
unlike fractions (p. 30)
raising a fraction to higher terms (p. 30)
least common denominator (p. 31)
prime number (p. 31)
composite number (p. 31)
equal fractions (p. 39)
cross products (p. 39)

Chapter 2 Review Problems

Identify the types of fractions (answers to all the problems are at the back of the book).

1. $\dfrac{10}{\frac{4}{5}}$
2. $\dfrac{20}{21}$
3. $\dfrac{1}{12}$
4. $\dfrac{72}{5}$
5. $22\dfrac{11}{23}$
6. $\dfrac{\frac{13}{20}}{\frac{11}{14}}$
7. $\dfrac{78}{11}$
8. $4\dfrac{2}{33}$

Convert the following mixed numbers to improper fractions:

9. $4\dfrac{5}{6}$
10. $7\dfrac{3}{8}$
11. $13\dfrac{5}{11}$
12. $49\dfrac{1}{7}$

Convert the following improper fractions to mixed numbers:

13. $\dfrac{7}{3}$
14. $\dfrac{22}{7}$
15. $\dfrac{41}{18}$
16. $\dfrac{98}{45}$

Reduce the following fractions to lowest terms:

17. $\dfrac{2}{8}$
18. $\dfrac{11}{33}$
19. $\dfrac{25}{125}$
20. $\dfrac{7}{63}$
21. $\dfrac{35}{105}$
22. $\dfrac{16}{96}$

Decide whether the following fractions are equal or not. If they are equal, write Yes; if not, write No.

23. Is $\dfrac{2}{3} = \dfrac{6}{12}$?
24. Is $\dfrac{7}{27} = \dfrac{1}{9}$?
25. Is $\dfrac{3}{21} = \dfrac{2}{7}$?
26. Is $\dfrac{25}{30} = \dfrac{5}{6}$?
27. Is $\dfrac{3}{4} = \dfrac{9}{12}$?
28. Is $\dfrac{22}{77} = \dfrac{24}{84}$?
29. Is $\dfrac{7}{28} = \dfrac{1}{4}$?
30. Is $\dfrac{20}{10} = \dfrac{4}{5}$?
31. Is $\dfrac{20}{10} = \dfrac{10}{5}$?
32. Is $\dfrac{23}{39} = \dfrac{27}{42}$?

Arrange the following fractions in order of size, least to greatest:

33. $\dfrac{3}{4}, \dfrac{1}{2}, \dfrac{3}{10}$
34. $\dfrac{1}{8}, \dfrac{3}{16}, \dfrac{5}{32}$
35. $\dfrac{7}{8}, \dfrac{3}{4}, \dfrac{3}{16}$
36. $\dfrac{7}{12}, \dfrac{5}{16}, \dfrac{1}{2}$
37. $\dfrac{18}{64}, \dfrac{1}{8}, \dfrac{5}{16}$
38. $\dfrac{10}{27}, \dfrac{1}{9}, \dfrac{2}{3}$
39. $\dfrac{1}{2}, \dfrac{1}{3}, \dfrac{3}{4}$
40. $\dfrac{75}{100}, \dfrac{21}{25}, \dfrac{3}{5}$
41. $\dfrac{3}{8}, \dfrac{1}{16}, \dfrac{1}{2}$
42. $\dfrac{1}{3}, \dfrac{8}{9}, \dfrac{3}{11}$

Arrange the following fractions in order of size, greatest to least:

43. $\dfrac{2}{5}, \dfrac{1}{5}, \dfrac{7}{10}$
44. $\dfrac{2}{5}, \dfrac{3}{10}, \dfrac{4}{15}$
45. $\dfrac{1}{2}, \dfrac{2}{3}, \dfrac{1}{24}$
46. $\dfrac{1}{7}, \dfrac{5}{21}, \dfrac{2}{3}$
47. $\dfrac{7}{8}, \dfrac{5}{6}, \dfrac{3}{4}$
48. $\dfrac{3}{7}, \dfrac{5}{14}, \dfrac{1}{2}$
49. $\dfrac{17}{30}, \dfrac{5}{10}, \dfrac{6}{15}$
50. $\dfrac{11}{100}, \dfrac{1}{10}, \dfrac{2}{3}$
51. $\dfrac{7}{12}, \dfrac{3}{4}, \dfrac{1}{2}$
52. $\dfrac{3}{5}, \dfrac{21}{25}, \dfrac{82}{100}$

Add the following fractions (reduce answer to lowest terms and express as a mixed number):

53. $\dfrac{19}{24} + \dfrac{3}{24} + \dfrac{5}{24}$
54. $\dfrac{3}{5} + \dfrac{4}{5} + \dfrac{2}{5}$
55. $\dfrac{2}{13} + \dfrac{5}{13} + \dfrac{10}{13}$
56. $\dfrac{7}{18} + \dfrac{17}{18}$
57. $\dfrac{15}{21} + \dfrac{13}{21} + \dfrac{8}{21}$
58. $\dfrac{12}{16} + \dfrac{3}{16} + \dfrac{9}{16}$

Raise the following fractions to higher terms, using the denominator or numerator given:

59. $\dfrac{3}{5} = \dfrac{21}{d}$
60. $\dfrac{11}{12} = \dfrac{n}{156}$
61. $\dfrac{57}{125} = \dfrac{n}{625}$
62. $\dfrac{7}{15} = \dfrac{28}{d}$

Find the least common denominator of the following fractions:

63. $\dfrac{3}{4}, \dfrac{1}{2}, \dfrac{2}{3}$
64. $\dfrac{5}{6}, \dfrac{6}{7}, \dfrac{7}{8}$
65. $\dfrac{14}{15}, \dfrac{15}{16}, \dfrac{16}{17}$
66. $\dfrac{9}{11}, \dfrac{1}{12}, \dfrac{4}{5}$

Add the following fractions:

67. $\dfrac{13}{14} + \dfrac{2}{11} + \dfrac{5}{9}$
68. $\dfrac{9}{10} + \dfrac{99}{100} + \dfrac{6}{7}$
69. $\dfrac{5}{6} + \dfrac{6}{7} + \dfrac{17}{48}$
70. $\dfrac{1}{11} + \dfrac{1}{22} + \dfrac{1}{16}$

Add the following mixed numbers:

71. $6\dfrac{1}{5} + 1\dfrac{3}{5}$
72. $7\dfrac{3}{5} + 9\dfrac{2}{3}$
73. $14\dfrac{1}{3} + 15\dfrac{1}{5}$
74. $11\dfrac{2}{3} + 11\dfrac{1}{2}$
75. $6\dfrac{5}{7} + 5\dfrac{5}{8}$
76. $221\dfrac{1}{3} + 171\dfrac{3}{12} + 42\dfrac{5}{7}$

Subtract the following problems:

77. $\dfrac{7}{9} - \dfrac{5}{9}$
78. $\dfrac{3}{4} - \dfrac{1}{3}$
79. $14\dfrac{12}{13} - 7\dfrac{1}{2}$
80. $43\dfrac{1}{8} - 3\dfrac{15}{58}$
81. $375\dfrac{13}{24} - 81\dfrac{15}{16}$
82. $42 - 33\dfrac{1}{3}$

Multiply the following problems:

83. $\dfrac{3}{7} \times \dfrac{5}{8}$
84. $\dfrac{14}{17} \times \dfrac{1}{2}$
85. $7\dfrac{3}{7} \times 11\dfrac{1}{4}$
86. $1\dfrac{1}{9} \times 3\dfrac{2}{7}$
87. $2\dfrac{5}{16} \times 7$
88. $4 \times 3\dfrac{4}{5}$

Divide the following problems:

89. $\dfrac{2}{7} \div \dfrac{25}{49}$
90. $\dfrac{1}{4} \div \dfrac{2}{17}$
91. $3\dfrac{15}{16} \div \dfrac{1}{8}$
92. $5\dfrac{1}{8} \div 2\dfrac{3}{4}$
93. $3\dfrac{1}{2} \div 4\dfrac{1}{4}$
94. $9\dfrac{1}{2} \div 4$

Solve the following word problems:

95. A diabetic uses Testape® to check his urine for sugar. On Monday he used 4⅓ inches, on Tuesday he used 3⅝ inches, and on Wednesday he used 2½ inches. How many inches of Testape® did he use?

96. Dr. Phillips, a dermatologist, has prescribed a special cream for her patients. To Patient A, she gives as a sample ⅔ of a jar, to Patient B she gives ¾ of a jar, and to Patient C she gives ⅛ of a jar. How much cream did Dr. Phillips give to her patients?

97. A patient is instructed by the doctor to take 1¼ tablets 3 times a day. If the patient takes this medication for 14 days, how many total tablets will have been taken by this patient?

98. An EKG (electrocardiogram) has been ordered by the physician. A total of 42 inches of EKG strip paper was run off. Of that, several small strips were included in the patient's chart: a 4½ inch strip, a 7⅛ inch strip, a 5¾ inch strip, and a 3⅚ inch strip. How much EKG strip paper was not included in the patient's chart?

99. The pharmacy makes I.V. additive labels for the I.V. fluid mixtures (intravenous fluids with drugs added) sent to the nursing stations. If each label is 4½ inches long and 111 labels were made in one day, how many inches of blank label material are left if the roll was 600 inches to begin with?

100. An order is written for Baby Franklin. The neonatologist wants Baby Franklin to have ¼ teaspoonful of medicine at breakfast, ⅔ teaspoonful at lunch, ½ teaspoonful at dinner, and ¼ teaspoonful after dinner. How many teaspoonsful of medicine did Baby Franklin receive?

3 Decimals

OBJECTIVES After studying this chapter, you should be able to:

1. Read and write decimals.
2. Change fractions into decimals and decimals into fractions.
3. Add, subtract, multiply, and divide decimals.
4. Round off decimals.

The first chapter explained the importance of lining up columns of numbers *before* adding, subtracting, multiplying, or dividing whole numbers; the ones column contains ones, the tens column contains tens, and so on. The position, or place, of numbers is equally important in problems that involve decimals.

> **Definition:** A *decimal* is a fraction with a denominator that is a power of 10. The decimal point is used to represent place value.

A decimal, therefore, is a fraction with a denominator that is a power of 10—for example, $3/10 = .3$, $125/1000 = .125$, and $6/1{,}000{,}000 = .000006$.

Section 3.1 Reading and Writing Decimals

When arranging decimals in columns, line up the numbers to the right of the decimal point as well as to the left.

> **Rule: Reading Decimals**
> *Step 1:* Find the decimal point.
> *Step 2:* Read the numbers to the left of the decimal point as whole numbers.
> *Step 3:* Read the numbers to the right as decimal fractions.
> *Step 4:* The decimal point is read as "and."

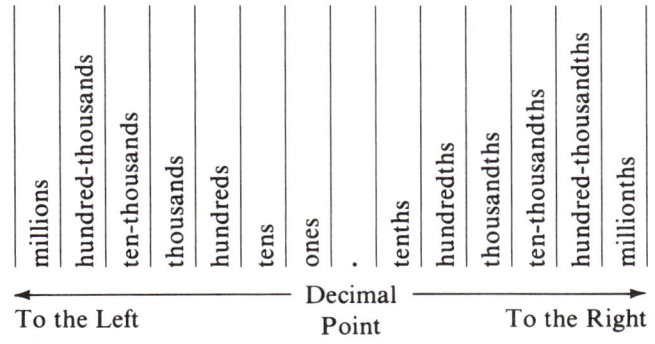

Figure 3-1.

Study Figure 3-1. Notice that, as you read from left to right, each place is $1/10$ as large as the value of the place to its left; that is,

$$\begin{aligned}10 &\text{ is one tenth of } 100,\\ 1 &\text{ is one tenth of } 10,\\ .1 &\text{ is one tenth of } 1,\\ .01 &\text{ is one tenth of } .1,\end{aligned}$$

and so on.

Example 1: Read the following decimal:

7.3

Solution:
Step 1: Locate the decimal point.

        7.3
        ↑

Step 2: Read the number to the left of the decimal point as a whole number.
Step 3: Read the number to the right of the decimal point as a decimal fraction.

ones	tenths
7 .	3

Step 4: The number 7.3 is read as "seven and three tenths." The decimal point is read as "and."

Example 2: Read the following decimal:

    4.125

Solution:
Step 1: Locate the decimal point.

    4.125
    ↑

Step 2: Read the number to the left of the decimal point as the whole number "four."
Step 3: Read the number to the right of the decimal point as the decimal fraction "one hundred twenty-five thousandths."

ones	tenths	hundredths	thousandths
4 .	1	2	5

Step 4: Read the number 4.125 as "four and one hundred twenty-five thousandths."

Example 3: Read the following decimal:

    0.00006

Solution:
Step 1: Locate the decimal point.

    0.00006
    ↑

Step 2: The number to the left of the decimal point is the whole number zero and is not read aloud.
Step 3: Read the number to the right of the decimal point as the decimal fraction "six hundred-thousandths."

Section 3.1 Reading and Writing Decimals

```
       |  |  |  |  |  |
       |  |  |  |  |  |
       |  |  |  |  | hundred-thousandths
       |  |  |  | ten-thousandths
       |  |  | thousandths
       |  | hundredths
       | tenths
       ones
     0 . 0  0  0  0  6
```

Step 4: Read the number 0.00006 as "six hundred-thousandths." (People in the medical professions often write the zero to the left of the decimal point for clarity. However, this does not change the way the decimal is read.)

Another popular way to read decimals is to say "point" instead of "and" for the decimal point. This reading is often used in hospital environments. For example, when the nurse has taken your body temperature, which is 98.6° F, he or she could read it as "ninety-eight *and* six tenths" or as "ninety-eight *point* six." Nurses usually read body temperatures using "point" to indicate the decimal point.

Using "point" instead of "and," you would read Examples 1, 2, and 3 as follows: 7.3 (seven point three), 4.125 (four point one two five), and 0.00006 (zero point zero zero zero zero six). Whether or not you use this approach in this course will be up to your instructor. It is included for reference only.

The next topic is writing decimals.

Example 4: Use numbers to write the following decimal: twenty-three and eight tenths.
Solution: 23.8

Example 5: Use numbers to write the following decimal: seventeen and thirty-nine hundredths.
Solution: 17.39

Example 6: Use numbers to write the following decimal: one hundred seventy-five thousandths.
Solution: 0.175

Example 7: Use numbers to write the following decimal: six ten-thousandths.
Solution: 0.0006

Example 8: Use numbers to write the following decimal: four hundred-thousandths.
Solution: 0.00004

Section 3.1 Readiness Review

Read the following decimals (see Examples 1, 2, & 3):

1. 37.3 = _____

2. 8.015 = _____

3. 0.0002 = _____

Use numbers to write the following decimals (see Examples 4, 5, 6, 7, & 8):

4. Eighteen and two hundredths = _____

5. Fourteen and one hundred thirty-six thousandths = _____

6. One hundred twenty-eight ten-thousandths = _____

Section readiness review answers (not given in order):

1. Thirty-seven and three tenths   3. Two ten-thousandths

5. 14.136   4. 18.02   6. 0.0128

2. Eight and fifteen thousandths

Section 3.1 Review Problems

Read the following decimals (answers to the odd-numbered problems are at the back of the book):

1. 2.5   2. 3.4   3. 18.7

4. 10.2   5. 8.14   6. 22.22

7. 1.123   8. 5.678   9. 12.001

10. 9.0005   11. 0.000083   12. 0.000001

Write the following decimals using numbers:

13. Three and six tenths

14. Five and five tenths

15. Nineteen and three tenths

16. Eleven and seven tenths

17. Four and twenty-one hundredths

18. Thirty-three and thirty-three hundredths

19. Two and three hundred forty-five thousandths

20. Nine and eight hundred one thousandths

21. Thirteen and thirteen thousandths

22. Six and two ten-thousandths

23. Eleven millionths

24. Seven hundred-thousandths

25. Five thousandths

Section 3.2 Changing Fractions to Decimals and Decimals to Fractions

Fractions can be changed into decimals, and vice versa. To change a fraction into a decimal, use the following rule:

Section 3.2 Changing Fractions to Decimals and Decimals to Fractions 67

> **Rule: Changing Fractions into Decimals**
> *Step 1:* Write the fraction (a/b) in division form ($b\overline{)a}$).
> *Step 2:* Place a decimal point after the numerator ($b\overline{)a.}$).
> *Step 3:* Insert zeros after the decimal point when required to complete the division problem.
> *Step 4:* The division may come out even, or it may continue until the desired number of decimal places is reached. (Stopping the division at a particular decimal place involves rounding off, which we will cover in a later section.)
> *Step 5:* Place the decimal point in the answer directly above the decimal point in the numerator.

Example 1: Change the fraction ¼ into a decimal.
Solution:
Step 1: Write the fraction in division form.

$$4\overline{)1}$$

Step 2: Place a decimal point after the numerator.

$$4\overline{)1.}$$

Step 3: Insert zeros after the decimal point when required to complete the division problem.

$$\begin{array}{r} 25 \\ 4\overline{)1.00} \\ \underline{8} \\ 20 \\ \underline{20} \end{array}$$

Step 4: The division comes out even.
Step 5: Move the decimal point up.

$$\begin{array}{r} .25 \\ 4\overline{)1.00} \\ \underline{8} \\ 20 \\ \underline{20} \end{array}$$

The answer is ¼ = .25 or 0.25.

Example 2: Change the fraction ⅓ into a decimal. (If the answer does not come out even, continue working to four decimal places.)
Solution:
Step 1: Write the fraction in division form.

$$3\overline{)1}$$

Step 2: Place a decimal point after the numerator.

$$3\overline{)1.}$$

Step 3: Put zeros after the decimal point to complete the division problem.

$$3 \overline{\smash{)}\begin{array}{r} 3333 \\ 1.0000 \end{array}}$$
$$\begin{array}{r} \underline{9} \\ 10 \\ \underline{9} \\ 10 \\ \underline{9} \\ 10 \\ \underline{9} \\ 1 \end{array}$$

Step 4: The division does not come out even. There is always a remainder of 1. In this case, we work the division to four decimal places.

Step 5: Move the decimal point up.

$$3 \overline{\smash{)}\begin{array}{r} .3333 \\ 1{.}0000 \end{array}}$$
$$\begin{array}{r} \underline{9} \\ 10 \\ \underline{9} \\ 10 \\ \underline{9} \\ 10 \\ \underline{9} \\ 1 \end{array}$$

The answer is $1/3 \approx .3333$ or 0.3333. The symbol $\approx$ stands for "is approximately equal to"; therefore, $1/3$ is approximately equal to 0.3333.

In the preceding examples, we have reviewed the process of changing fractions into decimals. Now we'll discuss how to change decimals into fractions. There are two ways to do this. The first method involves expressing the decimal in word form.

Rule: Changing Decimals into Fractions—Word Method
  Step 1: Express the decimal in words.
  Step 2: Write the words in the form of a fraction.
  Step 3: Reduce the fraction to lowest terms.

Example 3: Convert the decimal 0.78 to a fraction using the word method.
Solution:
  Step 1: Express the decimal in words. The decimal 0.78 is "seventy-eight hundredths."
  Step 2: Write "seventy-eight hundredths" in fraction form.

$$\text{seventy-eight hundredths} = \frac{78}{100}$$

Section 3.2 Changing Fractions to Decimals and Decimals to Fractions 69

Step 3: Reduce the fraction to lowest terms.

$$\frac{78}{100} = \frac{39}{50}$$

The decimal $0.78 = 39/50$.

Example 4: Convert the decimal 0.971 to a fraction using the word method.
Solution:
Step 1: Express the decimal in words. The decimal 0.971 is "nine hundred seventy-one thousandths."
Step 2: Write "nine hundred seventy-one thousandths" in fraction form.

$$\text{nine hundred seventy-one thousandths} = \frac{971}{1000}$$

Step 3: The fraction is already in lowest terms.

The second way to convert decimals to fractions involves counting the number of places to the right of the decimal point.

Rule: Changing Decimals into Fractions—Alternate Method
Step 1: Count the number of decimal places to the right of the decimal point. This number represents the number of zeros to be used in the denominator.
Step 2: Write the denominator as 1 followed by the number of zeros found in the step above.
Step 3: Remove the decimal point from the number you are converting to a fraction; this number becomes the numerator.
Step 4: Reduce fraction to lowest terms.

Example 5: Using the alternate method, convert the decimal 0.16 to a fraction.
Solution:
Step 1: Count the number of places to the right of the decimal point.

0.16 (two places)
↑↑

Step 2: There are two places; therefore, the fraction's denominator is 1 followed by two zeros (100).
Step 3: Remove the decimal point from the number you are converting to a fraction.

016

This number becomes the numerator. (The zero on the left is no longer needed. It is included for clarity when reading and writing decimals.)

$$\frac{16}{100}$$

70   Chapter 3 Decimals

Step 4:   Reduce the fraction to lowest terms.

$$\frac{16}{100} = \frac{4}{25}$$

The decimal $0.16 = 4/25$.

Example 6:   Using the alternate method, convert the decimal 0.125 to a fraction.
Solution:
  Step 1:   Count the number of places to the right of the decimal point.

  0.125   (three places)
  ↑↑↑

  Step 2:   There are three places; therefore, the fraction's denominator is 1 followed by three zeros (1000).
  Step 3:   Remove the decimal point.

  0125

  The original number becomes the numerator (extra zero on the left is no longer needed), the 1000 the denominator.

  $$\frac{125}{1000}$$

  Step 4:   Reduce the fraction to lowest terms.

  $$\frac{125}{1000} = \frac{1}{8}$$

  The decimal $0.125 = 1/8$.

Example 7:   Using the alternate method, convert the decimal 0.833 to a fraction.
Solution:
  Step 1:   Count the number of places to the right of the decimal point.

  0.833   (three places)
  ↑↑↑

  Step 2:   There are three places; therefore, the fraction's denominator is 1 followed by three zeros (1000).
  Step 3:   Remove the decimal point.

  0833

  The original number becomes the numerator, the 1000 the denominator.

  $$\frac{833}{1000}$$

Step 4: The fraction is already in lowest terms. The decimal $0.833 = {}^{833}/_{1000}$.

Section 3.2 Readiness Review

Change the following fractions into decimals (see Examples 1 & 2). (If an answer does not come out even, carry the division to four places.)

1. $\frac{3}{8} =$ _____   2. $\frac{1}{16} =$ _____   3. $\frac{5}{6} =$ _____

Convert the following decimals to fractions, using either method (see Examples 3–7). Reduce to lowest terms.

4. $0.23 =$ _____   5. $0.415 =$ _____   6. $0.777 =$ _____

Section readiness review answers (not given in order):

5. $\frac{83}{200}$ 1. 0.375   3. 0.8333

6. $\frac{777}{1000}$ 4. $\frac{23}{100}$ 2. 0.0625

Section 3.2 Review Problems

Convert the following decimals to fractions, using either method. Reduce answers to lowest terms (answers to the odd-numbered problems are at the back of the book):

1. 0.8   2. 0.5   3. 0.32
4. 0.44   5. 0.25   6. 0.15
7. 0.627   8. 0.871   9. 0.375
10. 0.125   11. 0.5625   12. 0.8125

Convert the following fractions to decimals. (If an answer does not come out even, carry the division out to four decimal places. Remember, you will learn how to round off in Section 3.4.)

13. $\frac{1}{5}$ 14. $\frac{1}{2}$ 15. $\frac{2}{5}$

16. $\frac{4}{5}$ 17. $\frac{3}{4}$ 18. $\frac{1}{4}$

19. $\frac{13}{16}$ 20. $\frac{7}{16}$ 21. $\frac{1}{6}$

22. $\frac{1}{3}$ 23. $\frac{11}{32}$ 24. $\frac{13}{32}$

25. $\frac{5}{12}$

72   Chapter 3 Decimals

Section 3.3 Addition and Subtraction of Decimals

Adding and subtracting decimals is done in much the same way as adding and subtracting whole numbers.

Rule: Adding Decimals
Step 1: Arrange the decimals to be added into columns so that the decimal points are lined up directly under one another.
Step 2: Add the decimals in the same way that whole numbers are added.
Step 3: Place a decimal point in the answer directly under the other decimal points.

Example 1:   Add 0.3 and 1.85.
Solution:
Step 1:   Arrange the decimals to be added into columns so that the decimal points are lined up directly under each other. Notice that the ones column is lined up, as well as the tenths and hundredths columns.

```
  0.3
  1.85
```

Step 2:   Add the decimals from right to left, starting with the hundredths column.

```
  0.3
  1.85
  2 15
```

Step 3:   Place a decimal point in the answer directly under the other decimal points.

```
  0.3
  1.85
  2.15
```

Therefore, 0.3 plus 1.85 is equal to 2.15 (0.3 + 1.85 = 2.15).

When adding long columns of decimals, it may be helpful to write zeros in the open spaces to keep the columns straight. Including extra zeros in these open spaces will not change the value of the decimals themselves.

Example 2:   Find the sum .4 + 21.7 + 3 + 4.26.
Solution:
Step 1:   Arrange the decimals into columns so that the decimal points are lined up directly under one another.

```
   .4
  21.7
   3.
   4.26
```

Step 2: Add from right to left. Write zeros in the open spaces to the right of the decimals to keep the columns straight.

```
 .40
21.70
 3.00
 4.26
─────
29 36
```

Step 3: Place a decimal point in the answer directly under the other decimal points.

```
 .40
21.70
 3.00
 4.26
─────
29.36
```

The sum is 29.36.

Example 3: Add the decimals 3.125 + 0.47 + 11.2 + 16.
Solution:
Step 1: Arrange the decimals in columns.

```
 3.125
 0.47
11.2
16.
─────
```

Step 2: Put zeros in the open spaces and add.

```
 3.125
 0.470
11.200
16.000
──────
30 795
```

Step 3: Bring the decimal point down.

```
 3.125
 0.470
11.200
16.000
──────
30.795
```

The sum is 30.795.

Example 4: For his 8-year-old patient, Dr. Arthur Foley II orders neomycin 0.5 g to be given several times before surgery. The doses of the drug are to be given at the following times:

11 A.M.	neomycin 0.5 g	11 P.M.	neomycin 0.5 g
3 P.M.	neomycin 0.5 g	3 A.M.	neomycin 0.5 g
7 P.M.	neomycin 0.5 g	7 A.M.	neomycin 0.5 g

How many grams of neomycin were given to this patient?

Solution: In this problem you must determine the total amount of neomycin given to the patient before surgery. (Remember, the explanation of grams can be found in Chapter 6—Metric System.) To find this amount, add the doses.

Step 1: Arrange the decimals into columns so that the decimal points are lined up directly under one another.

0.5 g	11 A.M.
0.5 g	3 P.M.
0.5 g	7 P.M.
0.5 g	11 P.M.
0.5 g	3 A.M.
0.5 g	7 A.M.

Step 2: Add the decimals.

```
  0.5 g
  0.5 g
  0.5 g
  0.5 g
  0.5 g
  0.5 g
 ─────
  3 0 g
```

Step 3: Place the decimal point in the answer under the other decimal points.

```
  0.5 g
  0.5 g
  0.5 g
  0.5 g
  0.5 g
  0.5 g
 ─────
  3.0 g
```

The patient received 3.0 g of neomycin before surgery.

The next topic in this section is the subtraction of decimals. The method used to subtract decimals is similar to the addition of decimals.

Rule: Subtracting Decimals

Step 1: Arrange the decimals to be subtracted into columns so that the decimal points line up directly under one another.

Step 2: Subtract the decimals in the same way that whole numbers are subtracted.

Step 3: Place a decimal point in the answer directly under the other decimal points.

Example 5: Subtract 34.75 from 181.8.

Section 3.3 Addition and Subtraction of Decimals

Solution:
Step 1: Arrange the decimals to be subtracted into columns so that the decimal points line up directly under each other.

$$\begin{array}{r} 181.8 \\ -34.75 \\ \hline \end{array}$$

Step 2: Subtract the decimals. Put a zero in the open space so you can use it in borrowing as well as to keep the columns straight.

$$\begin{array}{r} 181.80 \\ -34.75 \\ \hline 14705 \end{array}$$

Step 3: Place a decimal point in the answer directly under the other decimal points.

$$\begin{array}{r} 181.80 \\ -34.75 \\ \hline 147.05 \end{array}$$

The difference between 181.8 and 34.75 is 147.05 (181.8 − 34.75 = 147.05).

Example 6: Calculate 47.1 − 9.423.
Solution:
Step 1: Line up the decimals.

$$\begin{array}{r} 47.1 \\ -9.423 \\ \hline \end{array}$$

Step 2: Subtract the decimals. Put zeros in the open spaces.

$$\begin{array}{r} 47.100 \\ -9.423 \\ \hline 37677 \end{array}$$

Step 3: Place the decimal point in the answer.

$$\begin{array}{r} 47.100 \\ -9.423 \\ \hline 37.677 \end{array}$$

The difference between 47.1 and 9.423 is 37.677 (47.1 − 9.423 = 37.677).

Example 7: Subtract 17.39 from 25.988.
Solution:
Step 1: Line up the decimals.

$$\begin{array}{r} 25.988 \\ -17.39 \\ \hline \end{array}$$

Step 2: Subtract the decimals. Put a zero in the open space.

$$\begin{array}{r} 25.988 \\ -\ 17.390 \\ \hline 8\ 598 \end{array}$$

Step 3: Place the decimal point in the answer.

$$\begin{array}{r} 25.988 \\ -\ 17.390 \\ \hline 8.598 \end{array}$$

The difference between 25.988 and 17.39 is 8.598 (25.988 − 17.39 = 8.598).

Example 8: Mr. Henderson was admitted to the hospital as a last resort for weight reduction. The physician wired Mr. Henderson's jaws shut and thus restricted him to a liquid diet. Mr. Henderson's weight on admission was 308 pounds. When he was discharged, he weighed 191.5 pounds. How many pounds did Mr. Henderson lose?

Solution: The problem tells us that on admission Mr. Henderson weighed 308 pounds and on discharge he weighed 191.5 pounds. A reasonable estimate of the difference might be 120 pounds (310 − 190 = 120 pounds).

Step 1: Line up the decimals.

$$\begin{array}{r} 308.\ \text{lb} \\ -\ 191.5\ \text{lb} \end{array}$$

Step 2: Subtract the decimals. Put a zero in the open space.

$$\begin{array}{r} 308.0\ \text{lb} \\ -\ 191.5\ \text{lb} \\ \hline 116\ 5 \end{array}$$

Step 3: Place the decimal point in the answer.

$$\begin{array}{r} 308.0\ \text{lb} \\ -\ 191.5\ \text{lb} \\ \hline 116.5\ \text{lb} \end{array}$$

Comparing the answer, 116.5 pounds, with the estimate of 120 pounds, we see that they are close.

Section 3.3 Readiness Review

Add the following decimals (see Examples 1–3):

1.	74.35	2.	246.138	3.	5.3
	0.2		64.		36.06
	14.9		29.5		8.009
	3.				91.1005

Section 3.3 Addition and Subtraction of Decimals 77

Subtract the following decimals (see Examples 5–7):

4. 248
   − 51.2

5. 7171.304
   −   35.24

6. 152.306
   − 10.3577

Section readiness review answers (not given in order):

5. 7136.064   4. 196.8   1. 92.45

3. 140.4695   2. 339.638   6. 141.9483

Section 3.3 Review Problems

Add the following decimals (answers to the odd-numbered problems are at the back of the book):

1. 13.06
    74.005
    11.573
     6.03

2. 0.05
    0.006
    0.206
    0.3

3. 9.7
    10.04
    17.0006

4. 1560.40207
       6.8305
      42.74
        .005

5. 201.044
    190.040
    1202.
      11.1

6. 181.0403
      8.6
    406.760
      5.09009
     14.

7.   78.001
      0.04
      7.1
    137.0065
       .38008

8.   0.000435
    40.14801
     7.83
     9.50002

9.   2.601
     0.090317
    79.5302
     1.70063

Subtract the following decimals:

10.   7.8
    − 4.18

11.   5.3
    − 2.41

12. 85.0029
    − 1.96

13. 307.4001
    − 25.246

14. 502.3008
    − 14.111

15.   0.838
    − 0.090257

16. 14
    − 9.0187

17. 163
    − 42.001

Solve the following word problems:

18. A bottle contains 30 g of cocaine flakes. Every time a pharmacist removes cocaine to compound a prescription, the amount removed is noted on the bottle as follows:

Date	Beginning total	Amount removed	Ending total	Initial
5/30	30 g	7.1 g	22.9 g	MKM
6/7	22.9 g	5.75 g		MKM

What was the ending total on June 7?

19. Dr. Fetz, a urologist, wants to see if Mr. Bargello's kidneys are working properly. After fluids have been withheld from him for 12 hours, Mr. Bargello is given 1500 mℓ of water to drink. Then his urine is collected over a four-hour period, and the specific gravity of the urine is measured (specific-gravity measurements can show the physician how well the kidneys are concentrating or diluting urine). The measurements are as follows:

    Specific gravity
    Reading 1 (SG) = 1.025
    Reading 2 (SG) = 1.03
    Reading 3 (SG) = 1.031

    Find the difference between the specific-gravity measurements of Readings 1 and 2, Readings 2 and 3, and Readings 1 and 3.

20. A baby is taken to the emergency room. The father states that the infant is suffering from fever, vomiting, and diarrhea. The emergency-room technician takes the baby's temperature rectally three times over a four-hour period. The temperatures are:

    5:00 P.M.   104° F
    6:30 P.M.   102.2° F
    8:55 P.M.   97.7° F

    Find the difference between the baby's temperature at 5:00 P.M. and at 8:55 P.M.

21. A patient requiring thyroid-replacement therapy is given 0.2 mg of Synthroid® daily; however, she soon shows overdosage symptoms. The doctor gradually reduces the dose to 0.025 mg of Synthroid® daily. Find the difference between the original dose of Synthroid® and the current dose.

22. A 3-year-old child is given 0.6 mℓ of Tylenol® drops to treat a fever. The Tylenol® drops come in a 15-mℓ bottle with a dropper. How many milliliters of Tylenol® are left in the bottle?

23. Ruth Harris, a nurse on the third floor, must give her patient, Mr. Frink, his Decadron® tablets. The doctor performed brain surgery on Mr. Frink three days ago and has prescribed the drug to reduce the swelling in the brain. The order reads:

    Give Decadron®

    First day of therapy
    3 mg after breakfast
    3 mg at 3 P.M.
    3 mg at bedtime

    Second day of therapy
    2.5 mg after breakfast
    2.5 mg at 3 P.M.
    1.5 mg at bedtime

    Third day of therapy
    1.5 mg after breakfast
    1.5 mg at 3 P.M.
    1.0 mg at bedtime

Fourth day of therapy
1.0 mg after breakfast
0.75 mg at 3 P.M.
0.75 mg at bedtime

Fifth day of therapy
0.75 mg after breakfast
0.5 mg at 3 P.M.
0.5 mg at bedtime

Sixth day of therapy
0.5 mg after breakfast
0.5 mg at 3 P.M.
0.25 mg at bedtime

Seventh day of therapy
0.25 mg after breakfast
0.25 mg at 3 P.M.
0.25 mg at bedtime

Ms. Harris monitors the patient's daily doses of Decadron®. What is the total number of milligrams of Decadron® given to the patient during the entire seven days of therapy?

24. A child has just consumed a bottle of baby aspirin. Life-support measures are started, and blood is drawn to determine how much aspirin is in the child's blood. If the child's blood contains 35.25 mg of aspirin per 100 mℓ of blood and the toxic level is 30 mg per 100 mℓ, how far above the toxic level is the concentration of aspirin in the child's blood?

25. Char Davis complains of severe allergies. Dr. Kevin decides to place her on six days of Celestone® treatment. The directions are as follows:

Day 1	3.6 mg
Day 2	3.0 mg
Day 3	2.4 mg
Day 4	1.8 mg
Day 5	1.2 mg
Day 6	0.6 mg

How many milligrams of Celestone® does Char take over the six-day period?

Section 3.4 Rounding Off Decimals

In Section 3.2, we discussed how to convert fractions into decimals through division. You'll recall that some fractions divide evenly, with a remainder of 0; such fractions are represented by a terminating decimal—for example,

$$\frac{13}{80} = 0.1625$$

But there are also many fractions that, when converted through division to a decimal, never lead to a 0 remainder and are thus represented by a decimal that never ends—for example,

$$\frac{93}{148} = 0.62837837837837\ldots$$

(The three dots indicate that the decimal repeats in the same pattern forever.) In either case, we usually shorten the decimal by rounding it off to the degree of accuracy we require; that is, we choose an approximation of the actual number.

> **Rule: Rounding Off Decimals**
> *Step 1:* First, identify the place to which you're to round off the decimal. The problem will usually tell you to round off the decimal to the nearest hundred, ten, one, tenth, hundredth, or thousandth, and so on.
> *Step 2:* Once you know where you're to round off your answer, locate that place in the answer and circle it.
> *Step 3:* If the digit to the right of the circled number is less than 5, drop that digit and any others that may follow it. If the digit to the right of the circled number is 5 or greater, increase the circled number by 1 and drop all digits to the right of it.

Example 1: Round off 35.247 to the nearest hundredth (that is, to two decimal places).
Solution:
Steps 1 and 2: Identify the place to which you're to round off the decimal. Because the decimal 35.247 is to be rounded off to two decimal places, circle the 4.

35.2④7

Step 3: Because 7 is greater than 5, increase the ④ to 5 and drop the 7.

The decimal 35.247 rounded off to the nearest hundredth is 35.25.

Example 2: Round off 17.3521 to the nearest thousandth (three decimal places).
Solution:
Steps 1 and 2: Identify the decimal place. Since 17.3521 is to be rounded off to three decimal places, circle the 2.

17.35②1

Step 3: The next digit, 1, is less than 5, so it can be dropped.

The decimal 17.3521 rounded off to three decimal places is 17.352.

Example 3: Round off 8.17 to the nearest tenth.
Solution:
Steps 1 and 2: Identify the place to which you're to round off the decimal. Because 8.17 is to be rounded off to the nearest tenth, circle the 1, which is in the tenths place.

8.①7

Step 3: Because 7 is greater than 5, increase the ① to 2 and drop the 7.

8.17 is rounded off to 8.2.

Example 4: Round off 3.00456 to the nearest thousandth.
Solution:
Steps 1 and 2: Identify the decimal place you're to round off to and circle it.

3.00④56

Section 3.4 Rounding Off Decimals 81

Step 3: The digit to the right of the ④ is 5 or greater. Increase the circled number by 1 and drop the 5 and 6.

The decimal 3.00456 is rounded off to 3.005.

Example 5: Round off 3467.25 to the nearest one.
Solution:
Steps 1 and 2: Circle the number in the ones place.

346⑦.25

Step 3: Because the number to the right of the ⑦ is less than 5, drop both the 2 and the 5.

The decimal 3467.25 rounded off to the nearest one is 3467.

Example 6: Round off 0.986 to the nearest tenth.
Solution:
Steps 1 and 2: Circle the number in the tenths place.

0.⑨86

Step 3: The number to the right of the ⑨ is greater than 5; therefore, increase the 9 by 1. You will notice that increasing the ⑨ by 1 causes it to become ⑩. According to the way decimals are defined, 10 tenths are 1 one. So, in the decimal 0.⑨86, increase the ⑨ to ⑩, drop the 8 and the 6, and write the ⑩ as 1.0.

The decimal 0.986 rounded off to the nearest tenth is 1.0.

Example 7: Round off 555.51 to the nearest ten.
Solution:
Steps 1 and 2: Circle the number in the tens place.

5⑤5.51

Step 3: Because the number to the right of ⑤ is 5 or greater, increase ⑤ by 1. The 5 to the left of the decimal is changed to a zero. The 5 and 1 to the right of the decimal are dropped.

The decimal 555.51 rounded off to the nearest ten is 560.

Example 8: Round off 45.6789 to the nearest tenth and to the nearest thousandth.

Solution: There are two answers to this problem.

Steps 1 and 2: First, circle the number in the tenths place.

45.⑥789

Step 3: The number to the right of the ⑥ is greater than 5, so increase ⑥ by 1 and drop the 7, 8, and 9; therefore, 45.6789 is rounded off to 45.7 (first answer).

Steps 1 and 2: To obtain the second answer, circle the number in the thousandths place.

45.67⑧9

82 Chapter 3 Decimals

Step 3: The number to the right of the ⑧ is greater than 5, so increase ⑧ to ⑨ and drop the last 9. The decimal 45.6789 rounded off to the nearest thousandth is 45.679 (second answer).

Example 9: Round off 596.8614 to the nearest tenth and to the nearest thousandth.
Solution:
Steps 1 and 2: First, circle the number in the tenths place.

596.⑧614

Step 3: The number to the right of the ⑧ is greater than 5, so increase ⑧ by 1 and drop the 6, 1, and 4. The decimal 596.8614 rounded off to the nearest tenth is 596.9.

Steps 1 and 2: Next, circle the number in the thousandths place.

596.86①4

Step 3: The number to the right of ① is less than 5, so drop the 4. The decimal 596.8614 rounded off to the nearest thousandth is 596.861.

Section 3.4 Readiness Review

Round off each of the following decimals to the place indicated (see Examples 1–7):

1. 8.75 to the nearest one = _____

2. 1.23 to one decimal place = _____

3. 14.816 to two decimal places = _____

4. 25.0110 to the nearest thousandth = _____

Round off each of the following decimals to the nearest tenth and to the nearest thousandth (see Examples 8 & 9):

5. 15.4321 = _____ and _____

6. 101.8345 = _____ and _____

Section readiness review answers (not given in order):

| 3. 14.82 | 4. 25.011 | 1. 9 |
| 5. 15.4; 15.432 | 6. 101.8; 101.835 | 2. 1.2 |

Section 3.4 Review Problems

Round off each of the following decimals to the place indicated (answers to the odd-numbered problems are at the back of the book):

1. 4.13 to the nearest tenth

2. 3.22 to the nearest tenth

3. 21.55 to one decimal place

4. 17.69 to one decimal place

5. 15.861 to the nearest hundredth

6. 37.103 to the nearest hundredth

7. 39.208 to two decimal places

8. 25.555 to two decimal places

9. 13.4082 to the nearest ten

10. 195.5311 to the nearest ten

11. 211.0066 to three decimal places

12. 3.1789 to three decimal places

13. 316.1421 to the nearest thousandth

14. 1001.0125 to three decimal places

15. 475.0123 to the nearest thousandth

Round off each of the following decimals to the nearest tenth and to the nearest thousandth:

16. 1.2346	17. 0.8453	18. 1.3576
19. 5.7162	20. 7.6445	21. 0.5006
22. 0.1302	23. 3.32549	24. 8.14532
25. 0.0045		

Section 3.5 Multiplication and Division of Decimals

Multiplication of decimals is like multiplication of whole numbers. When you multiply decimals, you must place the decimal point correctly in the answer. To determine where to place the decimal, use the following rule.

> **Rule: Multiplying Decimals**
> *Step 1:* Multiply the numbers as if they were whole numbers.
> *Step 2:* Count the *total* number of decimal places to the right of the decimal points in the two numbers you have just multiplied.
> *Step 3:* The total found in Step 2 will tell you how many decimal places the answer has; just count up to that number, starting from the right side of the answer, and put in the decimal point.

Example 1: Multiply 4.27 by 3.5.
Solution:
Step 1: Multiply the numbers as if they were whole numbers.

$$\begin{array}{r} 427 \\ \times\ 35 \\ \hline 2135 \\ 1281 \\ \hline 14945 \end{array}$$

84   Chapter 3 Decimals

Step 2:   Count the total number of decimal places to the right of the decimal points in the numbers being multiplied.

$$\begin{array}{r} 4.27 \quad \text{(two decimal places)} \\ \times \ \ 3.5 \quad \text{(one decimal place)} \\ \hline \text{(three decimal places in the answer)} \end{array}$$

Step 3:   Count three places from right to left and put in the decimal point.

$$\begin{array}{r} 4.27 \\ \times \ \ 3.5 \\ \hline 2\ 135 \\ 12\ 81\ \ \\ \hline 14.945 \end{array}$$

The product of 4.27 and 3.5 is 14.945 (4.27 × 3.5 = 14.945).

Example 2:   Multiply 0.019 by 0.04.
Solution:
Step 1:   Multiply the numbers as if they were whole numbers.

$$\begin{array}{r} 19 \\ \times \ \ 4 \\ \hline 76 \end{array}$$

Step 2:   Count the total number of decimal places to the right of the decimal points in the numbers being multiplied.

$$\begin{array}{r} 0.019 \quad \text{(three decimal places)} \\ \times \ \ 0.04 \quad \text{(two decimal places)} \\ \hline \text{(five decimal places in the answer)} \end{array}$$

Step 3:   Count five places from right to left (inserting zeros as needed) and insert the decimal point.

$$\begin{array}{r} 0.019 \\ \times \ \ 0.04 \\ \hline 0.00076 \end{array}$$

The product of 0.019 × 0.04 = 0.00076.

Example 3:   Dr. Garfield orders a loading dose of Lanoxin® 0.25 mg (a heart drug) every six hours for her patient. How many milligrams of Lanoxin® does the patient receive in 24 hours?

Solution: Read the problem carefully. It states that a dose of the drug is given every six hours, so the patient receives four doses in 24 hours. Each dose is 0.25 mg. To find the answer, you must multiply 4 by 0.25.

Section 3.5 Multiplication and Division of Decimals

Step 1: Multiply the numbers as if they were whole numbers.

$$\begin{array}{r} 25 \\ \times\ 4 \\ \hline 100 \end{array}$$

Step 2: Count the total number of decimal places to the right of the decimal points in the numbers being multiplied.

$$\begin{array}{r} 0.25\ \text{mg} \quad \text{(two decimal places)} \\ \times\quad 4 \quad \text{(no decimal places)} \\ \hline \text{(two decimal places in the answer)} \end{array}$$

Step 3: Count the two places from right to left and insert the decimal point.

$$\begin{array}{r} 0.25\ \text{mg} \\ \times\quad 4 \\ \hline 1.00\ \text{mg} \end{array}$$

The answer is 1.00 mg, which is read "1 milligram of Lanoxin®."

The study of decimal multiplication includes multiplication of decimals by powers of ten, which is described by the following definition and rule.

Definition: Multiplying by *powers of ten* is simply multiplying by 10, 100, 1000, and so on.

Rule: Multiplying Decimals by a Power of Ten
  Step 1: Count the number of zeros in the power-of-ten multiplier.
  Step 2: Move the decimal point in the multiplicand to the right as many places as there are zeros in Step 1.

Example 4: Multiply 3.451 by 100.
Solution:

Step 1: Count the number of zeros in the power-of-ten multiplier.

  100   (There are two zeros.)

Step 2: Move the decimal point in the multiplicand two decimal places to the right.

  $3.451 \times 100 = 3\,45.1$

There are two zeros in the multiplier, 100. The decimal point in the multiplicand is moved two places to the right. The product of $3.451 \times 100 = 345.1$.

Example 5: Multiply 1.7 × 1000.
Solution:
Step 1: Count the number of zeros in the power-of-ten multiplier.

1000 (There are three zeros.)

Step 2: Move the decimal point in the multiplicand three decimal places to the right.

1.700 × 1000 = 1 700.

Extra zeros are inserted as needed.

The answer is 1700.

Now, take some time to learn about dividing decimals. Division of decimals is like division of whole numbers.

Rule: Dividing Decimals
Step 1: Rewrite the division problem in the form used in Chapter 1—that is,

$$\text{DIVISOR} \overline{\smash{)}\text{DIVIDEND}}^{\text{QUOTIENT}}$$

Step 2: If the divisor is a whole number, nothing needs to be changed. If, however, the divisor is a decimal, you must change it to a whole number by moving the decimal point to the right.
Step 3: If you had to shift the decimal point to the right in the divisor, you must also move the decimal point in the dividend the same number of places to the right (insert zeros as necessary).
Step 4: Now that the decimal points in the divisor and the dividend have been taken care of, place a decimal point directly above the one in the dividend—where the quotient will be.
Step 5: Work the problem as if you were dividing whole numbers and don't worry about putting the decimal point in the answer (you have taken care of it already).

Example 6: Divide 1.25 by 5.
Solution:
Step 1: Rewrite the problem as

$5\overline{)1.25}$

Step 2: Because there is no decimal point in the divisor, leave it alone.
Step 3: Does not apply.
Step 4: Place the decimal point in the quotient directly above the decimal point in the dividend.

$5\overline{)1.25}^{\,.}$

Section 3.5 Multiplication and Division of Decimals

Step 5:   Solve the problem as if it were a whole-number division problem.

$$5\overline{\smash{)}1.25} \atop \underline{1\ 0} \atop 25 \atop \underline{25}$$

quotient: .25

The answer is 0.25 (1.25 ÷ 5 = 0.25).

Example 7:   Divide 3.9 by 1.3.
Solution:
Step 1:   Rewrite the problem as

$$1.3\overline{\smash{)}3.9}$$

Step 2:   Move the decimal point in the divisor to the right.

$$1.3\overline{}$$

Step 3:   Now, because you moved the decimal point in the divisor one place to the right, you must also move the decimal point in the dividend one place to the right.

$$1.3\overline{\smash{)}3.9}$$

Step 4:   Place the decimal point in the quotient directly above the decimal point in the dividend.

$$13\overline{\smash{)}39.}$$

Step 5:   Work the problem as if you were dividing whole numbers.

$$13\overline{\smash{)}39.}\atop \underline{39}$$

quotient: 3.

The answer is 3 (3.9 ÷ 1.3 = 3).

Example 8:   Divide 0.009 by 1.5.
Solution:
Step 1:   Rewrite the problem as

$$1.5\overline{\smash{)}0.009}$$

Step 2:   There is a decimal point in the divisor. It must be moved one place to the right.

$$1.5\overline{}$$

Step 3: The decimal point in the dividend must also be moved one place to the right.

$$1.5\overline{)0.0\;09}$$

Step 4: Place the decimal point in the quotient directly above the decimal point in the dividend.

$$15\overline{)\overset{.}{.}09}$$

Step 5: Work the problem as if you were dividing whole numbers (you will need to insert a zero).

$$15\overline{).090}^{\;.006}$$

The answer is 0.006 (0.009 ÷ 1.5 = 0.006).

Example 9: Divide 5.7 by 0.9.
Solution:
Step 1: Rewrite the problem as

$$0.9\overline{)5.7}$$

Step 2: Move the decimal point in the divisor one place to the right.

$$0.9\overline{)}$$

Step 3: Move the decimal point in the dividend also one place to the right.

$$0\;9.\overline{)5\;7.}$$

Step 4: Place the decimal point in the quotient.

$$9\overline{)57\overset{.}{!}}$$

Step 5: Solve the problem (insert zeros as needed).

$$
\begin{array}{r}
6.33 \\
9\overline{)57.00} \\
\underline{54} \\
3\;0 \\
\underline{2\;7} \\
30 \\
\underline{27} \\
3
\end{array}
$$

What has happened? The answer doesn't come out even. This is the time to apply your knowledge of rounding off. In order to round off, you must work the division far enough so that you can follow the rounding-off procedure and can round the answer to a decimal approximation that will suit your needs. In this problem, let's round off to the tenths place. (In actual practice, you may base your decision where to round a number off on the exactness of the equipment you're using; for example, if you're using a syringe that is marked off in tenths of a milliliter, you would measure a 0.88-mℓ dose as 0.9 mℓ, because 0.88 is closer to 0.9 than to 0.8.) (Remember, we will talk about milliliters in Chapter 6.)

To round 6.33 off to the nearest tenth, first circle the tenths position.

6.③3

Because 3 is less than 5, leave the ③ as it is and drop the 3 to the right. The answer is 6.3.

Rule: Dividing Decimals by a Power of Ten
Step 1: Count the number of zeros in the power-of-ten divisor.
Step 2: Move the decimal point in the dividend as many places to the left as there are zeros in the divisor.

Example 10: Divide 15.5 by 100.
Solution:
Step 1: Count the number of zeros in the power-of-ten divisor.

100   (There are two zeros.)

Step 2: Move the decimal point in the dividend two places to the left.

15.5 ÷ 100 = .15 5

The answer is 0.155 (15.5 ÷ 100 = 0.155).

Example 11: Divide 0.3 by 1000.
Solution:
Step 1: Count the number of zeros in the power-of-ten divisor.

1000   (There are three zeros.)

Step 2: Move the decimal point in the dividend three places to the left (insert zeros as necessary).

0.3 ÷ 1000 = .000 3

The answer is 0.0003 (0.3 ÷ 1000 = 0.0003).

Section 3.5 Readiness Review

Multiply the following decimals (see Examples 1, 2, 4, & 5):

1. 15.4
    × 0.7 = _____

2. 21.7
    × 0.83 = _____

3. 0.127
    × 0.04 = _____

4. 5.14 × 10 = _____

5. 17.76 × 1000 = _____

Divide the following decimals (see Examples 6–11). Round off answers to the nearest thousandth when necessary.

6. 29.5 ÷ 8 = _____

7. 12.54 ÷ 6 = _____

8. 72.72 ÷ 0.0909 = _____

9. 3.68 ÷ 0.2 = _____

10. 1.01 ÷ 0.6 = _____

11. 4.37 ÷ 10 = _____

12. 0.6 ÷ 1000 = _____

Section readiness review answers (not given in order):

1. 10.78	3. 0.00508	5. 17760	6. 3.688
7. 2.09	2. 18.011	4. 51.4	8. 800
10. 1.683	11. 0.437	12. 0.0006	9. 18.4

Section 3.5 Review Problems

Multiply the following decimals (answers to the odd-numbered problems are at the back of the book):

1. 63.9
    × 0.5

2. 12.7
    × 0.8

3. 92.1
    × 8.4

4. 16.93
    × 8.04

5. 63.9
    × 21.8

6. 0.032
    × 0.3

7. 0.058
    × 0.2

8. 0.0804
    × 3.5

9. 0.125
    × 6

10. 2.87 × 10 =

11. 555.22 × 1000 =

Divide the following decimals. Round off answers to the nearest thousandth when necessary.

12. $0.312 \div 2 =$ **13.** $2897.5 \div 25 =$ **14.** $0.66 \div 333 =$

15. $0.6\overline{)1.6} =$ **16.** $18.42 \div 100 =$ **17.** $1.937 \div 10 =$

Solve the following word problems:

18. A patient with a duodenal ulcer is placed on atropine 0.5-mg tablets. The patient is instructed to take one tablet with each meal (breakfast, lunch, and dinner) and one tablet at bedtime. If the patient stays on this therapy for 21 days, taking four tablets a day, how many total milligrams of atropine will the patient have taken?

19. John Andrews, a gout patient for some time, takes 1.5 g of Probenecid® daily. He has taken this drug for a year (365 days). How many grams of Probenecid® has he taken?

20. A teenager with acne sees her dermatologist, Dr. Semion, for treatment of the condition. Dr. Semion places the patient on tetracycline (broad-spectrum antibiotic) capsules 0.25 g twice daily. If the patient takes two capsules per day of tetracycline 0.25 g for 30 days, how many grams will the patient have received in all?

21. A patient with a history of congestive heart failure has been on Diuril® 0.5-g tablets, one per day, for a while. If the patient has taken a total of 17.5 g, how many days has he been on the medication?

22. Luteesha Monroe has been placed on 0.2 g of quinidine sulfate, three times daily, for her heart problem. If Luteesha has taken a total of 12.6 g, how many days has she been taking the quinidine?

23. Each of the family members of a hepatitis-A patient (inflammation of the liver) is to receive a gamma-globulin injection for protection against infection. If Martha, the patient's sister, weighs 50 pounds, and the dose of gamma globulin is 0.01 mℓ for each pound of body weight, how much of the drug will she receive?

24. A patient with a urinary-tract infection is given Gantrisin® 0.5 g, four times daily. If the patient has taken a total of 20 g, how many days has he been on the medication?

25. Edward Spence has been placed on a salt-restricted diet. His doctor has recommended a 1.25-g-salt-per-day diet. Mr. Spence has been on this diet for 133 days. How many grams of salt has he consumed in this period of time?

Chapter 3 Readiness Review

Write the following decimals in word form:

1. 4.37 = _____

2. 0.1501 = _____

3. 27.8 = _____

4. 120.025 = _____

Use numbers to write the following decimals:

5. Seven and thirty-two thousandths = _____

6. One hundred nine and five tenths = _____

7. Twenty-one and two hundred twelve thousandths = _____

8. Six hundred twenty-five ten-thousandths = _____

Change the following fractions into decimals. Round off uneven answers to the nearest hundredth.

9. $\frac{1}{2}$ = _____

10. $\frac{4}{15}$ = _____

11. $\frac{2}{7}$ = _____

Convert the following decimals to fractions, using either method discussed. Reduce answers to lowest terms.

12. 0.43 = _____

13. 0.126 = _____

14. 0.511 = _____

Add the following decimals:

15. 14.75 + 74.021 + 111.35 + .7 = _____

16. 0.510 + 0.07 + 0.3 + 0.1256 = _____

17. 4.3 + 15.72 + 22.0004 = _____

Subtract the following decimals:

18. 10.7 − 3.29 = _____

19. 73.0045 − 1.31 = _____

20. 8.146 − 0.2 = _____

Round off each of the following decimals to the place indicated:

21. 5.42 to the nearest tenth = _____

22. 68.8 to the nearest one = _____

23. 0.85 to the nearest tenth = _____

24. 334.7216 to the nearest thousandth = _____

Round off each of the following decimals to the nearest tenth and to the nearest thousandth:

25. 4.3721 = _____ and _____

26. 10.94379 = _____ and _____

Multiply the following decimals:

27. 14.3
   × 0.8 = _____

28. 25.5
   × 0.42 = _____

29. 0.131
   × 1.7 = _____

30. 9.137 × 100 = _____

Divide the following decimals. Round off uneven answers to the nearest thousandth.

31. 47.3 ÷ 6 = _____

32. 15.55 ÷ 5 = _____

33. 36.36 ÷ 0.0606 = _____

34. 5.87 ÷ 10 = _____

35. 0.7 ÷ 100 = _____

Chapter readiness review answers (not given in order):

33. 600	34. 0.587	35. 0.007
29. 0.2227	30. 913.7	31. 7.883
32. 3.11	25. 4.4; 4.372	26. 10.9; 10.944
27. 11.44	28. 10.710	21. 5.4
22. 69	23. 0.9	24. 334.722
17. 42.0204	18. 7.41	19. 71.6945
20. 7.946	13. $^{63}/_{500}$	14. $^{511}/_{1000}$
15. 200.821	16. 1.0056	9. 0.5
10. 0.27	11. 0.29	12. $^{43}/_{100}$
5. 7.032	6. 109.5	7. 21.212
8. 0.0625		

2. fifteen hundred one ten-thousandths

4. one hundred twenty and twenty-five thousandths

1. four and thirty-seven hundredths

3. twenty-seven and eight tenths

Chapter 3 Summary

Define these terms in your own words; then compare your definitions with the text.

Key Words

decimal   (p. 63)
decimal point   (p. 63)
power of ten   (p. 63)

place value   (p. 63)
decimal fraction   (p. 63)
rounding off   (p. 80)

Chapter 3 Review Problems

Read the following decimals (answers to all the problems are at the back of the book):

1. 2.4
2. 2.3
3. 17.8
4. 37.88
5. 15.26
6. 4.987
7. 4.0004
8. 0.001
9. 0.00375

Use numbers to write the following decimals:

10. four and eight tenths
11. twenty and one tenth
12. three and twenty-four hundredths
13. forty-four and forty-four hundredths
14. eight and five hundred sixty-seven thousandths
15. three and one hundred seven thousandths
16. fourteen and fourteen thousandths
17. seventy-five ten-thousandths
18. twelve millionths

Change the following decimals into fractions using either method discussed:

19. 0.4
20. 0.1
21. 0.65
22. 0.88
23. 0.22
24. 0.463
25. 0.625
26. 0.875
27. 0.6875

Convert the following fractions to decimals. Round off to four places when necessary.

28. $\frac{3}{5}$
29. $\frac{5}{10}$
30. $\frac{7}{10}$
31. $\frac{15}{16}$
32. $\frac{5}{16}$
33. $\frac{3}{16}$
34. $\frac{2}{3}$
35. $\frac{1}{11}$
36. $\frac{1}{32}$
37. $\frac{7}{32}$

Add the following decimals:

38. 2.1 + 7.604 + 3.0068
39. 145.32 + 2.906 + 5.32
40. 6.05 + 7.12 + 8.061 + 4.22
41. 9.05 + 17.003 + 5.043 + 36.57
42. 13.06 + 24.017 + 2.6 + 40.05

43. 251.093 + 1.8 + 40.01 + 16.006

44. 6.05 + 15.007 + 13.06 + 910.215 + 13

45. 7.21 + 3.09 + 71.024 + 3.8 + 10.001

46. 8.601 + 5.304 + 2.4 + 10.46 + 907.003

47. 44.2 + 56.37 + 54. + 16.032 + 49.1

48. 9.1 + .003 + .011 + 14.009

49. 14.37 + 6.0002 + 69. + 1.04

Subtract the following decimals:

50. 8.4
    − 6.05

51. 5.3
    − 1.69

52. 10.58
    − 8.7

53. 17.94
    − 5.63

54. 55.55
    − 9.38

55. 61.125
    − 17.087

56. 42.17
    − 13.898

57. 131.578
    − 18.06

58. 125.158
    − 18.01

Round off each of the following decimals to the place indicated:

59. 3.14 to the nearest one

60. 19.94 to the nearest one

61. 51.25 to one decimal place

62. 67.19 to one decimal place

63. 18.582 to the nearest hundredth

64. 41.683 to the nearest hundredth

65. 13.304 to two decimal places

66. 155.9134 to the nearest thousandth

67. 121.6060 to three decimal places

68. 5.1987 to three decimal places

Round off each of the following decimals to the nearest tenth and to the nearest thousandth:

69. 2.3164

70. 0.1543

71. 6.7151

72. 6.4456

73. 0.1203

74. 4.25394

Multiply the following decimals:

75. 41.2
    × 0.8

76. 1.5
    × 0.6

77. 12.8
    × 0.69

78. 42.15
    × 7.08

79. 72.7
    × 28.1

80. 0.053
    ×   0.2

81. 0.035
    ×   0.6

82. 0.0641
    ×   2.4

83. 0.375
    ×   4

84. 0.0045
   ×   31

85. 3.841 × 100 =

86. 4.25 × 100 =

Divide the following decimals. Round off answers to the nearest thousandth when necessary.

87. 0.127 ÷ 3 =

88. 13.46 ÷ 5 =

89. 1.42 ÷ 0.142 =

90. 0.9⟌.08 =

91. 0.21⟌0.7 =

92. 23.1 ÷ 1000 =

93. 0.6 ÷ 1000 =

94. 48.035 ÷ 10 =

Solve the following word problems:

95. A laboratory technician measures the diameter of several red blood cells (RBC). The recorded measurements are as follows:

    RBC 1 = 5.52 microns
    RBC 2 = 6.7 microns
    RBC 3 = 5.325 microns
    RBC 4 = 7.0 microns

Find the sum of the measurements of RBCs 1, 2, 3, and 4.

96. A patient has been admitted to the intensive care unit (ICU) because of an overdose of barbiturates (sleeping pills). The patient is in a coma, and the barbiturate level per 100 mℓ of blood is as follows:

    blood sample 1 = 1.5 mg
    blood sample 2 = 1.47 mg

Find the difference between the barbiturate levels in blood samples 1 and 2.

97. A 6-month-old baby measures 67.3 cm in length. This same baby, at birth, measured 50.8 cm. How much has the baby grown in six months?

98. A sample of cow's milk is compared with a sample of human milk.

Vitamins	Human milk	Cow's milk	Difference
A	500 IU	220 IU	280 IU
D	10 IU	4.41 IU	?
C	10.8 mg	1.4 mg	?

Find the difference between both the vitamin-D and the vitamin-C content in human and cow's milk. (IU is the abbreviation for international units, which is used as a measurement for some vitamins.)

99. Frank Richards was hospitalized for an acute asthma attack. After initial measures were started, Frank was maintained on a total of 6.0 g of aminophylline over a three-day period. The aminophylline comes in a 0.5-g container, and one container equals one dose. The doses were given at regular intervals. How many times per day was the drug administered?

100. Dr. JoAnn Leaf sees her patient, Mr. Lewis, in her office for treatment of hypertension (high blood pressure). She prescribes 0.25 mg of reserpine daily. Mr. Lewis will stay on the medication for 14 days and will be rechecked at the end of that time. How many total milligrams of reserpine will he take in 14 days?

4
Ratio, Proportion, and Dimensional Analysis

OBJECTIVES After studying this chapter, you should be able to:

1. Define ratio and proportion.
2. Solve a proportion problem for a missing term.
3. Use ratio and proportion skills to solve problems.
4. Use dimensional analysis (the units method) to solve unit problems and word problems.

Section 4.1 Ratio

> **Definition:** A *ratio* is a comparison of two quantities.

A ratio may be written in the form of a fraction or in ratio form. In the fractional form, one quantity is on the top and the other quantity is on the bottom. In the ratio form, the two quantities being compared are separated by a colon (colon notation).

$\dfrac{A}{B}$ and $A:B$ are both read as "the ratio of A to B."

> **Rule: Changing Fractional-Form Ratios into Ratio (Colon-Notation) Form**
> *Step 1:* The number on top (numerator) is written to the left of the colon; this is the first term of the ratio.
> *Step 2:* The number on the bottom (denominator) is written to the right of the colon; this is the second term of the ratio.

Example 1: Use colon notation to write the following fractional-form ratios:

a. $\dfrac{1}{12}$

b. $\dfrac{1}{1000}$

c. $\dfrac{4}{3}$

Solution: a. $\dfrac{1}{12}$

Step 1: The number on top is the first term of the ratio; write it to the left of the colon.

1:

Step 2: The number on the bottom is the second term of the ratio; write it to the right of the colon.

1:12

The ratio $1/12$ is written as 1:12. Read both forms as "the ratio of 1 to 12."

b. $\dfrac{1}{1000}$

Steps 1 & 2: Write 1, the first term, to the left of the colon and 1000, the second term, to the right of the colon.

1:1000

The ratio $1/1000$ is written as 1:1000 and is read as "the ratio of 1 to 1000."

Section 4.1 Ratio 99

c. $\frac{4}{3}$

Step 1: 4:
Step 2: 4:3

The ratio ⁴⁄₃ is written as 4:3 and is read as "the ratio of 4 to 3."

In an earlier chapter you learned how to reduce fractions to lowest terms. Because ratios can be written in fractional form, you will occasionally have to reduce them to lowest terms, too.

Rule: Reducing Ratios to Lowest Terms
Step 1: The ratio must be in fractional form.
Step 2: Divide both the numerator and the denominator by the largest number that will divide into both of them. (For a review of reducing fractions to lowest terms, see Section 2.1.)

Example 2: Reduce the following ratios to lowest terms:
 a. 32 to 16
 b. 125 to 80
 c. 11 to 6

Solution: **a.** 32 to 16
Step 1: Write the ratio in fractional form.

$$\frac{32}{16}$$

Step 2: Divide both the numerator and the denominator by 16.

$$\frac{32 \div 16}{16 \div 16} = \frac{2}{1}$$

The ratio is ²⁄₁, or 2:1.

b. 125 to 80
Step 1: Write the ratio in fractional form.

$$\frac{125}{80}$$

Step 2: Divide both the numerator and the denominator by 5.

$$\frac{125 \div 5}{80 \div 5} = \frac{25}{16}$$

The ratio is ²⁵⁄₁₆, or 25:16.

c. 11 to 6
Step 1: Write the ratio in fractional form.

$$\frac{11}{6}$$

Step 2: This ratio is already reduced to lowest terms. The ratio is ¹¹⁄₆, or 11:6.

Section 4.1 Readiness Review

Use colon notation to write the following fractional-form ratios (see Example 1):

1. $\frac{1}{7} =$ ___1:7___
2. $\frac{3}{5} =$ ___3:5___
3. $\frac{14}{8} =$ ___14:8___
4. $\frac{1}{2} =$ ___1:2___
5. $\frac{4}{1} =$ ___4:1___
6. $\frac{1}{100} =$ ___1:100___

Reduce the following ratios to lowest terms (see Example 2):

7. 22 to 100 = ___$\frac{22}{100} = \frac{11}{50}$___
8. 44 to 50 = ___$\frac{44}{50} \, \frac{22}{25}$___
9. 15 to 20 = ___$\frac{15}{20} = \frac{3}{4}$___
10. 27 to 4 = ___$\frac{27}{4}$___
11. 16 to 5 = ___$\frac{16}{5}$___
12. 3125 to 1000 = ___$\frac{3125}{1000} \, \frac{125}{40} \, \frac{25}{8}$___

Section readiness review answers (not given in order):

4. 1:2
7. $\frac{11}{50}$
8. $\frac{22}{25}$
10. $\frac{27}{4}$
1. 1:7
3. 14:8
2. 3:5
5. 4:1
6. 1:100
9. $\frac{3}{4}$
11. $\frac{16}{5}$
12. $\frac{25}{8}$

Section 4.1 Review Problems

Use colon notation to write the following fractional-form ratios (answers to the odd-numbered problems are at the back of the book):

1. $\frac{1}{6}$
2. $\frac{2}{4}$
3. $\frac{13}{7}$
4. $\frac{2}{1}$
5. $\frac{1}{4}$
6. $\frac{1}{10}$
7. $\frac{1}{10000}$
8. $\frac{7}{8}$
9. $\frac{4}{3}$
10. $\frac{33}{35}$

Reduce the following ratios to lowest terms:

11. 66 to 50
12. 34 to 20
13. 125 to 1000
14. 345 to 1000
15. 406 to 36
16. 805 to 35
17. 5 to 10
18. 8 to 2
19. 202 to 12

20. 34 to 3   **21.** 65 to 100   **22.** 4 to 28

23. 121 to 13   **24.** 17 to 20   **25.** 15 to 16

Section 4.2 Proportion

A ratio is a comparison between two quantities. A proportion shows that two ratios are equal.

> **Definition:** A *proportion* states that two ratios are equal.

There are two ways to express a proportion; (1) in fractional form

$$\frac{A}{B} = \frac{C}{D}$$

which is read "*A* is to *B* as *C* is to *D*," and (2) in colon-notation form

$$A:B::C:D$$

which is also read as "*A* is to *B* as *C* is to *D*." The first and fourth terms (the *outside* terms) of a proportion are the *extremes*. The second and third terms (the *inside* terms) are the *means*.

$$\underbrace{A : \overbrace{B :: C} : D}_{\text{EXTREMES (outside)}}^{\text{MEANS (inside)}}$$

The golden rule of solving proportion problems is:

> **Rule:** The product of the means equals the product of the extremes.

Be sure to memorize this rule. The product of the means *always* equals the product of the extremes; that is, if you cross multiply the proportion

$$\frac{A}{B} \diagdown\!\!\!\!= \diagup\!\!\!\!\frac{C}{D}$$

then

$$B \times C = A \times D$$

The product of the means, $B \times C$, will equal the product of the extremes, $A \times D$. Now, it stands to reason that, if you know all but one of the terms of a proportion, you can find the missing term. There are two ways to find it.

> **Rule: Method A—To Solve a Proportion When One Term is Missing**
> *Step 1A:* Write the two ratios and the proportion in the colon-notation form. Use an alphabet letter to represent the unknown term.
> *Step 2A:* Identify the means and the extremes.
> *Step 3A:* Compute the product of the means and the product of the extremes, and set them equal.
> *Step 4A:* To solve for the missing term, divide the numbers on each side of the equals sign by the number in front of the unknown term.

> **Rule: Method B—To Solve a Proportion When One Term is Missing**
> *Step 1B:* Write the two ratios of the proportion in fractional form, and set them equal. Use an alphabet letter to represent the unknown term.
> *Step 2B:* Cross multiply the terms of the proportion.
> *Step 3B:* To solve for the missing term, divide the numbers on each side of the equals sign by the number in front of the unknown term.

Example 1: Dr. Dome has prescribed vitamin supplements for her patient, Mr. Schmidt. Each morning he takes 2 green vitamins and 3 red vitamins. During the last few weeks, he has taken 63 red vitamins. How many green vitamins did he take during this time?

Solution: First, solve this problem using Method A; then use Method B to solve it.

METHOD A

Step 1A: Write the two ratios and the proportion in colon form.

$$2 \text{ green} : 3 \text{ red} :: N \text{ green} : 63 \text{ red}$$
$$2 : 3 :: N : 63$$

Step 2A: Identify the means and the extremes.

$$\underbrace{2 : \overbrace{3 :: N} : 63}_{\text{EXTREMES}}^{\text{MEANS}}$$

Step 3A: Compute the product of the means and the product of the extremes, and set them equal.

$$3 \times N = 2 \times 63$$
$$3 \times N = 126$$

Step 4A: Divide the numbers on each side of the equals sign by the number in front of the N.

$$\frac{3 \times N}{3} = \frac{126}{3}$$
$$N = 42$$

The patient took 42 green vitamins.

METHOD B

Step 1B: Write the two ratios of the proportion in fraction form, and set them equal.

$$\frac{2 \text{ green}}{3 \text{ red}} = \frac{N \text{ green}}{63 \text{ red}}$$

$$\frac{2}{3} = \frac{N}{63}$$

Step 2B: Cross multiply the terms of the proportion.

$$\frac{2}{3} \diagdown = \diagup \frac{N}{63}$$

$$3 \times N = 2 \times 63$$
$$3 \times N = 126$$

Step 3B: Divide the numbers on each side of the equals sign by the number in front of the missing term.

$$\frac{3 \times N}{3} = \frac{126}{3}$$
$$N = 42$$

The patient took 42 green vitamins.

Example 2: Solve each of the following proportions for the unknown:

a. $3 : x :: 6 : 10$ b. $S : \frac{1}{2} :: 4 : 20$

c. $5 : 7 :: K : 28$ d. $0.125 : 0.375 :: 2 : z$

Solution: a. $3 : x :: 6 : 10$

Step 1A: The problem is already in colon form.

Step 2A: Identify the means and the extremes.

```
       MEANS
       ┌──┐
    3 : x :: 6 : 10
    └──────────┘
      EXTREMES
```

Step 3A: Compute the product of the means and the product of the extremes, and set them equal.

$$6 \cdot x = 10 \cdot 3$$
$$6 \cdot x = 30$$

Step 4A: Divide the numbers on both sides of the equals sign by 6.

$$\frac{6 \cdot x}{6} = \frac{30}{6}$$
$$x = 5$$

b. $S:\frac{1}{2} :: 4:20$

Step 1A: The problem is already in colon form.

Step 2A: Identify the means and the extremes.

$$\text{MEANS}$$
$$S:\underbrace{\overbrace{\frac{1}{2} :: 4}}_{}:20$$
$$\text{EXTREMES}$$

Step 3A: Multiply the means and the extremes, and set them equal.

$$4 \times \frac{1}{2} = 20 \times S$$
$$2 = 20 \times S$$

Step 4A: Divide the numbers on both sides of the equals sign by 20.

$$\frac{2}{20} = \frac{20 \times S}{20}$$
$$\frac{1}{10} = S$$

c. $5:7 :: K:28$

Step 1A: The problem is already in colon form.

Step 2A: Identify the means and the extremes.

$$\text{MEANS}$$
$$5:\underbrace{\overbrace{7 :: K}}_{}:28$$
$$\text{EXTREMES}$$

Step 3A: Multiply the means and the extremes, and set them equal.

$$7 \times K = 5 \times 28$$
$$7 \times K = 140$$

Step 4A: Divide the numbers on both sides of the equals sign by 7.

$$\frac{7 \times K}{7} = \frac{140}{7}$$
$$K = 20$$

d. $0.125:0.375 :: 2:z$

Step 1A: The problem is already in colon form.

Step 2A: Identify the means and the extremes.

$$\text{MEANS}$$
$$0.125 : 0.375 :: 2 : z$$
$$\text{EXTREMES}$$

Step 3A: Multiply the means and the extremes, and set them equal.

$$0.375 \times 2 = 0.125 \times z$$
$$0.750 = 0.125 \times z$$

Step 4A: To simplify this problem, move the decimal points three places to the right.

$$750 = 125 \times z$$

Divide the numbers on both sides of the equals sign by 125.

$$\frac{750}{125} = \frac{125 \times z}{125}$$
$$6 = z$$

Example 3: Solve each of the following proportions for the unknown:

    a. $\dfrac{1}{3} = \dfrac{m}{9}$ **b.** $\dfrac{7}{12} = \dfrac{14}{Q}$

    c. $\dfrac{T}{10} = \dfrac{5}{25}$ **d.** $\dfrac{4}{B} = \dfrac{13}{39}$

Solution: **a.** $\dfrac{1}{3} = \dfrac{m}{9}$

Step 1B: The problem is already in fractional form.
Step 2B: Cross multiply the terms of the proportion.

$$\frac{1}{3} = \frac{m}{9}$$
$$3 \times m = 1 \times 9$$
$$3 \times m = 9$$

Step 3B: Divide the numbers on both sides of the equals sign by 3.

$$\frac{3 \times m}{3} = \frac{9}{3}$$
$$m = 3$$

    b. $\dfrac{7}{12} = \dfrac{14}{Q}$

Step 1B: The problem is already in fractional form.

106   Chapter 4 Ratio, Proportion, and Dimensional Analysis

Step 2B: Cross multiply the terms of the proportion.

$$\frac{7}{12} = \frac{14}{Q}$$

$$12 \times 14 = 7 \times Q$$
$$168 = 7 \times Q$$

Step 3B: Divide both sides by 7.

$$\frac{168}{7} = \frac{7 \times Q}{7}$$

$$24 = Q$$

c. $\dfrac{T}{10} = \dfrac{5}{25}$

Step 1B: The problem is already in fractional form.
Step 2B: Cross multiply.

$$\frac{T}{10} = \frac{5}{25}$$

$$10 \times 5 = T \times 25$$
$$50 = T \times 25$$

Step 3B: Divide both sides by 25.

$$\frac{50}{25} = \frac{T \times 25}{25}$$

$$2 = T$$

d. $\dfrac{4}{B} = \dfrac{13}{39}$

Step 1B: The problem is already in fractional form.
Step 2B: Cross multiply.

$$\frac{4}{B} = \frac{13}{39}$$

$$B \times 13 = 4 \times 39$$
$$B \times 13 = 156$$

Step 3B: Divide the numbers on both sides of the equals sign by 13.

$$\frac{B \times 13}{13} = \frac{156}{13}$$

$$B = 12$$

Section 4.2 Readiness Review

Solve each of the following proportions for the unknown:

1. $7:49 :: 3:R$ $R =$ _____   (see Example 2)

2. $4:B :: 5:10$ $B =$ _____   (see Example 2)

3. $\frac{2}{3} : 4 :: C : 12$ $C =$ _____   (see Example 2)

4. $\frac{5}{7} = \frac{D}{21}$ $D =$ _____   (see Example 3)

5. $\frac{9}{99} = \frac{5}{T}$ $T =$ _____   (see Example 3)

6. $\frac{E}{0.3} = \frac{0.4}{1.2}$ $E =$ _____   (see Example 3)

Section readiness review answers (not given in order):

3. 2   6. 0.1   5. 55   4. 15   2. 8   1. 21

Section 4.2 Review Problems

Solve each of the following proportions for the unknown (answers to the odd-numbered problems are at the back of the book):

1. $1 : 2 :: 4 : N$

2. $\frac{1}{2} : 4 :: \frac{1}{3} : X$

3. $0.35 : 4 :: 0.70 : G$

4. $14 : 0.11 :: H : 0.22$

5. $\frac{9}{12} : I :: \frac{1}{3} : 15$

6. $J : \frac{1}{10} :: 4 : \frac{1}{100}$

7. $4 : 11 :: K : 33$

8. $5 : L :: 1 : 3$

9. $\frac{M}{10} = \frac{100}{100}$

10. $\frac{5\frac{1}{3}}{16} = \frac{N}{20}$

11. $\frac{41}{P} = \frac{20.5}{1}$

12. $\frac{Q}{3.75} = \frac{4}{7}$

13. $\frac{1}{2} = \frac{5}{R}$

14. $\frac{3}{17} = \frac{6}{S}$

15. $\frac{5}{20} = \frac{T}{10}$

16. $\frac{1}{10} = \frac{a}{100}$

17. $\frac{9}{V} = \frac{27}{60}$

18. $\frac{11}{w} = \frac{22}{4}$

19. $\frac{x}{1} = \frac{0.5}{15}$

20. $\frac{y}{0.2} = \frac{1}{4}$

21. On Mondays in the clinic the ratio of nurses' aides to R.N.s (Registered Nurses) scheduled to work is 5 to 2. Last Monday there were 25 nurses' aides working. How many R.N.s were on duty?

22. In your hospital's pharmacy, the ratio of pharmacists to technicians is 1 to 3. If there are six pharmacists scheduled to work, how many technicians will be working?

108   Chapter 4 Ratio, Proportion, and Dimensional Analysis

23. Mrs. Lee, a diabetic, takes 5 units of one type of insulin, insulin *x*, mixed with 20 units of another type of insulin, insulin *y*. (We'll cover units in Chapter 12. You don't have to know about them to work this problem.) Over a certain period of time Mrs. Lee used 1000 units of insulin *y*. How many units of insulin *x* did she use?

24. Towne Hospital has discovered, by checking records, that, for every 200 live births, 8 babies are twins (four sets of twins). If the hospital has 700 live births during the year, how many sets of twins can be expected?

25. Burn patient A requires ¾ of a jarful of burn cream at each dressing change. Burn patient B requires ⅓ of a jarful at each dressing change. Both patients' dressings are changed the same number of times. If burn patient A uses a total of three jarsful daily, how many jarsful of burn cream will burn patient B use per day?

Section 4.3 Dimensional Analysis

When solving mathematical problems, you must not mix up the units (or dimensions) in the problem. This rule is crucial for nurses to remember when they are performing dosage calculations. For example, if you need to determine a drug dosage in milligrams, and the information you have to find the dosage is given in units *other* than milligrams, you need to know how to convert all units to milligrams to solve your problem. Dimensional analysis will allow you to convert units.

> **Definition:** Changing from one unit to another *within* one system of measurement is called *conversion*.

> **Definition:** Units of length, volume, weight, and so on are referred to as *dimensional quantities*.

A dimensional quantity will have a numerical value that depends on the measurement system and the type of units used. For example, 1 hour may be written as 60 minutes, or as 3600 seconds; 1 yard may be written as 3 feet, or as 36 inches; 100 centimeters may be written as 1 meter; and so on.

> **Definition:** Dimensional analysis (the units method) is a cancellation process whereby dimensional quantities are converted from one system or type of units to another.

> **Rule: Dimensional Analysis**
> *Step 1:* Write down the dimensional quantity that you want to convert.
> *Step 2:* Consult a conversion table to find the conversion factor to use to change the dimensional quantity to the units you want. (Sometimes the conversion takes several steps, and so you may need to refer to the table several times.)
> *Step 3:* Write the conversion factor in the form of a fraction that is equal to 1 (multiplying by one doesn't change any of the values).
> *Step 4:* Cancel out like units and multiply the dimensional quantity you wrote down in Step 1 by the conversion-factor fraction from Step 3. (You may wish to review cancelling in Section 2.6.)

Example 1: Convert 3 hours into minutes.
Solution:

Step 1: Write down the dimensional quantity that you want to convert.

3 hours

Step 2: Consult Table 4-1 on page 110 to find a conversion factor to change hours to minutes. The table states that there are 60 minutes in 1 hour; this is your conversion factor.

Step 3: Write the conversion factor (60 minutes = 1 hour) as a fraction that is equal to 1.

$$\frac{60 \text{ minutes}}{1 \text{ hour}} = 1$$

Step 4: Now, after cancelling out like units (hours), multiply the dimensional quantity (3 hours) by the conversion factor (60 minutes) to obtain the answer.

$$3 \cancel{\text{ hours}} \times \frac{60 \text{ minutes}}{1 \cancel{\text{ hour}}} = 180 \text{ minutes}$$

Because the hours cancel out, you are left with minutes. 3 hours is the same as 180 minutes.

To apply the units method to "real-world" problems, you will first have to learn conversion factors. These factors allow you to convert from one type of unit to another type within the same system. The conversion problems in this section all involve changes of units within one system, such as pints to quarts, quarts to gallons, and so on. Conversion factors will also allow you to convert units in one system to units of another system—that is, metric to household, apothecary to metric, and so on. We'll study the latter type of conversion in Chapter 8.

The following problems include conversion tables with unit abbreviations. Get used to using the abbreviations while you work the problems.

110 Chapter 4 Ratio, Proportion, and Dimensional Analysis

Table 4-1. Commonly Used Unit-Conversion Table for Units of Time

60 seconds (sec) = 1 minute (min)
60 minutes (min) = 1 hour (hr)
24 hours (hr) = 1 day (d)
7 days (d) = 1 week (wk)

The next example involves more than one conversion. You must first find out how many days are in a week; then find out how many hours are in a day.

Example 2: Convert 2 weeks into hours.
Solution:
Step 1: Write down the dimensional quantity that you want to convert.

2 wk

Step 2: Find your conversion factors. Table 4-1 states that there are 7 days in a week and 24 hours in a day.

Step 3: Write the first conversion factor as a fraction that is equal to 1. (Use unit abbreviations.)

$$\frac{7 \text{ d}}{1 \text{ wk}} = 1$$

Then write the second conversion factor as a fraction that is equal to 1. (Put days on the bottom of the fraction.)

$$\frac{24 \text{ hr}}{1 \text{ d}} = 1$$

Step 4: Now, after cancelling out like units, multiply the dimensional quantity you are converting by the two conversion factors.

$$2 \text{ wk} \times \frac{7 \text{ d}}{1 \text{ wk}} \times \frac{24 \text{ hr}}{1 \text{ d}} =$$

You are left with

$$2 \times \frac{7}{1} \times \frac{24 \text{ hr}}{1} = 336 \text{ hr}$$

There are 336 hours in 2 weeks.

Table 4-2. Commonly Used Unit-Conversion Table for Units of Weight

16 ounces (oz) = 1 pound (lb)
2000 pounds (lb) = 1 ton (t)

Example 3: Convert ½ ton to ounces.
Solution:
Step 1: Write down

$$\frac{1}{2} t$$

Step 2: Table 4-2 tells you that 1 ton is equal to 2000 pounds and 1 pound is equal to 16 ounces. These are your conversion factors.

Step 3: Write 1 t = 2000 lb as a fraction that is equal to 1. (Put tons on the bottom.)

$$\frac{2000 \text{ lb}}{1 \text{ t}} = 1$$

Write 1 lb = 16 oz as a fraction equal to 1. (Put pounds on the bottom.)

$$\frac{16 \text{ oz}}{1 \text{ lb}} = 1$$

Step 4: After cancelling out like units, multiply the dimensional quantity you're converting by the two conversion factors.

$$\frac{1 \,\cancel{t}}{2} \times \frac{2000 \,\cancel{\text{lb}}}{1 \,\cancel{t}} \times \frac{16 \text{ oz}}{1 \,\cancel{\text{lb}}} =$$

You are left with

$$\frac{1}{2} \times \frac{2000}{1} \times \frac{16 \text{ oz}}{1} = 16{,}000 \text{ oz}$$

There are 16,000 ounces in ½ ton.

Table 4-3. Commonly Used Unit-Conversion Table for Units of Measure

8 fluid ounces (fl oz) = 1 cup (c)
2 cups (c) = 1 pint (pt)
2 pints (pt) = 1 quart (qt)
4 quarts (qt) = 1 gallon (gal)

Example 4: Convert 1 gallon into pints.
Solution:
Step 1: Write down

1 gal

Step 2: Table 4-3 tells you that there are 4 quarts in 1 gallon and 2 pints in 1 quart.
Step 3: Write 4 qt = 1 gal as a fraction equal to 1. (Put gallons on the bottom.)

$$\frac{4 \text{ qt}}{1 \text{ gal}} = 1$$

Write 1 quart = 2 pints as a fraction equal to 1. (Put quarts on the bottom.)

$$\frac{2 \text{ pt}}{1 \text{ qt}} = 1$$

Step 4: After cancelling out like units, multiply the dimensional quantity you're converting by the conversion factors.

$$1 \cancel{\text{ gal}} \times \frac{4 \cancel{\text{ qt}}}{1 \cancel{\text{ gal}}} \times \frac{2 \text{ pt}}{1 \cancel{\text{ qt}}} =$$

You are left with

$$1 \times \frac{4}{1} \times \frac{2 \text{ pt}}{1} = 8 \text{ pt}$$

There are 8 pints in 1 gallon.

The preceding examples show the approach to take in setting up unit-conversion problems. As you continue your study of mathematics, you will use dimensional analysis to set up, as well as check, complex problems. If you use the correct conversion factors and cancel out like units before you work through the problem, the answer should be in the units you want. The processes of multiplication, division, and so on that are involved in these problems will become routine.

Section 4.3 Readiness Review

Solve the following problems. Refer to the conversion tables when necessary.

1. Convert 1 week into hours.   _____   (see Examples 1 & 2)
2. Convert 1 day into minutes.   _____   (see Examples 1 & 2)
3. Convert 2 tons into ounces.   _____   (see Example 3)
4. Convert 1 gallon into cups.   _____   (see Example 4)

Section readiness review answers (not given in order):

3. 64,000 oz   4. 16 c

1. 168 hr   2. 1440 min

Section 4.3 Review Problems

Solve the following problems. You may refer to the unit-conversion tables in this section (answers to the odd-numbered problems are at the back of the book).

1. Convert 30 days into hours.
2. Convert 1 week into hours.
3. Convert 0.1 week into seconds.

4. Convert 1 day into minutes.
5. Convert 0.4 week into minutes.
6. Convert 5 pounds into ounces.
7. Convert 8/10 gallon into cups.
8. Convert 1 gallon into fluid ounces.
9. Convert 1 quart into fluid ounces.
10. Convert 0.15 quart into fluid ounces.
11. Convert 3.6 pounds into ounces.
12. Convert 5/8 gallon into cups.
13. Convert 7 quarts into fluid ounces.
14. Convert 6.5 quarts into fluid ounces.
15. Convert 3 hours into seconds.
16. Convert 11/12 hour into seconds.
17. Convert 11/12 hour into minutes.
18. Convert 8 days into minutes.
19. Convert 0.5 week into days.
20. Convert 0.5 week into minutes.

Chapter 4 Readiness Review

Use colon notation to write the following fractional-form ratios:

1. $\frac{1}{8} = $ _____
2. $\frac{3}{6} = $ _____
3. $\frac{14}{9} = $ _____
4. $\frac{5}{2} = $ _____
5. $\frac{1}{12} = $ _____
6. $\frac{7}{49} = $ _____

Reduce the following fraction-form ratios to lowest terms and then write them in colon form.

7. 44 to 50 = _____
8. 103 to 12 = _____
9. 1000 to 10 = _____
10. 32 to 6 = _____
11. 34 to 112 = _____
12. 18 to 48 = _____

Solve each of the following proportions for the unknown:

13. $5:6 :: 2:T$ $T = $ _____
14. $\frac{2}{3}:7 :: \frac{1}{6}:R$ $R = $ _____
15. $\frac{2}{8} = \frac{11}{P}$ $P = $ _____
16. $\frac{5}{25} = \frac{z}{100}$ $z = $ _____
17. $\frac{x}{2} = \frac{1.5}{15}$ $x = $ _____
18. $S:16 :: 2:9$ $S = $ _____

Solve the following problems. Refer to the conversion tables in Section 4.3 when necessary.

19. Convert 4 weeks into seconds. _____
20. Convert 3 weeks into minutes. _____
21. Convert 0.45 gallon into cups. _____
22. Convert 0.27 pint into fluid ounces. _____
23. Convert ¾ ton into pounds. _____
24. Convert 10 tons into ounces. _____

Chapter readiness review answers (not given in order):

24. 320,000 oz	23. 1500 lb	22. 4.32 fl oz	21. 7.2 c
20. 30,240 min	19. 2,419,200 sec	18. $S = 3\frac{5}{9}$	17. $x = 0.2$
16. $z = 20$	15. $P = 44$	14. $R = 1\frac{3}{4}$	13. $T = 2\frac{2}{5}$
12. $\frac{3}{8}$	11. $\frac{17}{56}$	10. $\frac{16}{3}$	9. $\frac{100}{1}$
8. $\frac{103}{12}$	7. $\frac{22}{25}$	6. $7:49$	5. $1:12$
4. $5:2$	3. $14:9$	2. $3:6$	1. $1:8$

Chapter 4 Summary

Define the following terms in your own words; then compare your definitions with the text.

Key Words

ratio (p. 98)
colon (p. 98)
proportion (p. 101)
means (p. 101)
extremes (p. 101)
cross multiply (p. 101)
units (p. 108)
units method (p. 108)
dimensional quantities (p. 108)
dimensional analysis (p. 108)

Chapter 4 Review Problems

Use colon notation to write the following fractional-form ratios (answers to all the problems are at the back of the book):

1. $\dfrac{1}{5}$
2. $\dfrac{2}{8}$
3. $\dfrac{13}{10}$
4. $\dfrac{12}{144}$
5. $\dfrac{3}{13}$
6. $\dfrac{10}{1}$
7. $\dfrac{100}{1}$
8. $\dfrac{1}{200}$
9. $\dfrac{1}{130}$
10. $\dfrac{1}{15,000}$

Reduce the following ratios to lowest terms:

11. 24 to 50   **12.** 16 to 32   **13.** 72 to 148

14. 126 to 376   **15.** 45 to 5   **16.** 68 to 14

17. 4 to 100   **18.** 23 to 14

Solve the following proportions:

19. $3:7 :: 9:B$

20. $\frac{5}{16}:20 :: C:30$

21. $\frac{1}{3}:12 :: \frac{1}{8}:D$

22. $1.25:8 :: 3.75:G$

23. $H:\frac{1}{10} :: \frac{1}{100}:\frac{1}{1000}$

24. $\frac{7}{15} = \frac{L}{25}$

25. $\frac{M}{0.65} = \frac{2}{1.3}$

26. $\frac{6}{8} = \frac{10}{N}$

27. $\frac{4}{18} = \frac{7}{P}$

28. $\frac{6}{21} = \frac{Q}{42}$

29. $\frac{10}{1} = \frac{R}{100}$

30. $\frac{18}{7} = \frac{6}{S}$

31. $\frac{3}{T} = \frac{1}{9}$

32. $\frac{U}{5} = \frac{0.1}{10}$

33. $\frac{V}{0.3} = \frac{1}{18}$

Solve the following word problems:

34. Father Flanigan was in the hospital recently. After numerous tests, the physician ordered the following medication for him: 9 green tablets daily and 3 white tablets daily. When Father Flanigan was discharged, he had taken a total of 24 white tablets. How many green tablets did he take while he was in the hospital?

35. Mary, the pharmacist, notices that, for every 3 aspirin suppositories dispensed, she also dispenses 2 acetaminophen (aspirin substitute) suppositories. It's time to order more of both. If Mary decides to order 144 aspirin suppositories, how many acetaminophen suppositories should she order to maintain a ratio of $3:2$?

36. It is the policy of a particular teaching hospital to enroll students from different health-care areas. The ratio of students is as follows: 7 medical students : 13 nursing students : 5 pharmacy students : 2 inhalation-therapy students.

   a. If there are 77 medical students in this program, how many nursing students are in it?

   b. If there are 85 pharmacy students, how many inhalation-therapy students are there?

   c. If there are 39 nursing students in the program during the summer, how many medical students are involved in it then?

Solve the following problems. You may refer to the unit-conversion tables in Section 4.3.

37. Convert 1 week into seconds.
38. Convert ¾ week into minutes.
39. Convert 8 weeks into minutes.
40. Convert ½ day into seconds.
41. Convert ⅕ ton into ounces.
42. Convert ½ gallon into pints.
43. Convert 3 gallons into cups.
44. Convert 6 quarts into cups.
45. Convert ½ pint into fluid ounces.
46. Convert 0.2 quart into fluid ounces.
47. Convert ¼ gallon into pints.
48. Convert 0.75 quart into fluid ounces.

5
Percent and Preparation of Solutions

OBJECTIVES   After studying this chapter, you should be able to:

1. Convert decimals to percents and percents to decimals.
2. Convert fractions to percents and percents to fractions.
3. Use the percent-proportion formula.
4. Work problems that involve percentage strength and ratio strength.

Chapter 5 Percent and Preparation of Solutions

Percents are commonly used in medical dosage calculations. An order for "one-percent hydrocortisone cream" has a definite meaning. By ordering this medication in terms of percent, the physician assures that a patient will receive the same strength of hydrocortisone in a cream base no matter where the patient is treated. Percent, percentage solutions, and ratio strength all have specific meanings.

Definition: A *percent*, written %, means per hundred, or parts of a hundred.

Section 5.1 Converting Decimals to Percents and Percents to Decimals

---

Rule: Converting Decimals to Percents
Step 1: Move the decimal point two places to the right (insert zeros when needed).
Step 2: Write the percent sign, %, after the number.

Example 1: Convert the following decimals to percents:

    a. 0.15   **b.** 0.375

    c. 3.29   **d.** 0.4

Solution: **a.** 0.15

Step 1: Move the decimal point two places to the right.

$$0.15. = 15$$

Step 2: Attach the percent sign.

15%

b. 0.375

Step 1: $0.37.5 = 37.5$

Step 2: 37.5%

c. 3.29

Step 1: $3.29. = 329$

Step 2: 329%

d. 0.4

Step 1: $0.40.$ (Insert a zero after the 4.)

Step 2: 40%

Section 5.1 Converting Decimals to Percents and Percents to Decimals

> **Rule: Converting Percents to Decimals**
> *Step 1:* Remove the percent sign.
> *Step 2:* Move the decimal point two places to the left (insert zeros when needed).

Example 2: Convert the following percents to decimals:

a. 51%   b. 16.4%

c. 3%   d. 205%

e. 0.5%

Solution: a. 51%
Step 1: 51
Step 2: .51 = 0.51

b. 16.4%
Step 1: 16.4
Step 2: .164 = 0.164

c. 3%
Step 1: 3
Step 2: .03 = 0.03 (Insert a zero before the 3.)

d. 205%
Step 1: 205
Step 2: 2.05 = 2.05

e. 0.5%
Step 1: 0.5
Step 2: .005 = 0.005

Section 5.1 Readiness Review

Convert the following decimals to percents (see Example 1):

1. 0.16 = _16%_   2. 3.51 = _351%_
3. 0.8 = _80%_   4. 0.448 = _44.8%_

Convert the following percents to decimals (see Example 2):

5. 52% = _.52_   6. 66.79% = _.6679_
7. .2% = _.002_   8. 307% = _3.07_

120   Chapter 5 Percent and Preparation of Solutions

Section readiness review answers (not given in order):

8. 3.07	**3.** 80%	**2.** 351%
1. 16%	**6.** 0.6679	**5.** 0.52
4. 44.8%	**7.** 0.002	

Section 5.1 Review Problems

Convert the following decimals to percents (answers to the odd-numbered problems are at the back of the book):

1. 0.18
2. 0.359
3. 0.125
4. 0.75
5. 0.627
6. 0.302
7. 8.17
8. 2.22
9. 0.1
10. 0.3

Convert the following percents to decimals:

11. 31%
12. 44%
13. 27.5%
14. 98.3%
15. 4%
16. .6%
17. 102%
18. 789%
19. 55.2%
20. 14%

Section 5.2 Converting Fractions to Percents and Percents to Fractions

When you convert fractions to percents, keep in mind that 100% = 1; thus, when you multiply a number by 100%, you change the form of the number without changing its value.

Rule: Converting Fractions to Percents
Step 1: Multiply by 100.
Step 2: Change improper fractions into mixed numbers whenever possible.
Step 3: Write the percent sign, %, after the number.

Example 1: Convert the following fractions to percents:

a. $\frac{5}{8}$

b. $8\frac{1}{4}$

c. $\frac{2}{10}$

d. $\frac{1}{2}$

Solution: **a.** $\dfrac{5}{8}$

Step 1: $\dfrac{5}{\underset{2}{\cancel{8}}} \times \dfrac{\overset{25}{\cancel{100}}}{1} = \dfrac{125}{2}$

Step 2: $\dfrac{125}{2} = 62\dfrac{1}{2}$

Step 3: $62\dfrac{1}{2}\%$

b. $8\dfrac{1}{4}$

Step 1: $8\dfrac{1}{4} \times \dfrac{100}{1} = \dfrac{33}{\underset{1}{\cancel{4}}} \times \dfrac{\overset{25}{\cancel{100}}}{1} = 825$

Step 2:   (not needed)
Step 3:   825%

c. $\dfrac{2}{10}$

Step 1: $\dfrac{2}{\underset{1}{\cancel{10}}} \times \dfrac{\overset{10}{\cancel{100}}}{1} = 20$

Step 2:   (not needed)
Step 3:   20%

d. $\dfrac{1}{2}$

Step 1: $\dfrac{1}{\underset{1}{\cancel{2}}} \times \dfrac{\overset{50}{\cancel{100}}}{1} = 50$

Step 2:   (not needed)
Step 3:   50%

Rule: Converting Percents to Fractions
   Step 1: Remove the percent sign.
   Step 2: Divide by 100.
   Step 3: Reduce answer to lowest possible terms.

Example 2:   Convert the following percents to fractions:

    a. 25%   **b.** 5%

    c. $3\dfrac{1}{8}\%$ **d.** $\dfrac{1}{4}\%$

    e. 225%

Solution: **a.** 25%
Step 1:   25
Step 2: $\dfrac{25}{100} = \dfrac{1}{4}$

Step 3: $\frac{1}{4}$ (Answer is already in lowest terms.)

b. 5%

Step 1: 5

Step 2: $\frac{5}{100} = \frac{1}{20}$

Step 3: $\frac{1}{20}$

c. $3\frac{1}{8}\%$

Step 1: $3\frac{1}{8}$

Step 2: $\frac{3\frac{1}{8}}{100} = \frac{\frac{25}{8}}{100} = \frac{25}{8} \div \frac{100}{1} = \frac{\overset{1}{\cancel{25}}}{8} \times \frac{1}{\underset{4}{\cancel{100}}} = \frac{1}{32}$

Step 3: $\frac{1}{32}$

d. $\frac{1}{4}\%$

Step 1: $\frac{1}{4}$

Step 2: $\frac{\frac{1}{4}}{100} = \frac{1}{4} \div \frac{100}{1} = \frac{1}{4} \times \frac{1}{100} = \frac{1}{400}$

Step 3: $\frac{1}{400}$

e. 225%

Step 1: 225

Step 2: $\frac{225}{100} = \frac{9}{4}$

Step 3: $\frac{9}{4} = 2\frac{1}{4}$

Section 5.2 Readiness Review

Convert the following fractions to percents (see Example 1):

1. $\frac{2}{3} =$ _____
2. $\frac{3}{10} =$ _____
3. $6\frac{1}{6} =$ _____
4. $\frac{4}{5} =$ _____
5. $21\frac{1}{2} =$ _____

Convert the following percents to fractions (see Example 2):

6. 24% = _____
7. 3% = _____
8. $10\frac{1}{10}\% =$ _____
9. $\frac{1}{5}\% =$ _____
10. 355% = _____

Section 5.3 Percent-Proportion Formula   123

Section readiness review answers (not given in order):

8. $\dfrac{101}{1000}$ 9. $\dfrac{1}{500}$ 7. $\dfrac{3}{100}$ 1. $66\dfrac{2}{3}\%$

3. $616\dfrac{2}{3}\%$ 2. 30% 4. 80% 6. $\dfrac{6}{25}$

5. 2150% 10. $3\dfrac{11}{20}$

Section 5.2 Review Problems

Convert the following fractions to percents (answers to the odd-numbered problems are at the back of the book):

1. $\dfrac{12}{13}$ 2. $\dfrac{6}{7}$ 3. $7\dfrac{1}{15}$

4. $2\dfrac{1}{2}$ 5. $9\dfrac{2}{3}$ 6. $\dfrac{7}{10}$

7. $\dfrac{1}{10}$ 8. $\dfrac{3}{4}$ 9. $\dfrac{5}{6}$

10. $\dfrac{12}{17}$ 11. $\dfrac{11}{15}$ 12. $\dfrac{13}{14}$

Convert the following percents to fractions:

13. 37% 14. 60% 15. 43%

16. 18% 17. 6% 18. 7%

19. $4\dfrac{1}{6}\%$ 20. $2\dfrac{1}{2}\%$ 21. $17\dfrac{11}{12}\%$

22. $\dfrac{1}{8}\%$ 23. $\dfrac{1}{6}\%$ 24. $\dfrac{3}{5}\%$

25. 370%

Section 5.3 Percent-Proportion Formula

Percent statements can be broken down into three sections. It is necessary to identify each section in order to solve percent problems. The three sections are: rate, base, and part.

> **Definition:** The *rate* is the number with a % sign or the word *percent* after it. It is symbolized by *R*.

> **Definition:** The *base* is commonly referred to as the "starting point." It is symbolized by *B*.

> **Definition:** The *part* (sometimes called *percentage*) is the result of multiplying the base by the rate. It is symbolized by *P*.

These three sections can be related by the use of the percent-proportion formula.

$$\frac{\text{Part}}{\text{Base}} = \frac{\text{Rate}}{100}$$

or

$$\frac{P}{B} = \frac{R}{100}$$

Consider the following problem:

What is 35% of 250?

To solve this problem, you should first identify each section and then apply the percent-proportion formula.

$$\underset{R}{35\%} \text{ of } \underset{B}{250} \text{ is equal to } \underset{P}{?}$$

We want to find *P*. Using the percent-proportion formula

$$\frac{P}{B} = \frac{R}{100}$$

we see that

$$\frac{P}{250} = \frac{35}{100}$$

Now cross multiply the proportion to solve for *P*.

$$250 \times 35 = P \times 100$$
$$8750 = P \times 100$$
$$\frac{8750}{100} = \frac{P \times \cancel{100}}{\cancel{100}}$$
$$87.5 = P$$

Therefore, 35% of 250 is equal to 87.5.

> **Rule: Finding Base, Rate, or Part**
> *Step 1:* Identify the sections of the percent problem.
> *Step 2:* Apply the percent-proportion formula.
> *Step 3:* Solve for the unknown in the proportion.

Section 5.3 Percent-Proportion Formula   125

Finding Part:
Example 1:   What is 20% of 18?
Solution:
Step 1:   20% of 18 is equal to ?
            ↑   ↑   ↑
            R   B   P

Step 2: $\dfrac{P}{B} = \dfrac{R}{100}$

$\dfrac{P}{18} = \dfrac{20}{100}$

Step 3:   Solve the proportion, with P as the unknown.

$$18 \times 20 = P \times 100$$
$$360 = P \times 100$$
$$\dfrac{360}{100} = \dfrac{P \times \cancel{100}^{1}}{\cancel{100}_{1}}$$
$$3.6 = P$$

Therefore, 20% of 18 is equal to 3.6.

Finding Base:
Example 2:   6 is 15% of what number?
Solution:
Step 1:   6 is 15% of ?
            ↑   ↑   ↑
            P   R   B

Step 2: $\dfrac{P}{B} = \dfrac{R}{100}$

$\dfrac{6}{B} = \dfrac{15}{100}$

Step 3: $B \times 15 = 6 \times 100$
$B \times 15 = 600$

$$\dfrac{B \times \cancel{15}^{1}}{\cancel{15}_{1}} = \dfrac{\cancel{600}^{40}}{\cancel{15}_{1}}$$

$$B = 40$$

Therefore, 6 is 15% of 40.

Finding Rate:
Example 3:   300 is what percent of 750?
Solution:
Step 1:   300 is what percent of 750?
            ↑   ↑   ↑
            P   R   B

Step 2: $\dfrac{P}{B} = \dfrac{R}{100}$

$\dfrac{300}{750} = \dfrac{R}{100}$

Step 3: $750 \times R = 300 \times 100$
$750 \times R = 30{,}000$

$$\frac{\overset{1}{\cancel{750}} \times R}{\underset{1}{\cancel{750}}} = \frac{\overset{40}{\cancel{30,000}}}{\underset{1}{\cancel{750}}}$$

$$R = 40$$

Therefore, 300 is 40% of 750.

Section 5.3 Readiness Review

Solve the following percent problems:

1. What is 35% of 10?   (see Example 1)
2. What is 20% of 24?   (see Example 1)
3. 28 is 7% of what number?   (see Example 2)
4. 0.5 is 40% of what number?   (see Example 2)
5. 40 is what percent of 20?   (see Example 3)
6. 110 is what percent of 275?   (see Example 3)

Section readiness review answers (not given in order):

6. 40%   3. 400   2. 4.8   1. 3.5   5. 200%   4. 1.25

Section 5.3 Review Problems

(Answers to the odd-numbered problems are at the back of the book.)

1. What is 52% of 85?   2. What is 190% of 40?
3. What is 3% of 61?   4. What is 1.4% of 12?
5. ¼ is 80% of what number?   6. 32 is 49% of what number?
7. 49 is 49% of what number?   8. 64 is 98% of what number?
9. 10 is what percent of 144?   10. 36 is what percent of 1?
11. 800 is what percent of 155?   12. 17 is what percent of 60?
13. Tagamet®, a drug that is now used to treat duodenal ulcers, underwent exhaustive studies prior to its release. In one study, 9 out of 11 patients' duodenal ulcers healed after being treated with Tagamet®. What is the percent of patients with healed duodenal ulcers?
14. The incidence of a particular side effect of Tagamet®, dizziness and rash, is 1%. If 3045 patients use the drug, what number are expected to experience this side effect?
15. Herpes encephalitis, a serious viral infection that involves the brain, has a death rate of 70%. A new drug, Ara-A®, has reduced this rate. In one study of 18 Herpes-encephalitis patients taking Ara-A®, 5 patients died. What is the percent mortality in the study group?

Section 5.4 Preparation of Solutions — Percentage Strength

As a nurse you will be required to apply external preparations such as soaks, irrigations, topical solutions, and so on. In many hospitals, the actual preparation will be done in the pharmacy, but there will be times when the nurse must assume this responsibility. It is for this reason that every nurse should become familiar with the preparation of solutions. Also, calculations involving solution preparation appear on most state board exams for nursing.

> **Definition:** *External solutions* are preparations that are applied to the outer surfaces of the body.

External solutions may be prepared from powders, crystals, tablets, stock solutions (highly concentrated solutions), and so on. The preparation of solutions involves solving proportions; therefore, problems involving solution preparation should appear familiar to you. Just be sure that, if you begin working the problem in the metric system, you end up with metric units; if you begin with the apothecaries' system, end up with apothecaries' units.

The strength of an external solution may be expressed as a percentage strength, a ratio strength, or by a specified amount (for example, "this preparation contains 5 mg of drug per 1 mℓ of solution"). Usually solution-preparation problems involving specified amounts can be easily calculated. Therefore, the rest of this chapter will cover problems involving percentage and ratio strength.

> **Definition:** *Percentage-strength solutions* tell how much of a drug (solute) is dissolved in a given amount of dissolving medium (solvent).

According to the preceding definition, a 5% solution contains 5 parts of a drug dissolved in approximately 95 parts of solvent to make a total of 100 parts of solution.[1] Therefore, "100 mℓ of 5% sodium chloride solution" means that 5 g of pure sodium chloride are diluted with enough water to give a final volume of 100 mℓ. Because the sodium chloride, the pure drug, is in dry form, grams are used to represent the amount. If the pure drug is in liquid form, milliliters or cubic centimeters are used.

> Use *grams* to represent the amount of pure drug in dry form, for example, tablets, crystals, powders, and so on, that is used in preparing an external solution.

---

[1] The concept of percentage strength is really more complicated than this statement indicates. However, it is not the purpose of this text to present this topic in great detail; rather, its purpose is to **give the** nursing student the basic information needed to prepare external solutions.

> Use *milliliters* or *cubic centimeters* to represent the amount of pure drug in liquid form that is used in preparing an external solution.

To solve percentage-strength solution problems, we use the following formula:

> Amount of pure drug = Strength of solution × Amount of solution

Strength, in this formula, refers to the strength of the final solution—that is, the strength of the solution the physician wants the patient to receive;

Amount of solution is the final volume of solution ordered that will contain the prescribed amount of pure drug;

Amount of pure drug is the quantity of drug in pure (100%) form, either dry or liquid, that is needed to make the solution as ordered.

Rule: Preparing Solutions—Percentage Strength

Step 1: Identify the different parts of the problem—that is, strength, amount of solution, amount of pure drug, and so on.

Step 2: Write the percentage strength in fractional form—for example, (dry form*) 5% sodium chloride solution is written as

$$\frac{5 \text{ g}}{100 \text{ m}\ell}$$

and (liquid form*) 10% acetic acid solution is written as

$$\frac{10 \text{ m}\ell}{100 \text{ m}\ell}$$

Step 3: Fill in the following formula

$$\text{Amount of pure drug} = \text{Strength of solution} \times \text{Amount of solution}$$

with information from the problem. Let *x* represent the unknown. Cancel units and solve for the unknown quantity.

Step 4: Explain how the actual solution should be made.

*The problem will usually stipulate whether the pure drug is in dry or liquid form.

Example 1: Solve the following problem:

Dr. Roisman orders a warm 0.9% saline gargle for his patient, who has a sore throat. He instructs the nurse to prepare a 75-mℓ solution. How many grams of pure (100%) sodium chloride are needed? How is this solution prepared?

Solution:

Step 1: Identify the different parts of the problem: 0.9% is the *strength;* 75 mℓ is the *amount of solution;* the *amount of pure drug* is what we're looking for.

Section 5.4 Preparation of Solutions — Percentage Strength 129

Step 2: Write the percentage strength in fractional form: 0.9% sodium chloride solution is written as

$$\frac{0.9 \text{ g}}{100 \text{ m}\ell}$$

You know the sodium chloride is in dry form before mixing, because the problem asks "How many *grams* of . . . needed?")

Step 3: Amount of pure drug = Strength of solution × Amount of solution

$$x = \frac{0.9 \text{ g}}{\underset{4}{\cancel{100 \text{ m}\ell}}} \cdot \overset{3}{\cancel{75 \text{ m}\ell}}$$

$$x = \frac{2.7 \text{ g}}{4} = 0.675 \text{ g}$$

Step 4: To prepare this solution, weigh 0.675 g of sodium chloride. Place this pure drug in a 100-mℓ graduate (see Figure 6-3 on page 147 for an illustration of graduates, which are containers used for measuring). Fill the graduate with warm tap water to the 75-mℓ mark. Stir well. Please note that you don't measure 75 mℓ of warm tap water first. If you did, and you then added the sodium chloride, the volume of the final solution would be *more* than 75 mℓ; that is, the volume of water *plus* the volume of water the salt displaces would equal more than 75 mℓ. This would be incorrect and not what the doctor ordered.

Example 2: Solve the following problem:
Ordered: Vinegar 1% solution in water: 250 mℓ
Directions: Use as a vaginal douche once daily, in the morning, for three days as directed by the physician.
Available: Bottle of vinegar

How much vinegar is required in this solution? How is the solution prepared?

Solution:
Step 1: 1% is the *strength;* 250 mℓ is the *amount of solution;* the *amount of pure drug* is what we're looking for.

Step 2: Write the percentage strength in the fractional form: 1% vinegar solution is written as

$$\frac{1 \text{ m}\ell}{100 \text{ m}\ell}$$

(Because vinegar is a liquid, milliliters are used.)

Step 3: Amount of pure drug = Strength of solution × Amount of solution

$$x = \frac{1 \text{ m}\ell}{\underset{4}{\cancel{100 \text{ m}\ell}}} \cdot \overset{10}{\cancel{250 \text{ m}\ell}}$$

$$x = \frac{10 \text{ m}\ell}{4} = 2.5 \text{ m}\ell$$

Step 4: To prepare this solution, measure 2.5 mℓ of vinegar. Place this amount in a 250-mℓ graduate. Fill the graduate with tap water to the 250-mℓ mark. Stir well.

Section 5.4 Readiness Review

Solve the following problems:

1. A patient is to receive a daily application of a 5% solution. The physician wants the entire contents of a 250-mℓ container to be used for each application. How many grams of pure drug (100%) are required in each solution? Using distilled water, how would you prepare this solution? (See Example 1.)

2. *Ordered:* Neomycin 1% irrigation in normal saline[2]: 1000 mℓ
   Directions: Use as directed.
   Available: Neomycin sulfate powder
   How many grams of neomycin sulfate are required? Using normal saline, how would you prepare this solution? (See Example 2.)

Section readiness review answers (not given in order):

2. 10 g of neomycin sulfate. Weigh 10 g of neomycin sulfate powder and place it in a 1000-mℓ graduate. Fill the graduate with normal saline to the 1000-mℓ mark. Stir well. 1. 12.5 g of drug. Weigh 12.5 g of drug and place it in a 250-mℓ graduate. Fill the graduate with distilled water to the 250-mℓ mark. Stir well.

Section 5.4 Review Problems

Solve the following problems (answers to the odd-numbered problems are at the back of the book):

1. A patient is to receive a daily application of a 3% solution. The physician wants the entire contents of the 125-mℓ container to be used for each application. How many grams of pure drug (100%) are required? Using distilled water, how would you prepare this solution?

2. *Ordered:* Gentian violet 1% in water: 7 mℓ
   Directions: Apply to affected area three to four times daily.
   Available: Gentian violet crystals

   How many grams of gentian violet are required? How is this solution prepared?

3. *Ordered:* Gentian violet 1% in rubbing alcohol: 15 mℓ
   Directions: Apply between toes three times daily.
   Available: Gentian violet crystals

   How many grams of gentian violet are required? How is this solution prepared?

4. *Ordered:* Hydrocortisone 0.5% in lotion: 60 mℓ
   Directions: Apply to rash twice daily as needed.
   Available: Hydrocortisone powder

   How many grams of hydrocortisone are required? How is this solution prepared? (Hydrocortisone powder is to be dissolved in lotion.)

---

[2]Normal saline is a 0.9% solution of sodium chloride.

5. *Ordered:* Neomycin 0.5% irrigation in normal saline: 1500 mℓ
Directions: UD (use as directed)
Available: Neomycin sulfate powder

How many grams of neomycin sulfate powder are required? How is this solution prepared? (Neomycin sulfate powder is to be dissolved in normal saline.)

Section 5.5 Preparation of Solutions — Ratio Strength and Dilution of Stock Solutions

> **Definition:** A *ratio-strength solution* tells you the strength of the solution expressed as a ratio; that is, a 1 : 200 solution contains 1 part of drug dissolved in approximately 199 parts of solvent to make a total of 200 parts of solution.

If the pure drug is in dry form, the ratio 1 : 200 means 1 gram of pure drug dissolved in every 200 mℓ of solution. When the pure drug is in liquid form, the ratio 1 : 200 means 1 milliliter of pure drug dissolved in every 200 mℓ of solution.

> **Rule: Preparing Solutions—Ratio Strength**
> *Step 1:* Identify the different parts of the problem—that is, strength, amount of solution, amount of pure drug, and so on.
> *Step 2:* Write the ratio strength in fractional form—for example, a 1 : 200 ratio strength is written as
>
> $$\frac{1 \text{ g of pure drug}}{200 \text{ m}\ell} \quad \text{(dry form)}$$
>
> or
>
> $$\frac{1 \text{ m}\ell \text{ of pure drug}}{200 \text{ m}\ell} \quad \text{(liquid form)}$$
>
> *Step 3:* Fill in the following formula
>
> $$\text{Amount of pure drug} = \text{Strength of solution} \times \text{Amount of solution}$$
>
> with information from the problem. Let *x* represent the unknown. Cancel units and solve for the unknown quantity.
> *Step 4:* Explain how the actual solution should be made.

Example 1: Solve the following problem:
Ordered: Potassium permanganate 1 : 2000 solution: 500 mℓ
Directions: Use as directed.
Available: Potassium permanganate crystals solution.

How many grams of potassium permanganate crystals are required? How is the solution prepared? (The solvent is distilled water.)

Solution:

Step 1: 1 : 2000 is the *strength;* 500 mℓ is the *amount of solution;* the *amount of pure drug* is what we're looking for.

Step 2: Write the ratio strength in fractional form: 1 : 2000 potassium permanganate solution is written as

$$\frac{1 \text{ g}}{2000 \text{ m}\ell}$$

Step 3: Amount of pure drug = Strength of solution × Amount of solution

$$x = \frac{1 \text{ g}}{\cancel{2000} \text{ m}\ell_4} \cdot \cancel{500}^{1} \text{ m}\ell$$

$$x = \frac{1 \text{ g}}{4} = 0.25 \text{ g}$$

Step 4: To prepare this solution, weigh 0.25 g of potassium permanganate crystals. Place them in a 500-mℓ graduate and fill with distilled water to the 500-mℓ mark. Stir well.

Up to now, the pure drugs in our problems have been 100% pure. However, after a pure drug is dissolved in a solution, you no longer have 100% pure drug and solution. What you have is known as a stock solution.

Definition: A *stock solution* is a mixture containing pure drug dissolved in a liquid.

The strength of a stock solution is indicated on the label in either percent or ratio form. Before you learn how to dilute stock solutions, you need to become familiar with several terms:

Amount of stock is the volume or quantity of stock solution that is required to make the solution as ordered;

Amount of prescribed is the final volume that will contain the drug as ordered;

Strength of prescribed is the final strength or final concentration of the solution the physician wants the patient to receive;

Strength of stock is the labeled strength of the stock solution in the form of a ratio or percentage.

Stock solutions are diluted to reduce the solution strength to the level prescribed by the doctor. If the order is for a solution stronger or more concentrated than the stock solution on hand, then a new solution must be prepared. To dilute stock solutions, use the following formula:

$$\text{Amount of stock} = \frac{\text{Amount of prescribed} \times \text{Strength of prescribed}}{\text{Strength of stock}}$$

Section 5.5 Preparation of Solutions — Ratio Strength and Dilution of Stock Solutions

> **Rule: Diluting Stock Solutions**
> *Step 1:* Identify the different parts of the problem—that is, amount of stock, amount of prescribed, strength of prescribed, and strength of stock solution.
> *Step 2:* Write the *strength of prescribed* and the *strength of stock* in fractional form.
> *Step 3:* Fill in the following formula
>
> $$\frac{\text{Amount of}}{\text{stock}} = \frac{\text{Amount of prescribed} \times \text{Strength of prescribed}}{\text{Strength of stock}}$$
>
> with information from the problem. Let x represent the unknown. Cancel units and solve for the unknown quantity.
> *Step 4:* Explain how the actual solution should be made.

Example 2: Solve the following problem:
Ordered: Potassium permanganate 1 : 2000 solution: 500 mℓ
Directions: Use as directed
Available: Potassium permanganate 4% stock solution

How many milliliters of the stock solution are required? How is this solution prepared? (The solvent is distilled water.)

Solution:

Step 1: *Amount prescribed* is 500 mℓ; *strength of prescribed* is 1 : 2000; *strength of stock* is 4%; *amount of stock* is what we're looking for.

Step 2: Strength prescribed is 1 : 2000, written as

$$\frac{1 \text{ mℓ}}{2000 \text{ mℓ}}$$

Strength of stock is 4%, written as

$$\frac{4 \text{ mℓ}}{100 \text{ mℓ}}$$

Step 3: Amount of stock $= \dfrac{\text{Amount of prescribed} \times \text{Strength of prescribed}}{\text{Strength of stock}}$

$$x = \frac{500 \text{ mℓ} \cdot \dfrac{1 \text{ mℓ}}{2000 \text{ mℓ}}}{\dfrac{4 \text{ mℓ}}{100 \text{ mℓ}}}$$

$$x = 500 \text{ mℓ} \cdot \frac{1}{2000} \cdot \frac{100}{4}$$

$$x = 6.25 \text{ mℓ}$$

Step 4: To prepare this solution, measure 6.25 mℓ of potassium permanganate stock solution. Place this in a 500-mℓ graduate and fill with distilled water to the 500-mℓ mark. Stir well.

134 Chapter 5 Percent and Preparation of Solutions

Example 3: Solve the following problem:
Prepare 300 mℓ of a 1:100 solution from a 3:10 stock solution. How many milliliters of stock solution will be required? How will you prepare this solution? (The solvent is distilled water.)

Solution:

Step 1: *Amount prescribed* is 300 mℓ; *strength prescribed* is 1:100; *strength of stock* is 3:10; *amount of stock* is what we're looking for.

Step 2: Strength prescribed is 1:100, written as

$$\frac{1 \text{ mℓ}}{100 \text{ mℓ}}$$

Strength of stock is 3:10, written as

$$\frac{3 \text{ mℓ}}{10 \text{ mℓ}}$$

Step 3: Amount of stock $= \dfrac{\text{Amount of prescribed} \times \text{Strength of prescribed}}{\text{Strength of stock}}$

$$x = \frac{300 \text{ mℓ} \cdot \dfrac{1 \text{ mℓ}}{100 \text{ mℓ}}}{\dfrac{3 \text{ mℓ}}{10 \text{ mℓ}}}$$

$$x = 300 \text{ mℓ} \cdot \frac{1}{100} \cdot \frac{10}{3}$$

$$x = 10 \text{ mℓ}$$

Step 4: To prepare this solution, measure 10 mℓ of stock solution. Place this in a 500-mℓ graduate and fill with distilled water to the 300-mℓ mark. Stir well.

Section 5.5 Readiness Review

Solve the following problems (see Examples 1 & 2):

1. *Ordered:* Potassium permanganate 1:100: 480 mℓ
   Directions: Use as directed.
   Available: Potassium permanganate crystals

   How many grams of potassium permanganate crystals are required? How is this solution prepared? (Solvent: distilled water)

2. Prepare 222 mℓ of a 1:2 solution from a 60% stock solution. How many milliliters of stock solution will be required? How will you prepare this solution? (Solvent: tap water)

Section readiness review answers (not given in order):

2. 185 mℓ. To prepare this solution, measure 185 mℓ of stock solution. Place this in a 250-mℓ graduate and fill with tap water to the 222-mℓ mark. Stir well. **1.** 4.8 g. To prepare this solution, weigh 4.8 g of potassium permanganate crystals. Place this in a 500-mℓ graduate and fill with distilled water to the 480-mℓ mark. Stir well.

Section 5.5 Review Problems

Solve the following problems (answers to the odd-numbered problems are at the back of the book):

1. *Ordered:* Potassium permanganate 1 : 5000 solution: 75 mℓ
   Directions: Apply sparingly, twice daily as directed.
   Available: Potassium permanganate crystals

   How many grams of potassium permanganate crystals are required? How is this solution prepared? (Solvent: distilled water)

2. Prepare 28 mℓ of a 4 : 700 solution from a 1 : 100 stock solution. How many milliliters of stock solution will be required? How will you prepare this solution? (Solvent: distilled water)

3. Prepare 125 mℓ of a 1 : 10,000 solution from a 1 : 200 stock solution. How many milliliters of stock solution will be required? How will you prepare this solution? (Solvent: alcohol)

4. Prepare 750 mℓ of a 1 : 2000 solution from a 7 : 50 stock solution. How many milliliters of stock solution will be required? How will you prepare this solution? (Solvent: tap water)

5. Prepare 250 mℓ of a 1 : 5000 solution from a 25% stock solution. How many milliliters of stock solution will be required? How will you prepare this solution? (Solvent: tap water)

Chapter 5 Readiness Review

Convert the following decimals to percents:

1. 0.13
2. 0.674
3. 5.26
4. 7.89
5. 0.2
6. 0.9

Convert the following percents to decimals:

7. 26%
8. 68%
9. 41.3%
10. 18.5%
11. 7%
12. 452%

Convert the following fractions to percents:

13. $\frac{10}{17}$
14. $6\frac{1}{4}$
15. $4\frac{4}{5}$
16. $\frac{5}{8}$
17. $\frac{1}{6}$
18. $\frac{23}{25}$

Convert the following percents to fractions:

19. 19%
20. 15%
21. 1%
22. 8%
23. $14\frac{2}{3}$%
24. $\frac{9}{10}$%

25. What is 15% of 68?
26. What is 75% of 45?
27. 64 is 80% of what number?
28. 27 is 100% of what number?
29. 24 is what percent of 60?
30. 56 is what percent of 14?

Solve the following problems:

31. The doctor has ordered a mild 0.5% silver nitrate solution for his patient, Mrs. Fong. He orders 15 mℓ of this solution to be prepared. How much of a 25% stock solution of silver nitrate is required to make the 0.5% solution? How is this solution prepared? (Use distilled water as the solvent.)

32. *Ordered:* Magnesium sulfate 10% solution: 750 mℓ
    Directions: Soak left foot twice daily, in the morning and evening, for 15 min.
    Available: Magnesium sulfate (Epsom salts) crystals

    How many grams of solute are required? How is this solution prepared? (Solvent: distilled water)

33. *Ordered:* Potassium permanganate 1 : 20 stock solution: 125 mℓ
    Available: Potassium permanganate crystals

    How many grams of the potassium permanganate crystals are required? How is this solution prepared? (Solvent: distilled water)

34. Prepare 50 mℓ of a 1 : 100 solution from a 12% stock solution. How many milliliters of stock solution will be required? How will you prepare this solution? (Solvent: tap water)

Chapter readiness review answers (not given in order):

34. 4.17 mℓ. To prepare this solution, measure 4.17 mℓ (realistically, you would measure 4 mℓ or 4.2 mℓ with the equipment available) of stock solution. Place this in a 100-mℓ graduate and fill with tap water to the 50-mℓ mark. Stir well.

33. 6.25 g. To prepare this solution, weigh 6.25 g of potassium permanganate crystals. Place this in a 250-mℓ graduate and fill with distilled water to the 125-mℓ mark. Stir well.

32. 75 g. To prepare this solution, weigh 75 g of magnesium sulfate crystals. Place this in a 1000-mℓ graduate and fill with distilled water to the 750-mℓ mark. Stir well.

31. 0.3 mℓ. To prepare this solution, measure 0.3 mℓ of silver nitrate stock solution. Place this in a 50-mℓ graduate and fill with distilled water to the 15-mℓ mark. Stir well.

30. 400%	29. 40%	28. 27	27. 80
26. 33.75	25. 10.2	24. $9/1000$	23. $11/75$
22. $2/25$	21. $1/100$	20. $3/20$	19. $19/100$
18. 92%	17. $16\,2/3\%$	16. $62\,1/2\%$	15. 480%
14. 625%	13. $58\,14/17\%$	12. 4.52	11. 0.07
10. 0.185	9. 0.413	8. 0.68	7. 0.26
6. 90%	5. 20%	4. 789%	3. 526%
2. 67.4%	1. 13%		

Chapter 5 Summary

Define each item in your own words; then compare your definitions with the text.

Key Words

percent   (p. 118)
percent-proportion formula
    (p. 124)
external solutions   (p. 127)
percentage strength   (p. 127)
stock solutions   (p. 132)

Chapter 5 Review Problems

Convert the following decimals to percents (answers to all the problems are at the back of the book):

1. 0.55
2. 0.23
3. 0.138
4. 0.123
5. 6.75
6. 1.54
7. 4.28
8. 3.72
9. 0.6
10. 0.5

Convert the following percents to decimals:

11. 36%
12. 20%
13. 81.5%
14. 5%
15. 7.3%
16. 14.4%
17. .8%
18. 201%
19. 123%
20. 565%

Convert the following fractions to percents:

21. $\frac{9}{11}$
22. $\frac{1}{8}$
23. $\frac{5}{6}$
24. $3\frac{1}{2}$
25. $7\frac{1}{4}$
26. $8\frac{1}{3}$
27. $\frac{3}{8}$
28. $\frac{9}{10}$
29. $\frac{3}{5}$
30. $\frac{13}{20}$

Convert the following percents to fractions:

31. 75%
32. 51%
33. 4%
34. 2%
35. $5\frac{2}{5}$%
36. $7\frac{3}{8}$%
37. $\frac{4}{7}$%
38. $\frac{2}{3}$%
39. $12\frac{3}{4}$%
40. $\frac{8}{9}$%

41. What is 150% of 14?
42. What is 9% of 81?
43. What is 80% of 75?
44. ⅓ is 50% of what number?
45. 72 is 10% of what number?
46. 125 is 750% of what number?
47. 130 is 25% of what number?
48. 950 is what percent of 275?
49. 13 is what percent of 156?
50. 49 is what percent of 70?

Solve the following problems:

51. Clinoril® is a drug used to treat arthritis. According to one study, the most common side effects—abdominal pain, nausea, and constipation—occurred in 17% of the 25,000 patients treated. How many patients suffered from abdominal pain, nausea, and constipation?

52. Patients who suffer from "traveler's diarrhea" may be able to prevent it by using doxycycline. In one study, only 1 traveler out of 18 developed diarrhea while being treated. What percent of the test group developed diarrhea?

53. Dr. Day ordered a 0.9% saline irrigation for her patient. She instructs the nurse to prepare a 500-mℓ solution using distilled water. How many grams of pure (100%) sodium chloride are needed? How is this solution prepared?

54. *Ordered:* Magnesium sulfate 5% solution: 112 mℓ
    Directions: Soak index finger daily, in the morning.
    Available: Magnesium sulfate (Epsom salts) crystals

    How many grams of solute are required? How is this solution prepared? (Solvent: distilled water)

55. *Ordered:* Hydrocortisone 1% in lotion: 120 mℓ
    Directions: Apply sparingly to affected area, every eight hours as needed.
    Available: Hydrocortisone powder

    How many grams of solute are required? How is this solution prepared? (Solvent: lotion)

56. *Ordered:* Potassium permanganate 1:150 solution: 55 mℓ
    Directions: Use as directed.
    Available: Potassium permanganate crystals

    How many grams of potassium permanganate crystals are required? How is this solution prepared? (Solvent: distilled water)

57. Prepare 1000 mℓ of a 1:5000 solution from a 1:20 stock solution. How many milliliters of stock solution will be required? How will you prepare this solution? (Solvent: distilled water)

58. Prepare 45 mℓ of a 1:1500 stock solution from a 1:80 stock solution. How many milliliters of stock solution will be required? How will you prepare this solution? (Solvent: alcohol)

59. Prepare 175 mℓ of a 1:1000 solution from a 17% stock solution. How many milliliters of stock solution will be required? How will you prepare this solution? (Solvent: alcohol)

60. Prepare 76 mℓ of a 1:30 solution from a 76% stock solution. How many milliliters of stock solution will be required? How will you prepare this solution? (Solvent: distilled water)

Two
Pharmacy Skills

6
Metric System: Weight and Measure

OBJECTIVES   After studying this chapter, you should be able to:

1. Identify the metric prefixes and the unit values they represent.
2. Work problems that involve metric measurements of length, volume, and weight.
3. Convert temperatures from the Fahrenheit thermometric scale to the Celsius scale and vice versa.

Section 6.1 Units of Length

> **Definition:** The *metric system* is a system of weight and measure based on powers of 10.

The metric system was developed by the French in the late 1700s. The United States accepted the metric system as a legal standard of measure in 1866. Recently, an updated, modern metric system has been developed that includes more accurate definitions of the basic metric units. This system is called the International System of Units, which is abbreviated SI, for Système International d'Unités.

The field of medical science uses SI constantly. It is for this reason that nursing students must become familiar with it and feel confident working metric problems. This textbook will stress the use of SI.

In the metric system, a prefix comes before a unit to denote a power of 10. Table 6-1 provides a table of metric prefixes.

Table 6-1. Metric Prefixes

Prefix		Denotes
Latin:		
milli-	=	0.001 unit
centi-	=	0.01 unit
deci-	=	0.1 unit
Greek:		
deka-	=	10 units
hecto-	=	100 units
kilo-	=	1000 units

Table 6-2 illustrates the units of metric length.

Table 6-2. Metric Lengths

1000 millimeters	=	1 meter
100 centimeters	=	1 meter
10 decimeters	=	1 meter
10 meters	=	1 dekameter
100 meters	=	1 hectometer
1000 meters	=	1 kilometer

The facts given in Table 6-2 lead to several interrelationships, which are illustrated in Table 6-3.

Table 6-3. Useful Metric Lengths

10 millimeters	=	1 centimeter (cm)
10 centimeters	=	1 decimeter (dm)
10 decimeters	=	1 meter (m)
10 meters	=	1 dekameter (dam)
10 dekameters	=	1 hectometer (hm)
10 hectometers	=	1 kilometer (km)

Some metric lengths are more commonly used than others. You will discover that the nursing sciences frequently use millimeters, centimeters, and meters. Occasionally, nurses also use kilometers and microns. A *micron* (μ) is a millionth of a meter (a thousandth of a millimeter).

1 micron = 0.000001 meter
1 meter = 1,000,000 μ

Before working metric problems, take time to familiarize yourself with the preceding tables. It may also be helpful to review Chapter 3, Decimals.

This chapter will discuss two methods for changing units of length in the metric system. One method involves dimensional analysis; the other uses ratio and proportion. Both techniques are popular, and your instructor may ask you to learn either one or the other, or both.

Section 6.1 Units of Length

The basic unit of length is the meter. It is commonly abbreviated as m.

> **Definition:** The SI definition of a *meter* is 1,650,763.73 wavelengths of the orange-red line of the spectrum of the isotope krypton-86 in a vacuum.

The preceding definition sounds quite complicated. However, you won't be expected to walk around the hospital defining the meter in terms of wavelengths of light. Just realize that the meter has a specific definition and that it can be reproduced anywhere in the world using the right equipment. Basically, a meter is somewhat longer than a yard (see Figure 6-1). Originally, it was defined as one ten-millionth of the distance from the earth's equator to the North Pole. But such a definition was useless to modern science, which demanded greater accuracy in measurements.

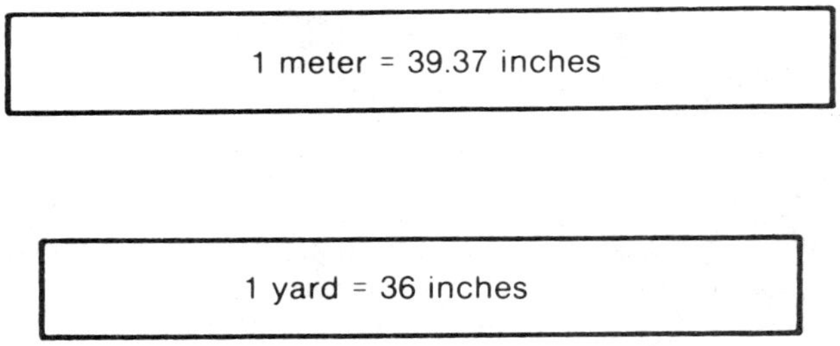

> 1 meter = 39.37 inches

> 1 yard = 36 inches

Figure 6-1. A meter and a yard

Section 6.1 Units of Length

By placing a metric prefix before the word *meter,* it is possible to express multiples of metric length.

Example 1: Solve the following problem:

3.5 m = _____ mm

The problem asks you to determine how many millimeters there are in 3.5 m. Let's see how this problem can be solved quickly.

$$3.5 \text{ m} \cdot \frac{1000 \text{ mm}}{1 \text{ m}} = 3500 \text{ mm} \quad \text{(method A)}$$

or

$$\frac{3.5}{x} = \frac{1}{1000}$$
$$x = 3500 \text{ mm} \quad \text{(method B)}$$

Example 1 has been solved using two different methods. Let's first study method A—dimensional analysis—then method B—ratio and proportion. We will first look at the rules for using method A; then we'll rework Example 1, slowly, step by step, using this method. Next we'll follow the same procedure using method B.

Rule: Changing Units of Length in the SI Metric System
Method A: Dimensional Analysis
Step 1A: Write down the given length.
Step 2A: Using dimensional analysis and known conversion factors, express 1 in fractional form(s) in order to obtain the desired unit and eliminate the given unit.
Step 3A: Cancel units and multiply the given length by the expression(s) for 1.

Use method A to solve the same problem:

3.5 m = _____ mm

Solution:

Step 1A: Write down the given length.

3.5 m

Step 2A: Express 1 in fractional form with millimeters on top and meters on the bottom. (This will cause meters to cancel out in Step 3A.)

$$\frac{1000 \text{ mm}}{1 \text{ m}} = 1$$

Step 3A: Cancel units and multiply the given length by the expression for 1.

$$3.5 \text{ m} \cdot \frac{1000 \text{ mm}}{1 \text{ m}} = 3500 \text{ mm}$$

Therefore, 3.5 m = 3500 mm.

144   Chapter 6 Metric System: Weight and Measure

> **Rule: Changing Units of Length in the SI Metric System**
> **Method B: Ratio and Proportion**
> *Step 1B:* Write the conversion as a ratio in fractional form; put the number you are converting *from* on top, and use x on the bottom to represent the number of units you are converting *to*.
> *Step 2B:* Write the fact(s) from Tables 6-2 and 6-3 that deal with both the units you started with and the units you want to convert to. Combine fact(s) when necessary. Write the fact(s) in fractional form.
> *Step 3B:* Write the ratios from Steps 1 and 2 as a proportion and cross multiply.
> *Step 4B:* To solve for the unknown quantity, divide the numbers on each side of the equals sign by the number on the same side as the x.

Now use method B to solve the same problem (3.5 m = _____ mm).

Solution:

Step 1B: Write the conversion as a ratio in fractional form, using x to represent the number of units you are converting to.

$$\frac{3.5}{x}$$

Step 2B: Because 1 m = 1000 mm, we write

$$\frac{1}{1000}$$

Step 3B: Write the two ratios as a proportion and cross multiply.

$$\frac{3.5}{x} = \frac{1}{1000}$$
$$x \cdot 1 = (3.5)(1000)$$
$$x \cdot 1 = 3500$$

Step 4B: To solve for the unknown quantity, divide the numbers on each side of the equals sign by the number on the same side as the x.

$$\frac{x \cdot 1}{1} = \frac{3500}{1}$$
$$x = 3500 \text{ mm}$$

Therefore, 3.5 m = 3500 mm.

The dimensional-analysis method can be used to check the ratio-and-proportion method, and vice versa.

Example 2: Use method B to solve the following problem:

$$4 \mu = \underline{\qquad} \text{ cm}$$

Solution:

Step 1B: Write the problem as a ratio in fractional form using x to represent the number of units you are converting to.

$$\frac{4}{x}$$

Section 6.1 Units of Length 145

Step 2B: Because 1,000,000 μ = 1 m = 100 cm, it follows that

$$1,000,000 \ \mu = 100 \text{ cm}$$

Now write this relationship in fractional ratio form.

$$\frac{1,000,000}{100}$$

Step 3B: Write the two ratios as a proportion and cross multiply.

$$\frac{4 \ \mu}{x \text{ cm}} = \frac{1,000,000 \ \mu}{100 \text{ cm}}$$
$$x \cdot 10,000 = (4)(1)$$
$$x \cdot 10,000 = 4$$

Step 4B: To solve for the unknown quantity, divide the numbers on each side of the equals sign by the number on the same side as the x.

$$\frac{x \cdot \cancel{10,000}}{\cancel{10,000}} = \frac{4}{10,000}$$

To divide by 10,000, move the decimal point four places to the left.

$$x = \frac{4}{10,000} = 0.0004 \text{ cm}$$

Therefore, 4 μ = 0.0004 cm.

Section 6.1 Readiness Review

Solve the following problems:

1. 18 μ = _____ m   (see Example 2)
2. 123 μ = _____ km   (see Example 2)
3. 9.89 mm = _____ cm   (see Example 2)
4. 0.06865 m = _____ cm   (see Example 1)
5. 4.157 m = _____ mm   (see Example 1)
6. 6.88 mm = _____ dm   (see Example 2)

Section readiness review answers (not given in order):

2. 0.000000123 km   4. 6.865 cm   6. 0.0688 dm

3. 0.989 cm   1. 0.000018 m   5. 4157 mm

Section 6.1 Review Problems

Solve the following problems (answers to the odd-numbered problems are at the back of the book):

1. 82 μ = _____ m   2. 50 μ = _____ m

3. 76 μ = _____ m   4. 2 μ = _____ mm

146   Chapter 6 Metric System: Weight and Measure

5. 4 μ = _____ mm
6. 8 μ = _____ mm
7. 0.3 μ = _____ cm
8. 0.61 μ = _____ cm   .000061
9. 0.5 μ = _____ cm
10. 300 μ = _____ km   .0000003
11. 141 μ = _____ km
12. 0.27 mm = _____ m   .00027
13. 657 mm = _____ m
14. 0.056 mm = _____ m
15. 4.77 mm = _____ cm
16. 1.79 mm = _____ cm
17. 32.81 mm = _____ km
18. 58.95 mm = _____ km
19. 8109 mm = _____ m
20. 1234 cm = _____ m
21. 509 cm = _____ m
22. 8.333 cm = _____ m
23. 6.25 cm = _____ m
24. 14 cm = _____ m
25. 503.94 cm = _____ km
26. 92 cm = _____ km
27. 4931 cm = _____ km
28. 502 cm = _____ km
29. 0.248 m = _____ cm
30. 0.05 m = _____ cm
31. 0.707 m = _____ cm
32. 0.125 m = _____ cm
33. 10.303 m = _____ mm
34. 8.11 m = _____ mm
35. 3.281 m = _____ mm
36. 26 m = _____ mm
37. 0.325 m = _____ μ
38. 0.001 m = _____ μ
39. 4.73 m = _____ cm
40. 0.5 m = _____ cm

Section 6.2 Units of Volume

In SI, the basic unit of volume is the cubic meter, m³. (The term *cubic meter* can be written as m × m × m or as m³.) In the medical community, the basic unit of volume is the liter. It is symbolized as ℓ. Even though the liter is not an official SI unit, it is defined in terms of metric units and is thus widely used throughout the world. This text will use the liter as the basic unit of volume.

> **Definition:** A *liter* is represented as the volume of water in a cube that measures 10 cm × 10 cm × 10 cm (see Figure 6-2).

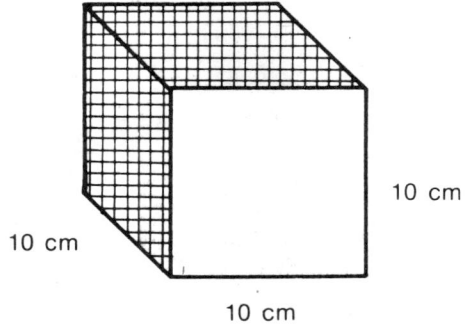

Figure 6-2. Representation of a liter

Again, the metric prefixes (see Table 6-1) are used to express the multiples of metric volume (see Table 6-4).

Table 6-4. Metric Volumes

1000 milliliters	= 1 liter
100 centiliters	= 1 liter
10 deciliters	= 1 liter
10 liters	= 1 dekaliter
100 liters	= 1 hectoliter
1000 liters	= 1 kiloliter

The facts given in Table 6-4 lead to several interrelationships, which are illustrated in Table 6-5.

Table 6-5. Useful Metric Volumes

10 milliliters	= 1 centiliter (cℓ)
10 centiliters	= 1 deciliter (dℓ)
10 deciliters	= 1 liter (ℓ)
10 liters	= 1 dekaliter (daℓ)
10 dekaliters	= 1 hectoliter (hℓ)
10 hectoliters	= 1 kiloliter (kℓ)
1 liter (ℓ) = 1000 mℓ = 1000 cc	

As stated in the definition, a liter is equal to the volume of fluid contained in a cube that has the dimensions 10 cm × 10 cm × 10 cm or 1000 cm³. A cubic centimeter (cm³) is the same volume as a milliliter (mℓ = 0.001 ℓ). A cubic centimeter (cm³) is commonly referred to as a "cc." In medicine the terms *cc* and *mℓ* are used interchangeably. Therefore:

$$1\ \ell = 1000\ \text{cc}$$
$$1\ \ell = 1000\ \text{m}\ell$$
$$\frac{1\ \ell}{1000\ \text{cc}} = \frac{1000\ \text{m}\ell}{1\ \ell} = \frac{1\ \text{m}\ell}{1\ \text{cc}}$$

Figure 6-3 shows some common examples of metric graduates, used to measure fluid volume. Graduates A, C, and D are of the conical variety. Graduate B is of the cylindrical type.

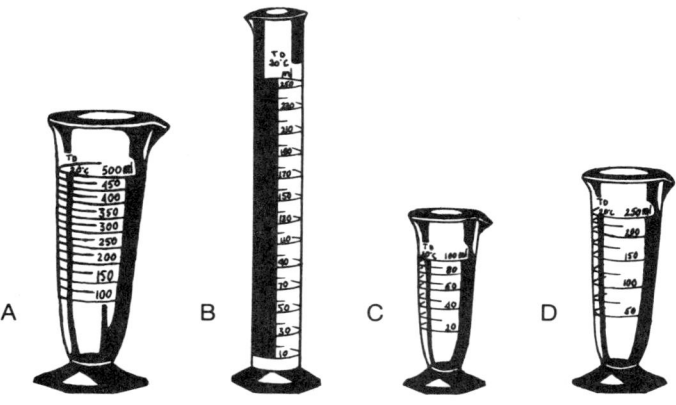

Figure 6-3. Examples of metric graduates

Chapter 6 Metric System: Weight and Measure

> **Rule: Changing Units of Volume in the Metric System**
> **Method A: Dimensional Analysis**
> *Step 1A:* Write down the given volume.
> *Step 2A:* Using dimensional analysis and known conversion factors, express 1 in fractional form(s) in order to obtain the desired unit and eliminate the given unit.
> *Step 3A:* Cancel units and multiply the given volume by the expression(s) for 1.

Example 1: Use method A to solve the following problem:

$$75 \text{ m}\ell = \underline{\hspace{1cm}} \ell$$

Solution:

Step 1A: Write down the given volume.

$$75 \text{ m}\ell$$

Step 2A: Express 1 in fractional form such that liters are on top and milliliters are on the bottom. (This will cause milliliters to cancel out in Step 3A.)

$$\frac{1 \ell}{1000 \text{ m}\ell} = 1$$

Step 3A: Cancel units and multiply the given volume by the expression for 1.

$$75 \cancel{\text{m}\ell} \cdot \frac{1 \ell}{1000 \cancel{\text{m}\ell}} = 0.075$$

Therefore, $75 \text{ m}\ell = 0.075 \ell$.

> **Rule: Changing Units of Volume in the Metric System**
> **Method B: Ratio and Proportion**
> *Step 1B:* Write the conversion as a ratio in fractional form; put the number you are converting *from* on top, and use x on the bottom to represent the number of units you are converting *to*.
> *Step 2B:* Write the fact(s) from Table 6-4, Metric Volumes, and/or Table 6-5, Useful Metric Volumes, that deal with both the units you started with and the units you want to convert to. Combine fact(s) when necessary. Write the fact(s) in fractional form.
> *Step 3B:* Write the ratios from Steps 1 and 2 as a proportion and cross multiply.
> *Step 4B:* To solve for the unknown quantity, divide the numbers on each side of the equals sign by the number on the same side as the x.

Now use method B to solve the same problem ($75 \text{ m}\ell = \underline{\hspace{1cm}} \ell$).

Section 6.2 Units of Volume   149

Solution:
Step 1B: Write the problem as a ratio in fractional form, using x to represent the unknown quantity.

$$\frac{75}{x}$$

Step 2B: Because 1000 mℓ = 1 ℓ, we write

$$\frac{1000}{1}$$

Step 3B: Write the two ratios as a proportion and cross multiply.

$$\frac{75}{x} = \frac{1000}{1}$$
$$x \cdot 1000 = 75 \cdot 1$$
$$x \cdot 1000 = 75$$

Step 4B: To solve for the unknown quantity, divide the numbers on each side of the equals sign by the number on the same side as the x.

$$\frac{x \cdot \cancel{1000}^{1}}{\cancel{1000}_{1}} = \frac{75}{1000}$$

To divide by 1000, move the decimal point three places to the left.

$$x = 0.075 \text{ ℓ}$$

Therefore, 75 mℓ = 0.075 ℓ.

Example 2:   Use method A to solve the following problem:

  4.8 ℓ = _____ mℓ

Solution:
Step 1A: Write down the given volume.

  4.8 ℓ

Step 2A: Express 1 in fractional form such that milliliters are on top and liters are on the bottom.

$$\frac{1000 \text{ mℓ}}{1 \text{ ℓ}} = 1$$

Step 3A: Cancel units and multiply the given volume by the expression for 1.

$$4.8 \cancel{\text{ ℓ}} \times \frac{1000 \text{ mℓ}}{1 \cancel{\text{ ℓ}}} = 4800 \text{ mℓ}$$

Therefore, 4.8 ℓ = 4800 mℓ.

Example 3:   Use method B to solve the following problem:

  375 cc = _____ ℓ

Solution:

Step 1B: Write the conversion as a ratio in fractional form, using x to represent the number of units you are converting to.

$$\frac{375}{x}$$

Step 2B: Because 1000 cc = 1 ℓ, we write

$$\frac{1000 \text{ cc}}{1 \text{ ℓ}}$$

Step 3B: Write the two ratios as a proportion and cross multiply.

$$\frac{375}{x} = \frac{1000}{1}$$

$$x \cdot 1000 = 375 \cdot 1$$

$$x \cdot 1000 = 375$$

Step 4B: Solve for the missing term.

$$\frac{x \cdot 1000}{1000} = \frac{375}{1000}$$

To divide by 1000, move the decimal point three places to the left.

$$x = 0.375 \text{ ℓ}$$

Therefore, 375 cc = 0.375 ℓ.

Section 6.2 Readiness Review

Solve the following problems:

1. 65 mℓ = _____ ℓ   (see Example 1)
2. 2.5 ℓ = _____ mℓ   (see Example 2)
3. 1000 mℓ = _____ ℓ   (see Example 1)
4. 0.41 ℓ = _____ cc   (see Example 3)
5. 500 mℓ = _____ cc   (see Example 3)
6. 17 mℓ = _____ ℓ   (see Example 1)

Section readiness review answers (not given in order):

1. 0.065 ℓ	3. 1 ℓ	6. 0.017 ℓ
2. 2500 mℓ	4. 410 cc	5. 500 cc

Section 6.2 Review Problems

Solve the following problems (answers to the odd-numbered problems are at the back of the book):

1. 75 mℓ = __.075__ ℓ
2. 100 ℓ = __100000__ mℓ
3. .05 ℓ = _____ mℓ
4. 19 mℓ = __.019__ ℓ
5. 480 mℓ = _____ ℓ
6. 4.68 mℓ = __.00468__ ℓ
7. 2 mℓ = _____ ℓ
8. 396 cℓ = __39.6__ dℓ
9. 350 mℓ = _____ ℓ
10. 2700 mℓ = __2.7__ ℓ
11. 3150 mℓ = _____ ℓ
12. 140 ℓ = __14__ daℓ
13. 110 daℓ = __11.0__ hℓ
14. 5 hℓ = __50__ daℓ
15. 13 hℓ = __1.3__ kℓ
16. 0.7 ℓ = __700__ mℓ
17. 0.4234 ℓ = __423.4__ mℓ
18. 2000 mℓ = __2__ ℓ
19. 0.4 mℓ = _____ ℓ
20. 18 mℓ = __.018__ ℓ
21. 1000 mℓ = _____ ℓ
22. 250 mℓ = __250__ cc
23. 317 mℓ = _____ cc
24. 24 mℓ = __24__ cc
25. 0.32 ℓ = _____ cc
26. 4.83 ℓ = __4830__ cc
27. 0.27 ℓ = _____ mℓ
28. 0.8 ℓ = __800__ mℓ
29. 0.42 ℓ = _____ cc
30. 0.33 ℓ = __330__ cc
31. 63 ℓ = _____ cc
32. 102 ℓ = __102000__ mℓ
33. 10 ℓ = _____ mℓ
34. 0.01 ℓ = __10__ mℓ
35. 0.101 ℓ = _____ mℓ
36. 1150 mℓ = __1.15__ ℓ
37. 25 ℓ = _____ cc
38. 0.03 ℓ = __30__ cc
39. 7.96 ℓ = _____ cc
40. 0.6619 ℓ = __661.9__ cc

Section 6.3 Units of Weight

The basic unit of weight in the metric system is the kilogram, symbolized kg. Technically speaking, the kilogram is a mass unit, not a weight unit. However, the distinction is beyond the scope of this text. We will use mass and weight to mean the same thing. (If you desire more information about mass and weight, consult any introductory physics text.)

> **Definition:** A *kilogram* is defined as the mass of the cylinder of platinum-iridium alloy that is kept in the Bureau of Standards near Paris.

The medical community, however, still favors the use of the gram as the basic unit of weight.

> **Definition:** A *gram* is defined as $1/1000$ of a kilogram.

The gram is symbolized by g. The gram used to be defined as the weight of distilled water contained in one cubic centimeter at 4° C and at sea level. (The Celsius thermometric scale is discussed in Section 6.4.) As it turned out, this definition was unsuitable to the needs of modern science, and the SI adopted the current kg standard and defined the gram in terms of it.

Once more, let's refer to the table of metric prefixes (Table 6-1) at the beginning of this chapter and express the multiples of metric weight (see Table 6-6).

Table 6-6. Metric Weights

1,000,000 micrograms	=	1 gram
1000 milligrams	=	1 gram
100 centigrams	=	1 gram
10 decigrams	=	1 gram
10 grams	=	1 dekagram
100 grams	=	1 hectogram
1000 grams	=	1 kilogram
0.000001 gram	=	1 microgram

The facts given in Table 6-6 lead to the interrelationships that are illustrated in Table 6-7.

Table 6-7. Useful Metric Weights

1000 micrograms (mcg)	=	1 milligram (mg)
10 milligrams	=	1 centigram (cg)
10 centigrams	=	1 decigram (dg)
10 decigrams	=	1 gram (g)
10 grams	=	1 dekagram (dag)
10 dekagrams	=	1 hectogram (hg)
10 hectograms	=	1 kilogram (kg)

Nursing sciences use the metric weights of micrograms, milligrams, grams, and kilograms most frequently (see Figure 6-4).

Figure 6-4. A set of metric weights

A connection can be demonstrated between the volume and the weight of water. One milliliter, also known as a cubic centimeter, of water weighs about one gram; that is,

$$1 \text{ g of water} = 1 \text{ m}\ell \text{ of water}$$

or

$$\begin{aligned} 1000 \text{ g of water} &= 1000 \text{ m}\ell \text{ of water} \\ &= 1 \text{ kg of water} \\ &= 1 \ \ell \text{ of water} \end{aligned}$$

The fact that one milliliter of water weighs about one gram will be obvious in some dosage situations. Remember, however, that one milliliter of a denser liquid may weigh more than one gram.

Rule: Changing Units of Weight in the Metric System
  Method A: Dimensional Analysis
Step 1A: Write down the given weight.
Step 2A: Using dimensional analysis and known conversion factors, express 1 in fractional form(s) in order to obtain the desired unit and eliminate the given unit.
Step 3A: Cancel units and multiply the given weight by the expression(s) for 1.

Example 1: Use method A to solve the following problem:

250 mg = _____ g

Solution:

Step 1A: Write down the given weight.

250 mg

Step 2A: Express 1 in fractional form such that grams are on top and milligrams are on the bottom. (This will cause milligrams to cancel out in Step 3A.)

$$\frac{1 \text{ g}}{1000 \text{ mg}} = 1$$

Step 3A: Cancel units and multiply the given weight by the expression for 1.

$$250 \text{ mg} \times \frac{1 \text{ g}}{1000 \text{ mg}} = 0.25 \text{ g}$$

Therefore, 250 mg = 0.25 g.

154   Chapter 6 Metric System: Weight and Measure

> **Rule: Changing Units of Weight in the Metric System**
> **Method B: Ratio and Proportion**
> *Step 1B:* Write the conversion as a ratio in the fractional form; put the number you are converting *from* on top, and use x to represent the number of units you are converting *to*.
> *Step 2B:* Write the fact(s) from Tables 6-6 and 6-7 on metric weight that deal with both the units you started with and the units you want to convert to. Combine fact(s) when necessary. Write the fact(s) in fractional form.
> *Step 3B:* Write the ratios from Steps 1 and 2 as a proportion and cross multiply.
> *Step 4B:* To solve for the unknown quantity, divide the numbers on each side of the equals sign by the number on the same side as the x.

Now use method B to solve the same problem (250 mg = _____ g).

Solution:

Step 1B: Write the conversion as a ratio in fractional form using x to represent the number of units you are converting to.

$$\frac{250}{x}$$

Step 2B: Because 1000 mg = 1 g, we write

$$\frac{1000}{1}$$

Step 3B: Write the two ratios as a proportion and cross multiply.

$$\frac{250}{x} = \frac{1000}{1}$$
$$x \cdot 1000 = 250 \cdot 1$$
$$x \cdot 1000 = 250$$

Step 4B: To solve for the unknown quantity, divide the numbers on each side of the equals sign by the number on the same side as the x.

$$\frac{x \cdot 1000}{1000} = \frac{250}{1000}$$

To divide by 1000, move the decimal point three places to the left.

$$x = 0.25$$

Therefore, 250 mg = 0.25 g.

Example 2: Use method A to solve the following problem:

5 mg = _____ mcg

Solution:

Step 1A: Write down the given weight.

5 mg

Section 6.3 Units of Weight 155

Step 2A: Express 1 in fractional form, such that micrograms are on the top and milligrams are on the bottom.

$$\frac{1000 \text{ mcg}}{1 \text{ mg}} = 1$$

Step 3A: Cancel units and multiply the given weight by the expression for 1.

$$\frac{5 \text{ mg}}{1} \times \frac{1000 \text{ mcg}}{1 \text{ mg}} = 5000 \text{ mcg}$$

So, we see that 5 mg = 5000 mcg.

Example 3: Use method B to solve the following problem:

1576 mℓ of water weighs _____ kg

Solution:

Step 1B: Write the conversion as a ratio in fractional form.

$$\frac{1576}{x}$$

Step 2B: Because 1000 mℓ of water weighs 1 kg, we write

$$\frac{1000}{1 \text{ kg}}$$

Step 3B: Write the two ratios as a proportion and cross multiply.

$$\frac{1576}{x} = \frac{1000}{1}$$
$$x \cdot 1000 = 1576 \cdot 1$$
$$x \cdot 1000 = 1576$$

Step 4B: Solve for the missing term.

$$\frac{x \cdot 1000}{1000} = \frac{1576}{1000}$$

To divide by 1000, move the decimal point three places to the left.

$$x = 1.576$$

Therefore, 1576 mℓ of water weighs 1.576 kg.

Section 6.3 Readiness Review

Solve the following problems:

1. 50 mg = _50000_ mcg   (see Example 2)
2. 193 mcg = _____ mg   (see Example 1)
3. 3.72 mg = _____ g   (see Example 1)

Chapter 6 Metric System: Weight and Measure

4. 14.17 g = _____ mg   (see Example 2)
5. 15 ml of water weighs _____ kg   (see Example 3)
6. 29 l of water weighs _____ kg   (see Example 3)

Section readiness review answers (not given in order):

6. 29 kg   5. 0.015 kg   4. 14,170 mg
3. 0.00372 g   2. 0.193 mg   1. 50,000 mcg

Section 6.3 Review Problems

Solve the following problems (answers to the odd-numbered problems are at the back of the book):

00.005

1. 80 mg = _____ mcg
2. 0.5 mg = 500 mcg
3. 0.37 mg = _____ mcg
4. 700 mcg = .0007 g
5. 1439 mcg = _____ g
6. 1745 mcg = .001745 g
7. 5 mcg = _____ g
8. 0.24 kg = 240 g
9. 0.05 kg = _____ g
10. 70 kg = 70 000 g
11. 1.4 kg = _____ g
12. 5002 mg = .005002 kg
13. 190 mg = _____ kg
14. 3500 mg = .0035 kg
15. 14 mg = _____ kg
16. 10000 mcg = .00001 kg
17. 30000 mcg = _____ kg
18. 15 mcg = .000015 g
19. 18 mcg = _____ g
20. 144 mg = .144 g
21. 63 mg = _____ g
22. 15.6 g = 15600 mg
23. 1052 kg = _____ g
24. 4.88 g = 4880 mg
25. 18.5 g = _____ mg
26. 303 dg = 30300 mg
27. 505 g = _____ dg
28. 18.16 g = 1.816 dag
29. 246 hg = _____ dag
30. 810 mg = .00081 kg
31. 182 mg = _____ kg
32. 222 mg = .222 g
33. 343 mg = _____ g
34. 63.8 kg = 63800 g
35. 30 ml of water weighs _____ kg

Know for test
36. 378 cc of water weighs _____ kg
37. 0.484 l of water weighs _____ g
38. 2.4 l of water weighs _____ kg
39. 25 ml of water weighs _____ kg
40. 0.3 l of water weighs _____ g

Section 6.4 Celsius Temperature

The basic SI unit for temperature is the Kelvin (symbolized K). A *Kelvin* is defined as $1/273.16$ of the triple point of water. The *triple point* is the temperature at which water can exist as a solid, a liquid, or in vapor form.

The Celsius scale is frequently used in medicine. According to this scale, the triple point of water is 0° C, and water boils at 100° C. A connection exists between the Kelvin scale and the Celsius scale: 1 K = 1° C. Therefore, water freezes at 273.16 K = 0° C, and water boils at 373.16 K = 100° C. The Kelvin and Celsius scales are both considered SI units. This text emphasizes the Celsius scale.

> On the Celsius temperature scale (symbolized ° C), water freezes at 0° C and boils at 100° C.

> On the Fahrenheit scale (symbolized ° F), the freezing point of water is 32° F, and the boiling point of water is 212° F.

The interval between freezing and boiling is 100 Celsius degrees, compared to 180 Fahrenheit degrees. Figure 6-5 compares the Celsius and Fahrenheit temperature scales.

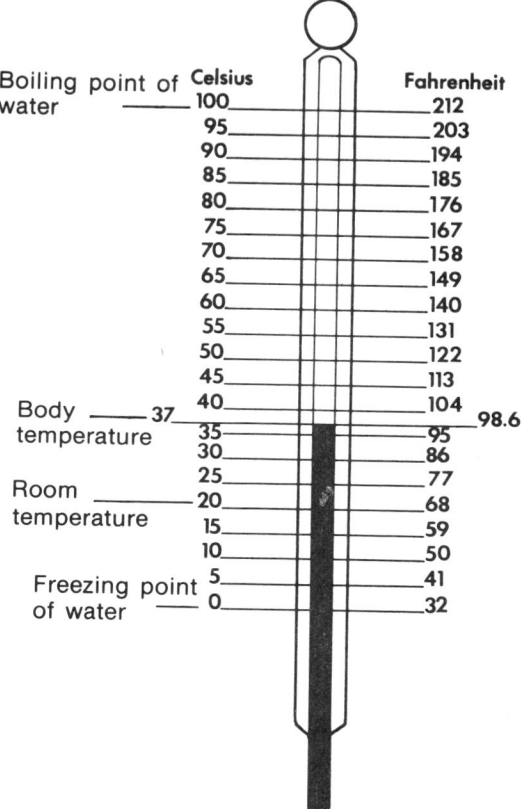

Figure 6-5. A comparison of the Celsius and Fahrenheit temperature scales

On the Celsius scale, the difference between the boiling point and the freezing point is 100° − 0° = 100°. The difference between the boiling and freezing points on the Fahrenheit scale is 212° − 32° = 180°. The ratio of Celsius to Fahrenheit degrees is 100 to 180, or 5 to 9 (5/9).

In many hospitals today, the reports of a patient's body temperature are given in terms of the Fahrenheit (° F) scale. There is, however, a trend toward routine use of the Celsius (° C) temperature scale; therefore, nursing students should become familiar with both temperature systems. (The term *centigrade* is sometimes used instead of Celsius.)

> **Rule: Converting from Fahrenheit to Celsius Temperature**
> *Step 1:* Subtract 32° from the Fahrenheit temperature.
> *Step 2:* Multiply that answer by 5/9.
> *Step 3:* Round off to the degree of accuracy required (in the nursing profession, this is normally the nearest tenth of a degree).
>
> $$C = \frac{5}{9}(F - 32)$$

Example 1: Convert 98.6° F to Celsius.
Solution:
Step 1: 98.6° − 32° = 66.6°
Step 2: C = 5/9 × 66.6 = 37
Step 3: Rounding off is not required. Therefore, 98.6° F = 37° C.

Example 2: Convert 70° F to Celsius.
Solution:
Step 1: 70° − 32° = 38°
Step 2: C = 5/9 × 38 = 21.11
Step 3: Rounded off to the nearest tenth, C = 21.1°. Therefore, 70° F = 21.1° C.

> **Rule: Converting from Celsius to Fahrenheit Temperature**
> *Step 1:* Multiply the Celsius temperature by 9/5.
> *Step 2:* Add 32° to the result found in Step 1.
> *Step 3:* Round off to the degree of accuracy required (in the nursing profession, this is normally to the nearest tenth of a degree).
>
> $$F = \frac{9}{5}C + 32°$$

Example 3: Convert 25° C to Fahrenheit.
Solution:
Step 1: 9/5 × 25° = 45°
Step 2: 45° + 32° = 77°
Step 3: Rounding off is not required. Therefore, 25° C = 77° F.

Example 4: Convert 73° C to Fahrenheit.
Solution:
Step 1: 9/5 × 73° = 131.4°
Step 2: 131.4° + 32° = 163.4°
Step 3: Rounding off is not required. Therefore, 73° C = 163.4° F.

Section 6.4 Readiness Review

Solve the following problems:

1. Convert 106° F to Celsius ___41.1°___ (see Examples 1 & 2)
2. Convert 75.6° F to Celsius ___24.2°___ (see Examples 1 & 2)
3. Convert 100° F to Celsius ___37.8°___ (see Examples 1 & 2)
4. Convert 2° C to Fahrenheit ___35.6°___ (see Examples 3 & 4)
5. Convert 41.2° C to Fahrenheit ___106.2___ (see Examples 3 & 4)
6. Convert 100° C to Fahrenheit ___212°___ (see Examples 3 & 4)

Section readiness review answers (not given in order):

3. 37.8° C	4. 35.6° F	6. 212° F
5. 106.2° F	1. 41.1° C	2. 24.2° C

Section 6.4 Review Problems

(Answers to the odd-numbered problems are at the back of the book.)

48.9° 1. Convert 120° F to Celsius   2. Convert 78.3° F to Celsius   25.7°
28°   3. Convert 82.4° F to Celsius   4. Convert 110° F to Celsius   43.3°
71.1° 5. Convert 160° F to Celsius   6. Convert 125.5° F to Celsius 51.9°
100°   7. Convert 212° F to Celsius   8. Convert 0° C to Fahrenheit   32°
68°   9. Convert 20° C to Fahrenheit   10. Convert 32.5° C to Fahrenheit 90.5°
110.1° 11. Convert 43.4° C to Fahrenheit 12. Convert 60.1° C to Fahrenheit 140.2
181.9° 13. Convert 83.3° C to Fahrenheit 14. Convert 90° C to Fahrenheit 194°
204.4 15. Convert 95.8° C to Fahrenheit

Chapter 6 Readiness Review

Solve the following problems:

1. 23 μ = _____ m
2. 0.25 mm = _____ m
3. 0.7578 m = _____ cm
4. 11.297 m = _____ mm

5. 25.37 mm = _____ cm
6. 750 μ = _____ m
7. 200 cℓ = _____ mℓ
8. 17 cℓ = _____ dℓ
9. 3000 ℓ = _____ daℓ
10. 1250 mℓ = _____ ℓ
11. 72 ℓ = _____ cℓ
12. 412 mℓ = _____ cc
13. 40 mg = _____ mcg
14. 0.01 kg = _____ g
15. 55 cg = _____ mg
16. 161.6 g = _____ mcg
17. 20 mℓ of water weighs _____ kg
18. 1.3 ℓ of water weighs _____ g
19. Convert 130° F to Celsius _____
20. Convert 50° F to Celsius _____
21. Convert 98.5° F to Celsius _____
22. Convert 4° C to Fahrenheit _____
23. Convert 30° C to Fahrenheit _____
24. Convert 73.2° C to Fahrenheit _____

Chapter readiness review answers (not given in order):

21. 36.9° C	22. 39.2° F	23. 86° F	24. 163.8° F
17. 0.02 kg	18. 1300 g	19. 54.4° C	20. 10° C
13. 40,000 mcg	14. 10 g	15. 550 mg	16. 161,600,000 mcg
9. 300 daℓ	10. 1.25 ℓ	11. 7200 cℓ	12. 412 cc
5. 2.537 cm	6. 0.00075 m	7. 2000 mℓ	8. 1.7 dℓ
1. 0.000023 m	2. 0.00025 m	3. 75.78 cm	4. 11297 mm

Chapter 6 Summary

Define each item in your own words, then compare your definitions with the text.

Key Words

metric system (p. 141)
milli- (p. 141)
centi- (p. 141)
deci- (p. 141)
deka- (p. 141)
hecto- (p. 141)
kilo- (p. 141)
SI (p. 141)
micron (p. 142)
meter (p. 142)
liter (p. 146)

cubic centimeter (p. 147)
metric graduates (p. 147)
gram (p. 152)
Kelvin (p. 157)
Celsius (p. 157)
temperature (p. 157)
Fahrenheit (p. 157)
freezing point (p. 157)
boiling point (p. 157)
triple point (p. 157)
centigrade (p. 158)

Chapter 6 Review Problems

Solve the following problems (answers to all the problems are at the back of the book):

1. 31 μ = .000031 m
2. 48 μ = .000048 m
3. 6 μ = .006 mm
4. 5 μ = .005 mm
5. 0.4 μ = .00004 cm
6. 0.35 μ = .000035 cm
7. 344 μ = .000344 m
8. 1000 μ = .001 m
9. 0.047 mm = .000047 m
10. 0.0389 mm = .0000389 m
11. 55.359 mm = .0000553 km
12. 19.95 mm = .00000199 km
13. 70.96 mm = .07096 m
14. 351 mm = .351 m
15. 603 cm = 6.03 m
16. 383.7 cm = 3.837 m
17. 6.87 cm = .0000687 km
18. 7.046 cm = .0000704 km
19. 2388 cm = 23.88 m
20. 5801 cm = 58.01 m
21. 2.989 m = 2989 mm
22. 9.125 m = 9125 mm
23. 3.72 mℓ = .372 cℓ
24. 20 mℓ = 2.0 cℓ
25. 481 mℓ = 4.81 dℓ
26. 15 mℓ = .015 ℓ
27. 0.3 ℓ = 300 mℓ
28. 1.4 mℓ = .0014 ℓ
29. 1.6 mℓ = .0016 ℓ
30. 17 mℓ = .017 ℓ
31. 300 mℓ = 300 cc
32. 75 mℓ = 75 cc
33. 3.55 ℓ = 3550 cc
34. 5.23 ℓ = 5230 mℓ
35. 27 ℓ = 27000 mℓ
36. 41 ℓ = 41000 mℓ
37. 18 ℓ = 18000 cc
38. 103 ℓ = 103000 cc
39. 0.001 ℓ = 1 cc
40. 0.011 ℓ = 11 cc
41. 0.5 ℓ = 500 mℓ
42. 0.45 ℓ = 450 mℓ
43. 30 mg = 30000 mcg
44. 0.2 mg = 200 mcg
45. 824 mcg = .000824 g
46. 7 mcg = .000007 g
47. 0.11 kg = 110 g
48. 75 kg = 75000 g
49. 2.9 kg = 2900 g
50. 4008 mg = .004008 kg
51. 2700 mg = .0027 kg
52. 20,000 mcg = .00002 kg
53. 40,000 mcg = .00004 kg
54. 16 mcg = .00000016 kg
55. 20 mcg = _____ kg
56. 50 cc of water weighs .05 kg
57. 423 cc of water weighs _____ kg
58. 0.6 ℓ of water weighs 600 g
59. 15 ℓ of water weighs 15 kg
60. Convert 32° F to Celsius
61. Convert 40° F to Celsius
62. Convert 60° F to Celsius

63. Convert 70° F to Celsius
64. Convert 81.1° F to Celsius
65. Convert 98.7° F to Celsius
66. Convert 150° F to Celsius
67. Convert 183.9° F to Celsius
68. Convert 91.3° C to Fahrenheit
69. Convert 50° C to Fahrenheit
70. Convert 6° C to Fahrenheit
71. Convert 10° C to Fahrenheit
72. Convert 37.5° C to Fahrenheit
73. Convert 60.1° C to Fahrenheit
74. Convert 74.5° C to Fahrenheit
75. Convert 88.7° C to Fahrenheit

7 Apothecaries' and Household Systems

OBJECTIVES After studying this chapter, you should be able to:

1. List the basic Roman numerals and their corresponding numbers, starting with $\overline{ss} = \frac{1}{2}$ and ending with $m = 1000$.
2. Use all the rules given in the text to combine Roman numerals.
3. Work problems that involve apothecaries' volume and weight.
4. Work problems that involve household fluid measure.

The apothecaries' and household systems of measurement are not as popular as the metric system. However, many physicians continue to use them when writing medication orders and prescriptions. It is important, therefore, that nurses become familiar with and have a working knowledge of all three systems.

Section 7.1 Roman Numerals

Roman numerals are commonly used to record quantities and drug strengths on prescription orders. In addition, the apothecaries' system of weight and measurement traditionally uses Roman numerals.

The Roman method of counting uses several letters of the alphabet to represent specific numbers. This text uses italicized "small" or lowercase letters (see Table 7-1).

Table 7-1. Roman Numerals

Roman Numeral	Number
$\overline{ss}$	½
i	1
v	5
x	10
l	50
c	100
d	500
m	1000

Example 1: Write each of the following as a Roman numeral:

a. 5 = _____   b. 1000 = _____

c. 50 = _____   d. 500 = _____

Solution: Referring to Table 7-1, you can see that:

a. 5 = *v*   b. 1000 = *m*

c. 50 = *l*   d. 500 = *d*

Example 2: Write each of the following as a whole number:

a. *c* = _____   b. *i* = _____

c. *x* = _____   d. *m* = _____

Solution: Once more, refer to Table 7-1.

a. *c* = 100   b. *i* = 1

c. *x* = 10   d. *m* = 1000

There are a few general rules for combining Roman numerals. If you are already familiar with these rules, review the examples and then work the problems.

> **Rule 1:**
> Arrange Roman numerals with the greatest-valued numeral on the left and the least-valued numeral on the right.

> **Rule 2:**
> When a numeral of lesser value follows a numeral of larger value, the lesser-valued numeral is added to the larger one.

Example 3: Write the number 6 as a Roman numeral.
Solution: The number 6 can be broken down to $5 + 1 = 6$. Table 7-1 indicates that the number 5 is written as *v* and the number 1 is written as *i*. The lesser-valued numeral, in this case, *i*, is added to the larger-valued numeral, *v*, to give *vi* (Rule 2). The Roman numerals *vi* are arranged in descending order of value from left to right (Rule 1).

Example 4: Write the number 155 as a Roman numeral.
Solution: The number 155 is broken down to $100 + 50 + 5$. The numbers 100, 50, and 5 are written as the Roman numbers *c*, *l*, and *v*, respectively. The lesser-valued numerals, *l* and *v*, are added to the larger-valued numerals, *c*, to give *clv* (Rule 2). The Roman numerals *clv* are arranged in descending order of value from left to right. They *cannot* be written as *lcv* or *vlc* and still have the same meaning (Rule 1).

> **Rule 3:**
> A Roman numeral may be repeated in sequence, but never more than three times. The one exception to this rule is never repeat the numerals *v*, *l*, and *d*, because, when doubled, their values may be expressed as *x*, *c*, and *m*.

Example 5: Write the number 3 as a Roman numeral.
Solution: The number 3 is broken down to $1 + 1 + 1$. The numbers 1, 1, and 1 are written as *i*, *i*, and *i*, respectively. The Roman numerals *iii* are repeated in sequence but not more than three times (Rule 3). The number 3 is written as *iii*.

Example 6: Write the number 262 as a Roman numeral.
Solution: The number 262 is broken down to $100 + 100 + 50 + 10 + 1 + 1$. The numbers 100, 100, 50, 10, 1, and 1 are written as the Roman numerals *c*, *c*, *l*, *x*, *i* and *i*, respectively. The Roman numerals *c* and *i* are repeated in sequence, but only twice (Rule 3). The Roman numerals *cclxii* are arranged in descending order of value from left to right (Rule 1).

> **Rule 4:**
> If Roman numerals do not occur in descending order of value—that is, if a Roman numeral of lesser value comes *before* a numeral of larger value—the lesser-valued numeral is subtracted from the larger one. (Examples: If *i* comes before *v* or *x*, subtract the *i* from *v* or *x;* if *x* comes before *l* or *c*, subtract the *x* from *l* or *c;* if *c* comes before *d* or *m*, subtract the *c* from *d* or *m*—that is, $iv = 4$, $ix = 9$, $xl = 40$, $xc = 90$, $cd = 400$, and $cm = 900$.) Exceptions to this rule are *il, ic, xd,* and *xm,* which *do not* exist. The numbers 49, 99, 490, and 990 are written as *xlix, xcix, cdxc,* and *cmxc,* respectively.

Example 7: Write the number 4 as a Roman numeral.
Solution: The number 4 is written as $5 - 1 = 4$. The numbers 5 and 1 are written as the Roman numerals *v* and *i*. Place the *i* before the larger-valued *v* and subtract it from the *v;* that is, $4 = iv$ (Rule 4).

Example 8: Write the number 90 as a Roman numeral.
Solution: The number 90 is written as $100 - 10 = 90$. The numbers 100 and 10 are written as the Roman numerals *c* and *x*. Place the *x* before the larger-valued *c* and subtract it from the *c* (Rule 4). Therefore, $90 = xc$.

> **Rule 5:**
> When a Roman numeral of lesser value appears *between* numerals of larger value, apply Rule 4. The difference is then added to the value of the first symbol.

Example 9: Write the number 14 as a Roman numeral.
Solution: The number 14 is written as $10 + 5 - 1$. The numbers 10, 5, and 1 are written as the Roman numerals *x, v,* and *i*. Place the *i* before the larger-valued *v* and subtract it from the *v* (Rule 4): the difference is 4, written as *iv,* which is added to the *x,* giving $xiv = 14$ (Rule 5).

Section 7.1 Readiness Review

Write each of the following as a Roman numeral (see Example 1):

1. $10 =$ _____
2. $500 =$ _____

Write each of the following as a whole number (see Example 2):

3. $v =$ _____
4. $c =$ _____

Write each of the following as a Roman numeral:

5. $51 =$ _____ (Rules 1 & 2)
6. $2\frac{1}{2} =$ _____ (Rules 1, 2, & 3)
7. $43 =$ _____ (Rules 3 & 4)
8. $591 =$ _____ (Rules 4 & 5)

Write each of the following as a whole number (Rules 1–5):

9. vii = _____   10. ccxxii = _____

11. mi = _____   12. mdccclxvi = _____

Section readiness review answers (not given in order):

2. d	4. 100	6. iiss	8. dxci
10. 222	12. 1866	1. x	3. 5
5. li	7. xliii	9. 7	11. 1001

Section 7.1 Review Problems

Write each of the following as a Roman numeral (answers to the odd-numbered problems are at the back of the book):

1. 1 = *i*   2. 1000 = M   3. 100 = C
4. 5 = v   5. 50 = *l*   6. ½ = ss
7. 10 = x   8. 500 = d

Write each of the following as a whole number:

9. v = 5   10. x = 10   11. l = 50
12. m = 1000   13. i = 1   14. c = 100
15. d = 500

Write each of the following as a Roman numeral:

16. 7 = vii   17. 52 = lii   18. 33 = xxxiii
19. 1½ = i ss   20. 9 = ix

Write each of the following as a whole number:

21. viii = 8   22. xlii = 42   23. cccxxi = 321
24. cix = 109   25. mmi = 2001

Section 7.2 Apothecaries' Fluid Measure

Section 7.1, Roman Numerals, referred to the apothecaries' system of measurement. Like the metric system, the apothecaries' system has basic units of weight and liquid measure (volume). In the apothecaries' system, Roman numerals are generally used to express amounts. However, sometimes the Roman numerals are written slightly differently—for example, *ii* is also written as īi, *vi* is written as v̄i, and so on. In the apothecaries' system, the unit of measure—weight or liquid measure—is written before the Roman numeral.

When you compare the apothecaries' system to the metric system, you will notice one major difference: apothecaries' units are grossly inaccurate compared to SI units. If you keep this in mind as you work through the next two sections, you will realize why the apothecaries' system is not as popular as the metric system.

The basic unit of fluid measure in the apothecaries' system is the minim. It is commonly abbreviated as ♏.

Definition: A *minim* can be thought of as the volume of water that weighs the same as a grain of wheat. Also, a minim is approximately equal to one drop of water.

Do not equate a minim to one drop of water when you need accurate measurement. In Section 7.4, you will see how important the size of the drop becomes in calculating dosages.

Table 7-2. Apothecaries' Fluid Measure (Volume)

60 minims (♏) = 1 fluidram (f ʒ or flʒ)
8 fluidrams (480 minims) = 1 fluidounce (f ℥ or fl℥)
16 fluidounces = 1 pint (pt or O)
2 pints (32 fluidounces) = 1 quart (qt)
4 quarts (8 pints) = 1 gallon (gal or C)

By combining the facts shown in Table 7-2, one can obtain useful relationships among apothecaries' units of fluid measure (see Table 7-3).

Table 7-3. Useful Apothecaries' Fluid Measure (Volume)

1 gal = 4 qt = 8 pt = 128 f ℥
1 qt = 2 pt = 32 f ℥
1 pt = 16 f ℥
1 f ℥ = 8 f ʒ = 480 ♏
1 f ʒ = 60 ♏

Before you start working problems involving the apothecaries' system, first review Section 7.1, Roman Numerals, and Tables 7-2 and 7-3. (See Figure 7-1 for a representation of several apothecaries' graduates.)

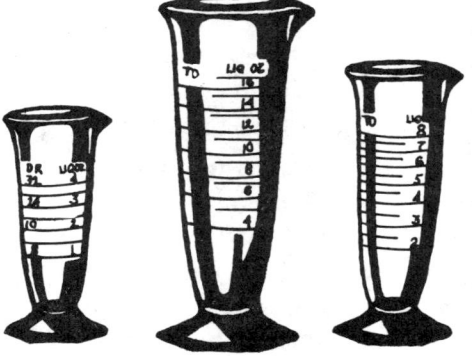

Figure 7-1. Apothecaries' graduates

Section 7.2 Apothecaries' Fluid Measure 169

> **Rule: Writing Apothecaries' Fluid Measures as if Read Aloud**
> *Step 1:* Determine what apothecaries' symbol is being used (if necessary, refer to Table 7-2, Apothecaries' Fluid Measure).
> *Step 2:* Change the Roman numeral into a whole number (or fraction, that is, $\overline{ss} = 1/2$).
> *Step 3:* Write the information from Steps 1 and 2 as if you were reading aloud.

Example 1: Write the following measurement as if read aloud: f ʒ ii
Solution:
Step 1: The apothecaries' symbol used in this problem is the fluidram.
Step 2: The Roman numeral may be rewritten as 2.
Step 3: f ʒ ii may be read aloud as "two fluidrams."

Example 2: Write the following measurement as if read aloud: ♏ xiv
Solution:
Step 1: The apothecaries' symbol used in the problem is the minim.
Step 2: The Roman numeral may be rewritten as 14.
Step 3: ♏ xiv may be read as "fourteen minims."

> **Rule: Writing Measures in the Apothecaries' System—Abbreviated Form**
> *Step 1:* Refer to Table 7-2, Apothecaries' Fluid Measure, and express the unit of measure in the abbreviated form.
> *Step 2:* Write the numerical portion of the measure as a Roman numeral.
> *Step 3:* Combine the abbreviation and the Roman numeral, placing the unit of measure before the Roman numeral.

Example 3: Write the following measure in the apothecaries' abbreviated form: 5½ fluidounces
Solution:
Step 1: The unit of measure, fluidounces, is expressed as f ℥
Step 2: Rewrite the numerical portion of the measure, 5½, as the Roman numeral $v\overline{ss}$.
Step 3: Combine the results of Steps 1 and 2: 5½ fluidounces may be expressed as f ℥ $v\overline{ss}$.

Example 4: Write the following measure in the apothecaries' abbreviated form: 8 quarts
Solution:
Step 1: The unit of measure, quarts, is expressed as qt.
Step 2: Rewrite the number 8 as the Roman numeral *viii*.
Step 3: Combine the results of Steps 1 and 2: 8 quarts may be expressed as qt *viii*.

170 Chapter 7 Apothecaries' and Household Systems

> **Rule: Changing from One Fluid Measure to Another in the Apothecaries' System**
> **Method A: Dimensional Analysis**
> *Step 1A:* Write down the given fluid measure (volume).
> *Step 2A:* Using dimensional analysis and known conversion factors, express 1 in fractional form(s) in order to obtain the desired unit and eliminate the given unit.
> *Step 3A:* Cancel units and multiply the given fluid measure by the expression(s) for 1.
> *Step 4A:* Rewrite the answer in the apothecaries' abbreviated form.

Example 5: Use method A to solve the following problem: 4 fluidounces = _____ fluidrams or f ʒ.

Solution:

Step 1A: Write down the given fluid measure.

4 fluidounces or 4 f ℥

Step 2A: Express 1 in fractional form such that fluidrams are on the top and fluidounces are on the bottom. The fact from Table 7-3 that relates fluidounces to fluidrams is

$$1 \text{ f ℥} = 8 \text{ f ʒ}$$

Therefore,

$$\frac{8 \text{ f ʒ}}{1 \text{ f ℥}} = 1$$

Step 3A: Cancel units and multiply the given fluid measure (volume) by the expression for 1.

$$\frac{4 \text{ f ℥}}{1} \times \frac{8 \text{ f ʒ}}{1 \text{ f ℥}} = 32 \text{ f ʒ}$$

Step 4A: 4 fluidounces = 32 fluidrams, or f ʒ *xxxii*.

> **Rule: Changing from One Fluid Measure to Another in the Apothecaries' System**
> **Method B: Ratio and Proportion**
> *Step 1B:* Write the conversion as a ratio in fractional form, using x to represent the number of units you are converting to.
> *Step 2B:* Write the fact(s) from Table 7-2 and 7-3 on apothecaries' fluid measure that deal with both the units you started with and the units you want to convert to. Combine fact(s) when necessary. Write the fact(s) in fractional form.
> *Step 3B:* Write the ratios from Steps 1 and 2 as a proportion and cross multiply. Be sure that the units on each side of the equation correspond to one another.
> *Step 4B:* To solve for the unknown quantity, divide the numbers on each side of the equals sign by the number on the same side as the x.
> *Step 5B:* Rewrite the answer in the apothecaries' abbreviated form.

Section 7.2 Apothecaries' Fluid Measure   171

Now use method B to solve the same problem (4 fluidounces = _____ fluidrams or f℥).

Solution:

Step 1B: Write the conversion as a ratio in fractional form, using x to represent the number of units you are converting to.

$$\frac{4\ \text{f}℥}{x\ \text{f}℥}$$

Step 2B: Because 1 f℥ = 8 f℥, we write

$$\frac{1\ \text{f}℥}{8\ \text{f}℥}$$

Step 3B: Write the two ratios as a proportion and cross multiply.

$$\frac{4\ \text{f}℥}{x\ \text{f}℥} = \frac{1\ \text{f}℥}{8\ \text{f}℥}$$
$$x \cdot 1 = (4)(8)$$
$$x \cdot 1 = 32$$

Step 4B: To solve for the unknown quantity, divide the numbers on each side of the equals sign by the number on the same side as the x.

$$\frac{x \times 1}{1} = \frac{32}{1}$$
$$x = 32\ \text{f}℥$$

Step 5B: 4 f℥ = 32 f℥, or f℥ *xxxii*.

Example 6: Use method A to solve the following problem: 20 fluidounces = _____ pints or pt.

Solution:

Step 1A: Write down the given volume.

20 fluidounces or 20 f℥

Step 2A: Express 1 in a fractional form such that pints are on the top and fluidounces are on the bottom. The fact from Table 7-3 that relates pints to fluidounces is

1 pt = 16 f℥

Therefore,

$$\frac{1\ \text{pt}}{16\ \text{f}℥} = 1$$

Step 3A: Cancel units and multiply the given volume by the expression for 1.

$$\frac{20\ \cancel{\text{f}℥}}{1} \times \frac{1\ \text{pt}}{16\ \cancel{\text{f}℥}} = 1\frac{1}{4}\ \text{pt}[1]$$

Therefore, 20 f℥ = 1¼ pints or pt.

---

[1] Because the only fraction that can be represented by a Roman numeral is ½ ($\overline{ss}$), any other fraction or any mixed number should be written with the apothecaries' unit following. (For example, 1¼ pints = 1¼ pt, 3⅜ fluidounces = 3⅜ f℥, and so on.)

Chapter 7 Apothecaries' and Household Systems

Section 7.2 Readiness Review

Write the following measurements as if read aloud (see Examples 1 & 2):

1. ♏ *xl* = _____
2. pt *xii* = _____

Write the following measures in the apothecaries' abbreviated form (see Examples 3 & 4):

3. 20 fluidrams = _____
4. ½ gallon = _____

Solve the following problems:

5. 240 minims = _____ fluidounces or f℥ _____ (Example 6)

6. 1¼ gallons = _____ fluidrams or f ʒ _____ (Example 5)

Section readiness review answers (not given in order):

3. f ʒ *xx*
6. 1280 fluidrams or f ʒ *mcclxxx*
1. forty minims
5. ½ fluidounce or f℥ $\overline{ss}$
2. twelve pints
4. gal $\overline{ss}$ or C $\overline{ss}$

Section 7.2 Review Problems

Write the following measurements as if read aloud (answers to the odd-numbered problems are at the back of the book):

1. ♏ *lx*
2. f ʒ *xi*
3. f℥ *xxv*
4. f℥ *xiv*
5. f℥ *iii*
6. O x$\overline{ss}$
7. qt *vii*
8. qt *ii*
9. gal ī$\overline{ss}$
10. C *xii*$\overline{ss}$

Write the following measures in the apothecaries' abbreviated form:

11. 360 minims
12. 25 minims
13. 15 fluidounces
14. 16 fluidounces
15. 2 pints
16. 2½ pints
17. 13 quarts
18. 6 quarts

Solve the following problems:

19. 480 minims = _____ fluidounces or f℥ _____

20. 3120 minims = _____ fluidounces or f℥ _____

21. 1024 fluidrams = _____ fluidounces or f℥ _____

22. 128 fluidrams = _____ fluidounces or f℥ _____

23. 128 fluidrams = _____ pints or pt _____

24. 48 fluidrams = _____ pints or pt _____

25. 256 fluidrams = _____ quarts or qt _____

26. 768 fluidrams = _____ quarts or qt _____

27. 512 fluidrams = _____ gallons or gal _____

Section 7.3 Apothecaries' Weight 173

28. 3584 fluidrams = _____ gallons or C _____
29. 8 fluidounces = __½__ pints or pt __ss__
30. 32 fluidounces = __2__ pints or O __ii__
31. 128 fluidounces = __1__ gallons or gal __i__
32. 256 fluidounces = __2__ gallons or C __ii__
33. 4 fluidounces = __1920__ minims or ♏ __MCMXX__
34. ⅔ fluidounce = __320__ minims or ♏ __CCCXX__
35. ¼ gallon = __256__ fluidrams or f ʒ __cclvi__
36. 2 gallons = __2048__ fluidrams or f ʒ __mmxlviii__
37. 1 pint = __128__ fluidrams or f ʒ __cxxviii__
38. ¼ pint = __32__ fluidrams or f ʒ __xxxii__
39. 5 fluidounces = __40__ fluidrams or f ʒ __xl__
40. ⅛ fluidounce = __1__ fluidrams or f ʒ __i__

Section 7.3 Apothecaries' Weight

The basic unit of weight in the apothecaries' system is the grain. It is commonly abbreviated as gr.

> An apothecaries' *grain* is traditionally defined as the average weight of a grain of wheat.

Like some other definitions you have studied so far, the traditional definition of *grain* is not precise. Additional apothecaries' weights are shown in Table 7-4.

Table 7-4. Apothecaries' Weights

60 grains = 1 dram (ʒ)
8 drams (480 grains) = 1 ounce (ʒ)
12 ounces (5760 grains) = 1 pound (℔)

Table 7-5 indicates the most useful apothecaries' weights.

Table 7-5. Useful Apothecaries' Weights

1 ℔ = 12 ʒ = 96 ʒ = 5760 gr
1 ʒ = 8 ʒ = 480 gr
1 ʒ = 60 gr

Because we normally think of 1 pound being equal to 16 ounces, you might wonder why an apothecaries' pound is equal to only 12 apothecaries' ounces (dry weight). The difference arises through the fact that the apothecaries' system is not the same as the English system, which we frequently use when we are grocery shopping. In the apothecaries' system, 1 ounce = 480 gr. In the English, or avoirdupois, system, the ounce = 437.5 gr. The avoirdupois grain and the apothecaries' grain are exactly the same; however, the apothecaries' ounce, by definition, is 42.5 gr heavier than the avoirdupois ounce.

As when working fluid-measure problems, you need to know Roman numerals to solve apothecaries' weight problems.

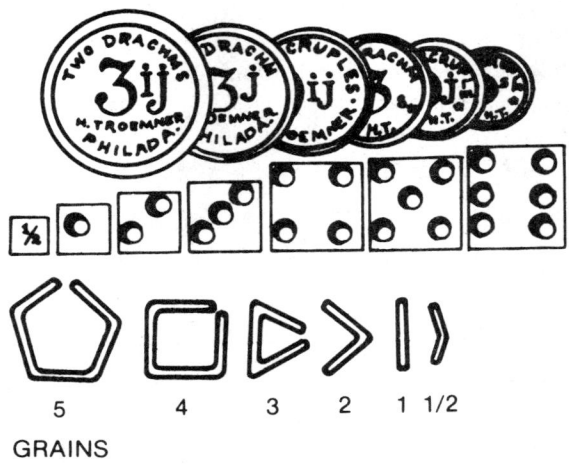

Figure 7-2. Apothecaries' weights

Figure 7-2 shows a set of apothecaries' weights. Note that, on the dram weights, dram is spelled drachm; this is just another form of the same word. The square-shaped weights are grain weights weighing ½ to 6 grains. The aluminum-wire weights are also commonly used to measure grains.

> **Rule: Writing Apothecaries' Weights as if Read Aloud**
> *Step 1:* Determine what apothecaries' symbol is being used (refer to Table 7-5, Apothecaries' Weights, if necessary).
> *Step 2:* Change the Roman numeral into a whole number (or fraction—that is, $\overline{ss} = \frac{1}{2}$).
> *Step 3:* Write the information from Steps 1 and 2 as if you were reading aloud.

Example 1: Write the following measurement as if read aloud:

ʒ *xii*

Solution:
Step 1: The apothecaries' symbol used in this problem is the dram.
Step 2: Rewrite the Roman numeral as 12.
Step 3: ʒ *xii* may be read aloud as "twelve drams."

Section 7.3 Apothecaries' Weight 175

Example 2: Write the following measurement as if read aloud:

℥ iiiss

Solution:
Step 1: The apothecaries' symbol used in this problem is the ounce.
Step 2: Rewrite the Roman numeral as 3½.
Step 3: ℥ iiiss may be read as "three and one-half ounces."

> **Rule: Writing Weights in the Apothecaries' System—Abbreviated Form**
> *Step 1:* Refer to Table 7-5, Apothecaries' Weights, and express the unit of measure in the abbreviated form.
> *Step 2:* Write the numerical portion of the measure as a Roman numeral.
> *Step 3:* Combine the information from Steps 1 and 2, placing the unit of weight before the Roman numeral.

Example 3: Write the following weight in the apothecaries' abbreviated form.

2 pounds

Solution:
Step 1: The unit of weight, pounds, is expressed as ℔.
Step 2: Rewrite the numerical portion of the problem, 2, as the Roman numeral *ii*.
Step 3: Combine "℔" with "*ii*": ℔ *ii*.

Example 4: Write the following weight in the apothecaries' abbreviated form:

$9\frac{1}{2}$ grains

Solution:
Step 1: Grains may be abbreviated as gr.
Step 2: 9½ may be rewritten as *ixss*.
Step 3: 9½ grains may be expressed as gr *ixss*.

> **Rule: Changing from One Weight to Another in the Apothecaries' System**
> **Method A: Dimensional Analysis**
> *Step 1A:* Write down the given weight.
> *Step 2A:* Using dimensional analysis and known conversion factors, express 1 in fractional form(s) in order to obtain the desired unit and eliminate the given unit.
> *Step 3A:* Cancel units and multiply the given weight by the expression(s) for 1.
> *Step 4A:* Rewrite the answer in the apothecaries' abbreviated form.

Example 5: Use method A to solve the following problem: 2½ ounces = _____ drams or ℥.

Solution:

Step 1A: Write down the given weight.

$$2\frac{1}{2} \text{ ounces or } 2\frac{1}{2} \; ℥$$

Step 2A: Express 1 in fractional form such that drams are on top and ounces are on the bottom. The fact in Table 7-5 that relates ounces to drams is 1 ℥ = 8 ℨ.

$$\frac{8 \; ℨ}{1 \; ℥} = 1$$

Step 3A: Cancel units and multiply the given weight by the expression for 1.

$$\frac{2\frac{1}{2} \; \cancel{℥}}{1} \times \frac{8 \; ℨ}{1 \; \cancel{℥}} =$$

$$\frac{5}{\cancel{2}} \times \frac{\cancelto{4}{8} \; ℨ}{1} = 20 \; ℨ$$

Step 4A: 2½ ounces = 20 drams, or ℨ xx.

Rule: Changing from One Weight to Another in the Apothecaries' System
   Method B: Ratio and Proportion

Step 1B: Write the conversion as a ratio in fractional form, using x to represent the number of units you are converting to.

Step 2B: Write the fact(s) from the Tables 7-4 and 7-5 on apothecaries' weight that deal with both the unit you started with and the units you want to convert to. Combine fact(s) when necessary. Write the fact(s) in fractional form.

Step 3B: Write the ratios from Steps 1 and 2 as a proportion and cross multiply. Be sure that the units on each side of the equation correspond to one another.

Step 4B: To solve for the unknown quantity, divide the numbers on each side of the equals sign by the number on the same side as the x.

Step 5B: Rewrite the answer in the apothecaries' abbreviated form.

Now use method B to solve the same problem (2½ ounces = _____ drams or ℨ).

Solution:

Step 1B: Write the conversion as a ratio in fractional form; x represents the unknown.

$$\frac{2\frac{1}{2} \; ℥}{x \; ℨ}$$

Section 7.3 Apothecaries' Weight

Step 2B: Because $1 \; ℥ = 8 \; ℨ$, we write

$$\frac{1 \; ℥}{8 \; ℨ}$$

Step 3B: Write the two ratios as a proportion and cross multiply.

$$\frac{2\frac{1}{2} \; ℥}{x \; ℨ} = \frac{1 \; ℥}{8 \; ℨ}$$

$$x \cdot 1 = 2\frac{1}{2} \cdot 8$$

$$x \cdot 1 = \frac{5}{\cancel{2}_1} \cdot \cancel{8}^4$$

$$x \cdot 1 = 20$$

Step 4B: Divide the numbers on each side of the equals sign by the number on the same side as x.

$$\frac{x \cdot 1}{1} = \frac{20}{1}$$

$$x = 20 \; ℨ$$

Step 5B: $2\frac{1}{2} \; ℥ = 20 \; ℨ$, or $ ℨ \; xx$.

Example 6: Use method B to solve the following problem: 90 grains = _____ drams or $ ℨ$.

Solution:

Step 1B: The conversion, in fractional form, is

$$\frac{90 \; gr}{x \; ℨ}$$

Step 2B: Because $60 \; gr = 1 \; ℨ$, we write

$$\frac{60 \; gr}{1 \; ℨ}$$

Step 3B: Write the ratios as a proportion and cross multiply.

$$\frac{90 \; gr}{x \; ℨ} = \frac{60 \; gr}{1 \; ℨ}$$

$$x \cdot 60 = 90 \cdot 1$$

$$x \cdot 60 = 90$$

Step 4B: Next, we divide:

$$\frac{x \cdot \cancel{60}^1}{\cancel{60}_1} = \frac{\cancel{90}^3}{\cancel{60}_2}$$

$$x = 1\frac{1}{2} \; ℨ$$

Step 5B: $90 \; gr = 1\frac{1}{2} \; ℨ$, or $ℨ \; \overline{iss}$

Chapter 7 Apothecaries' and Household Systems

Section 7.3 Readiness Review

Write the following measurements as if read aloud (see Examples 1 & 2):

1. gr ii _____
2. ℥ xix _____

Write the following weights in the apothecaries' abbreviated form (see Examples 3 & 4):

3. 14 drams _____
4. 2½ pounds _____

Solve the following problems:

5. 1 dram = _____ grains or gr _____

6. 12 ounces _____ drams or ℥ _____

Section readiness review answers (not given in order):

5. 60 grains or gr *lx*
6. 96 drams or ℥ *xcvi*
3. ℥ *xiv*
4. ℔ *iiss*
2. nineteen ounces
1. two grains

Section 7.3 Review Problems

Write the following measurements as if read aloud (answers to the odd-numbered problems are at the back of the book):

1. gr vi __6 grains__
2. ℥ xiii __13 drams__
3. ℥ xvi __16 drams__
4. ℥ ss __½ ounce__
5. ℔ xxv __25 lbs__
6. ℔ xxix __29 lbs__

Write the following weights in the apothecaries' abbreviated form:

7. 1½ grains __gr iss__
8. 27 drams __℥ xxvii__
9. 15 drams __℥ xv__
10. 4 ounces __℥ iv__
11. 2 pounds __℔ ii__
12. 50 pounds __℔ l__

Solve the following problems:

13. 90 grains = __1½__ drams or ℥ _____
14. 150 grains = __2½__ drams or ℥ _____
15. 720 grains = __1½__ ounces or ℥ _____
16. 960 grains = __2__ ounces or ℥ _____
17. 11,520 grains = __2__ pounds or ℔ _____
18. 57,600 grains = __10__ pounds or ℔ _____
19. 8 drams = __1__ ounces or ℥ _____
20. 16 drams = __2__ ounces or ℥ _____
21. 12 ounces = __1__ pounds or ℔ _____
22. 36 ounces = __3__ pounds or ℔ _____
23. ¼ pound = __1440__ grains or gr _____
24. 2/15 pound = __768__ grains or gr _____

25. 5 pounds = __480__ drams or ℨ _____
26. ¼ pound = __24__ drams or ℨ __XXIV__
27. 1 pound = __12__ ounces or ℨ _____
28. 1⅓ pounds = __16__ ounces or ℨ __XVI__
29. ⅓ ounce = __160__ grains or gr _____
30. 4¼ ounces = __2040__ grains or gr __MMXL__
31. 3¼ ounces = __26__ drams or ℨ _____
32. ⅔ ounce = __5⅓__ drams or ℨ _____   5.3
33. 10 drams = __600__ grains or gr _____
34. ⅓ ounce = __160__ grains or gr __CLX__
35. ¾ ounce = __360__ grains or gr _____
36. 16 drams = __2__ ounces or ℨ __ii__
37. 7 drams = __420__ grains or gr _____
38. ½ pound = __2880__ grains or gr __mmdccclxxx__
39. ⅔ pound = __64__ drams or ℨ _____
40. 4 drams = __½__ ounces or ℨ __ss__

Section 7.4 Household Fluid Measure

In your nursing practice, you will sometimes see directions for taking medication written like this: "Take 1 teaspoonful every eight hours," or "Take ½ tablespoonful . . . ." These directions are based on household fluid measurements.

> **Definition:** *Household fluid measure* uses utensils from cooking and eating as the basis for medication measurement.

Unlike the metric and apothecaries' systems, there is no real basic unit of fluid measure in the household system. It is extremely important to recognize that the household system is *not* accurate. It approximates those values that are commonly prescribed, and it is convenient for the patient to measure at home.

The smallest unit of household fluid measure is the drop, and the largest unit is the glassful (see Table 7-6).

Table 7-6. Household Fluid Measures

75 drops (gtt) = 1 teaspoonful (tsp)
3 teaspoonsful = 1 tablespoonful (tbsp)
2 tablespoonsful = 1 fluidounce (fℨ or flℨ)
8 fluidounces = 1 glassful

Chapter 7 Apothecaries' and Household Systems

Because household fluid measures are not truly accurate, it is often hard to establish their equivalents in the metric and apothecaries' systems. Table 7-7 is based on equivalent measures that have been established by custom.

Table 7-7. Household Fluid Measures—Equivalents

Household	Metric	Apothecaries'
1 drop (gtt)	0.06 mℓ	♏ i**
15 drops	1 mℓ	♏ xv*
75 drops	5 mℓ	♏ lxxv*
1 teaspoonful (tsp)	5 mℓ*	1⅓ f3
3 teaspoonsful	15 mℓ	f3 iv
1 tablespoonful (tbsp)	15 mℓ	f3 iv
1 tablespoonful (tbsp)	15 mℓ	f℥ ss
2 tablespoonsful	30 mℓ	f℥ i
1 glassful	240 mℓ	f℥ viii

*There still remains some controversy over the volume of a teaspoon. Some physicians prefer to use 4 mℓ as the volume equal to one teaspoonful. However, because the *United States Pharmacopeia* and the *National Formulary* (the sole references governing pharmacy measurements as well as medications) use 5 mℓ as the volume equal to one teaspoonful, we will, also. Therefore, the volume of one teaspoonful is 1⅓ fluidrams.

**One drop is not *always* equal to one minim; the paragraph following the table will explain why.

The next problem we must tackle is the concept that one drop is not *always* equal to one minim; for example, a drop of water is not equal to the volume of a drop of simple syrup. The volume of a drop of liquid depends on the liquid's density, temperature, viscosity (thickness), and so on. The volume of a drop of liquid also depends on the size of the dropper opening from which it was dropped. However, for the sake of simplicity, the volume of a drop used in problems throughout this text is equal to one minim, unless otherwise stated. Therefore:

1 drop (gtt) = 1 minim (♏)
1 drop (gtt) = 0.06 mℓ

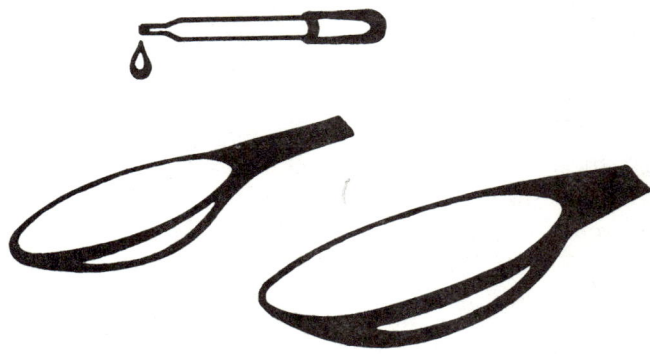

1 drop
1 teaspoonful
1 tablespoonful = 3 teaspoonsful

Figure 7-3. Some household fluid-measure devices

Section 7.4 Household Fluid Measure   181

Figure 7-3 shows several household fluid-measure devices and the approximations of the volume they hold.

Rule: Changing from Household Fluid Measure to Metric Volume
Method A: Dimensional Analysis

Step 1A: Write down the given volume.
Step 2A: Using dimensional analysis and known conversion factors, express 1 in fractional form(s) in order to obtain the desired unit and eliminate the given unit.
Step 3A: Cancel units and multiply the given volume by the expression(s) for 1.

Example 1: Use method A to solve the following problem: 5 drops = _____ milliliters.

Solution:

Step 1A: 5 gtt

Step 2A: The fact in Table 7-7 that relates drops to milliliters is 1 gtt = 0.06 mℓ.

$$\frac{0.06 \text{ m}\ell}{1 \text{ gtt}} = 1$$

Step 3A: Cancel units and multiply.

$$\frac{5 \text{ gtt}}{1} \times \frac{0.06 \text{ m}\ell}{1 \text{ gtt}} = 0.3 \text{ m}\ell$$

Therefore, 5 drops = 0.3 milliliters, or 5 gtt = 0.3 mℓ.

Rule: Changing from Household Fluid Measure to Metric Volume
Method B: Ratio and Proportion

Step 1B: Write the conversion as a ratio in fractional form, using x to represent the number of units you are converting to.
Step 2B: Write the fact(s) from the Table 7-7, Household Fluid Measures, that deal with both the units you started with and the units you want to convert to. Combine fact(s) when necessary. Write the fact(s) in fractional form.
Step 3B: Write the ratios from Steps 1 and 2 as a proportion and cross multiply. Be sure that the units on each side of the equation correspond to one another.
Step 4B: To solve for the unknown quantity, divide the numbers on each side of the equals sign by the number on the same side as the x.

Now use method B to solve the same problem (5 drops = _____ milliliters).

Solution:

Step 1B: Write the conversion as a ratio in fractional form.

$$\frac{5 \text{ gtt}}{x \text{ m}\ell}$$

Step 2B: Because 1 drop = 0.06 milliliter, we write

$$\frac{1 \text{ gtt}}{0.06 \text{ m}\ell}$$

Step 3B: Write the ratios as a proportion and cross multiply.

$$\frac{5 \text{ gtt}}{x \text{ m}\ell} = \frac{1 \text{ gtt}}{0.06 \text{ m}\ell}$$
$$x \cdot 1 = 5 \cdot 0.06$$
$$x \cdot 1 = 0.3$$

Step 4B: Next, we divide:

$$\frac{x \cdot 1}{1} = \frac{0.3}{1}$$
$$x = 0.3 \text{ m}\ell$$

Therefore, 5 drops = 0.3 milliliters, or 5 gtt = 0.3 mℓ.

Example 2: Use method A to solve the following problem: 4 tablespoonsful = _____ milliliters.

Solution:
Step 1A: 4 tbsp
Step 2A: The fact in Table 7-7 that relates tablespoonsful to milliliters is 1 tablespoonful = 15 milliliters.

$$\frac{15 \text{ m}\ell}{1 \text{ tbsp}} = 1$$

Step 3A: Cancel units and multiply.

$$\frac{4 \cancel{\text{ tbsp}}}{1} \times \frac{15 \text{ m}\ell}{1 \cancel{\text{ tbsp}}} = 60 \text{ m}\ell$$

Therefore, 4 tablespoonsful = 60 milliliters, or 4 tbsp = 60 mℓ.

By now, you should be used to using the dimensional analysis and ratio-and-proportion methods to solve problems. It is important to recognize that the ratio method requires memorizing many conversion factors. Dimensional analysis allows you to memorize fewer conversion factors. Whichever method you choose, you can use the other method as a check.

Rule: Changing from Metric Volume to Household Fluid Measure
Method A: Dimensional Analysis
Step 1A: Write down the given volume.
Step 2A: Using dimensional analysis and known conversion factors, express 1 in fractional form(s) in order to obtain the desired unit and eliminate the given unit.
Step 3A: Cancel units and multiply the given volume by the expression(s) for 1.

Section 7.4 Household Fluid Measure   183

Example 3: Use method A to solve the following problem: 45 milliliters = _____ tablespoonsful.

Solution:

Step 1A: 45 mℓ

Step 2A: The fact in Table 7-7 that relates milliliters to tablespoonsful is 15 milliliters = 1 tablespoonful.

$$\frac{1 \text{ tbsp}}{15 \text{ mℓ}} = 1$$

Step 3A: Cancel units and multiply.

$$\frac{\overset{3}{\cancel{45 \text{ mℓ}}}}{1} \cdot \frac{1 \text{ tbsp}}{\underset{1}{\cancel{15 \text{ mℓ}}}} = 3 \text{ tbsp}$$

Therefore, 45 milliliters = 3 tablespoonsful, or 45 mℓ = 3 tbsp.

Rule: Changing from Metric Volume to Household Fluid Measure
  Method B: Ratio and Proportion
  Step 1B: Write the conversion as a ratio in fractional form, using x to represent the number of units you are converting to.
  Step 2B: Write the fact(s) from Table 7-7, Household Fluid Measures, that deal with both the units you started with and the units you want to convert to. Combine fact(s) when necessary. Write the fact(s) in fractional form.
  Step 3B: Write the ratios from Steps 1 and 2 as a proportion and cross multiply. Be sure that the units on each side of the equation correspond to one another.
  Step 4B: To solve for the unknown quantity, divide the numbers on each side of the equals sign by the number on the same side as the x.

Now use method B to solve the same problem (45 milliliters = _____ tablespoonsful).

Solution:

Step 1B: Write the conversion as a ratio in fractional form.

$$\frac{45 \text{ mℓ}}{x \text{ tbsp}}$$

Step 2B: Because 15 milliliters = 1 tablespoonful, we write

$$\frac{15 \text{ mℓ}}{1 \text{ tbsp}}$$

Step 3B: Write the ratios as a proportion and cross multiply.

$$\frac{45 \text{ mℓ}}{x \text{ tbsp}} = \frac{15 \text{ mℓ}}{1 \text{ tbsp}}$$
$$x \cdot 15 = 45 \cdot 1$$
$$x \cdot 15 = 45$$

Step 4B: Next, we divide:

$$\frac{x \cdot \cancel{15}^1}{\cancel{15}_1} = \frac{\cancel{45}^3}{\cancel{15}_1}$$

$$x = 3 \text{ tbsp}$$

Therefore, 45 milliliters = 3 tablespoonsful, or 45 mℓ = 3 tbsp.

Example 4: Use method B to solve the following problem: 480 milliliters = _____ glassfuls.

Solution:

Step 1B: Write the conversion as a ratio in fractional form.

$$\frac{480 \text{ m}\ell}{x \text{ glassfuls}}$$

Step 2B: Because 240 milliliters = 1 glassful, we write

$$\frac{240 \text{ m}\ell}{1 \text{ glassful}}$$

Step 3B: Write the ratios as a proportion and cross multiply.

$$\frac{480 \text{ m}\ell}{x \text{ glassfuls}} = \frac{240 \text{ m}\ell}{1 \text{ glassful}}$$

$$x \cdot 240 = 480 \cdot 1$$
$$x \cdot 240 = 480$$

Step 4B: Now, divide:

$$\frac{x \cdot \cancel{240}^1}{\cancel{240}_1} = \frac{\cancel{480}^2}{\cancel{240}_1}$$

$$x = 2 \text{ glassfuls}$$

Therefore, 480 milliliters = 2 glassfuls, or 480 mℓ = 2 glassfuls.

Rule: Changing from Household Fluid Measure to Apothecaries' Fluid Measure
Method A: Dimensional Analysis

Step 1A: Write down the given volume.
Step 2A: Using dimensional analysis and known conversion factors, express 1 in fractional form(s) in order to obtain the desired unit and eliminate the given unit.
Step 3A: Cancel units and multiply the given volume by the expression(s) for 1.
Step 4A: Rewrite the answer in the apothecaries' abbreviated form.

Example 5: Use method A to solve the following problem: $1\frac{1}{2}$ tablespoonsful = _____ fluidrams.

Solution:
Step 1A: 1½ tbsp
Step 2A: The fact in Table 7-7 that relates tablespoonsful to fluidrams is 1 tablespoonful = 4 fluidrams.

$$\frac{4 \text{ f}\mathfrak{Z}}{1 \text{ tbsp}}$$

Step 3A: Cancel units and multiply.

$$\frac{1\frac{1}{2} \, \cancel{\text{tbsp}}}{1} \cdot \frac{4 \text{ f}\mathfrak{Z}}{1 \, \cancel{\text{tbsp}}} = 6 \text{ f}\mathfrak{Z}$$

Step 4A: 1½ tablespoonsful = 6 fluidrams, or f$\mathfrak{Z}$ vi.

Rule: Changing from Household Fluid Measure to Apothecaries' Fluid Measure
Method B: Ratio and Proportion

Step 1B: Write the conversion as a ratio in fractional form, using x to represent the number of units you are converting to.
Step 2B: Write the fact(s) from the Table 7-7, Household Fluid Measures, that deal with both the units you started with and the units you want to convert to. Combine fact(s) when necessary. Write the fact(s) in fractional form.
Step 3B: Write the ratios from Steps 1 and 2 as a proportion and cross multiply. Be sure that the units on each side of the equation correspond to one another.
Step 4B: To solve for the unknown quantity, divide the numbers on each side of the equals sign by the number on the same side as the x.
Step 5B: Rewrite the answer in the apothecaries' abbreviated form.

Now use method B to solve the same problem: 1½ tablespoonsful = _____ fluidrams.

Solution:
Step 1B: Write the conversion as a ratio in fractional form.

$$\frac{1\frac{1}{2} \text{ tbsp}}{x \text{ f}\mathfrak{Z}}$$

Step 2B: Because 1 tablespoonful = 4 fluidrams, we write

$$\frac{1 \text{ tbsp}}{4 \text{ f}\mathfrak{Z}}$$

Step 3B: Write the ratios as a proportion and cross multiply.

$$\frac{1\frac{1}{2} \text{ tbsp}}{x \text{ f}\mathfrak{Z}} = \frac{1 \text{ tbsp}}{4 \text{ f}\mathfrak{Z}}$$

$$x \cdot 1 = 1\frac{1}{2} \cdot 4$$

$$x \cdot 1 = \frac{3}{\cancel{2}_1} \cdot \cancel{4}^2$$

$$x \cdot 1 = 6$$

Step 4B: Next, we divide:

$$\frac{x \cdot 1}{1} = \frac{6}{1}$$

$$x = 6 \text{ fluidrams}$$

Therefore, 1½ tablespoonsful = 6 fluidrams.

Step 5B: 6 fluidrams = f℥ vi.

Example 6: Use method A to solve the following problem: 8 drops = _____ minims.
Solution:
Step 1A: 8 gtt
Step 2A: The fact in Table 7-7 that relates drops to minims is 1 drop = 1 minim.

$$\frac{1 \text{ m}}{1 \text{ gtt}} = 1$$

Step 3A: Cancel units and cross multiply.

$$\frac{8 \cancel{\text{ gtt}}}{1} \cdot \frac{1 \text{ m}}{1 \cancel{\text{ gtt}}} = 8$$

Therefore, 8 drops = 8 minims.

Step 4A: 8 minims = m viii.

> **Rule: Changing from Apothecaries' Fluid Measure to Household Fluid Measure**
> **Method A: Dimensional Analysis**
> *Step 1A:* Write down the given volume.
> *Step 2A:* Using dimensional analysis and known conversion factors, express 1 in fractional form(s) in order to obtain the desired unit and eliminate the given unit.
> *Step 3A:* Cancel units and multiply the given volume by the expression(s) for 1.

Example 7: Use method A to solve the following problem: 16 fluidounces = _____ glassfuls.
Solution:
Step 1A: 16 f℥
Step 2A: The fact in Table 7-7 that relates fluidounces to glassfuls is f℥ viii = 1 glassful, or 8 fluidounces = 1 glassful.

$$\frac{1 \text{ glassful}}{8 \text{ f℥}} = 1$$

Section 7.4 Household Fluid Measure

Step 3A: Cancel units and cross multiply.

$$\frac{\overset{2}{\cancel{16\,f\!\tilde{3}}}}{1} \times \frac{1 \text{ glassful}}{\underset{1}{\cancel{8\,f\!\tilde{3}}}} = 2 \text{ glassfuls}$$

Therefore, 16 fluidounces = 2 glassfuls, or 16 f$\tilde{3}$ = 2 glassfuls.

Rule: Changing from Apothecaries' Fluid Measure to Household Fluid Measure
Method B: Ratio and Proportion

Step 1B: Write the conversion as a ratio in fractional form, using x to represent the number of units you are converting to.
Step 2B: Write the fact(s) from the Table 7-7, Household Fluid Measures, that deal with both the units you started with and the units you want to convert to. Combine fact(s) when necessary. Write the fact(s) in fractional form.
Step 3B: Write the ratios from Steps 1 and 2 as a proportion and cross multiply. Be sure that the units on each side of the equation correspond to one another.
Step 4B: To solve for the unknown quantity, divide the numbers on each side of the equals sign by the number on the same side as the x.

Now use method B to solve the same problem (16 fluidounces = _____ glassfuls).

Solution:

Step 1B: Write the conversion as a ratio in fractional form.

$$\frac{16 \text{ f}\tilde{3}}{x \text{ glassfuls}}$$

Step 2B: Because 8 f$\tilde{3}$ = 1 glassful, we write

$$\frac{8 \text{ f}\tilde{3}}{1 \text{ glassful}}$$

Step 3B: Write the ratios as a proportion and cross multiply.

$$\frac{16 \text{ f}\tilde{3}}{x \text{ glassfuls}} = \frac{8 \text{ f}\tilde{3}}{1 \text{ glassful}}$$
$$x \cdot 8 = 16 \cdot 1$$
$$x \cdot 8 = 16$$

Step 4B: Now divide to solve for the unknown quantity.

$$\frac{x \cdot \overset{1}{\cancel{8}}}{\underset{1}{\cancel{8}}} = \frac{\overset{2}{\cancel{16}}}{\underset{1}{\cancel{8}}}$$
$$x = 2 \text{ glassfuls}$$

Therefore, 16 fluidounces = 2 glassfuls, or 16 f$\tilde{3}$ = 2 glassfuls.

188 Chapter 7 Apothecaries' and Household Systems

Example 8: Use method B to solve the following problem: 2 fluidounces = _____ teaspoonsful.

Solution:

Step 1B:
$$\frac{2\ f\!\!\!\!\!{\mathfrak{Z}}}{x\ \text{tsp}}$$

Step 2B: Because 1 fluidounce = 2 tbsp and 1 tbsp = 3 tsp, the following is true.

$$\frac{2\ \cancel{\text{tbsp}}}{1\ f\!\!\!\!\!{\mathfrak{Z}}} \times \frac{3\ \text{tsp}}{1\ \cancel{\text{tbsp}}} = \frac{6\ \text{tsp}}{1\ f\!\!\!\!\!{\mathfrak{Z}}}$$

Step 3B:
$$\frac{2\ f\!\!\!\!\!{\mathfrak{Z}}}{x\ \text{tsp}} = \frac{1\ f\!\!\!\!\!{\mathfrak{Z}}}{6\ \text{tsp}}$$
$$x \cdot 1 = 2 \cdot 6$$
$$x \cdot 1 = 12$$

Step 4B:
$$\frac{x \cdot 1}{1} = \frac{12}{1}$$
$$x = 12\ \text{teaspoonsful}$$

Therefore, 2 fluidounces = 12 teaspoonsful, or f$\mathfrak{Z}$ *ii* = 12 tsp.

Section 7.4 Readiness Review

Solve the following problems:

1. 4 gtt = __.24__ mℓ ✓ (see Examples 1 & 2)
2. 3 mℓ = __45__ gtt (see Examples 3 & 4)
3. 75 mℓ = __5__ tbsp ✓ (see Examples 3 & 4)
4. 5 tsp = __6⅔__ fluidrams or _____ f$\mathfrak{Z}$ (see Examples 5 & 6)
5. 2⅔ f$\mathfrak{Z}$ = __2__ tsp ✓ (see Examples 7 & 8)
6. 375 m̨ = __1⅔__ tbsp (see Examples 7 & 8)

Section readiness review answers (not given in order):

4. 6⅔ fluidrams or 6⅔ f$\mathfrak{Z}$ 3. 5 tbsp   1. 0.24 mℓ
2. 50 gtt   6. 1½ tbsp   5. 2 tsp

Section 7.4 Review Problems

Solve the following problems (answers to the odd-numbered problems are at the back of the book):

1. 8 gtt = __.48__ mℓ
2. 15 gtt = __9__ mℓ
3. 3 tsp = __15__ mℓ
4. 2½ tsp = __12.5__ mℓ
5. 1½ tbsp = __22½__ mℓ
6. 5 tbsp = __75__ mℓ
7. 6 glassfuls = __1440__ mℓ
8. 1½ glassfuls = __360__ mℓ
9. 10 mℓ = __166.⅔__ gtt
10. 0.5 mℓ = __8⅓__ gtt
11. 10 mℓ = __2__ tsp
12. 17.5 mℓ = __3½__ tsp

13. 30 mℓ = __2__ tbsp
14. 22.5 mℓ = _____ tbsp
15. 360 mℓ = __1½__ glassfuls
16. 180 mℓ = _____ glassfuls
17. 11 gtt = __11__ minims or ℳ _____
18. 3 gtt = __3__ minims or ℳ _____
19. 180 gtt = __2⅖__ tsp
20. 300 gtt = __4__ tsp
21. 4 tbsp = __16__ fluidrams or fʒ _____
22. 1 tbsp = __4__ fluidrams or fʒ _____
23. 2 tbsp = __1__ fluidounce or f℥ _____
24. 3 tbsp = __1½__ fluidounce or f℥ _____
25. 12 tsp = __2__ fluidounce or f℥ _____
26. 4 tsp = __1⅓ or ⅔__ fluidounce or f℥ _____
27. ½ glassful = __32__ fluidram or fʒ _____
28. ¼ glassful = __16__ fluidram or fʒ _____
29. ⅕ glassful = __12⅘__ fluidram or _____ fʒ
30. ⅛ glassful = __1__ fluidounce or f℥ _____
31. 3 tsp = __225__ minims or ℳ _____
32. 6 tsp = __450__ minims or ℳ _____
33. 10 tbsp = __2250__ minims or ℳ _____
34. 2 tbsp = __450__ minims or ℳ _____
35. ℳ i = __1__ gtt
36. ℳ iv = __4__ gtt
37. 1 tsp = __75__ gtt 83.3
38. 2 tsp = __150__ gtt
39. 1⅓ f℥ = __1__ tsp
40. f℥ viii = __6__ tsp
41. f℥ xx = __5__ tbsp
42. f℥ ss = __1__ tbsp
43. f℥ ii = __12__ tsp
44. f℥ iv = __24__ tsp
45. f℥ xii = __3/16__ glassful
46. f℥ xlviii = __¾__ glassful
47. f℥ xii = __1½__ glassful
48. f℥ vi = __¾__ glassful
49. ℳ xc = __1⅕__ tsp
50. ℳ cxx = __1⅗__ tsp

Chapter 7 Readiness Review

Write each of the following as a Roman numeral:

1. ½ = __ss̄__
2. 5 = _____
3. 1000 = __M__
4. 100 = __C__

Chapter 7 Apothecaries' and Household Systems

Write each of the following as a whole number:

5. c = 100
6. l = 50
7. d = 500
8. x = 10

Write the following numbers as Roman numerals:

9. 11 = xi
10. 55 = lv
11. 1101 = mci
12. 11½ = xi ss
13. 20 = xx
14. 23 = xxiii
15. 586 = dlxxxvi
16. 3½ = iii ss
17. 9 = ix
18. 40 = xl
19. 400 = cd
20. 9½ = ix ss
21. 19 = xix
22. 54 = liv
23. 140 = cxl
24. 19½ = xix ss

Write the following measurements as if read aloud:

25. ♏ c = 100 minims
26. qt iv = 4 quarts

Write the following measures in the apothecaries' abbreviated form:

27. 3 pints = pt iii
28. 17 fluidrams = f℥ xvii

Solve the following problems:

29. 30,720 minims = ½ gallon or gal ss
30. ½ fluidram = 30 minims or ♏ xxx

Write the following measurements as if read aloud:

31. gr iii = 3 grains
32. lb viii = 8 pounds

Write the following weights in the apothecaries' abbreviated form:

33. 2 grains = gr ii
34. ½ ounce = ℥ ss

Solve the following problems:

35. 30 grains = ½ drams or ℨ ss
36. ½ ounce = 4 drams or ℨ iv

Solve the following problems:

37. f℥ xii = 3 tbsp
38. f℥ viii = 1 glassfuls
39. 6 tsp = 2 tbsp
40. 5 tbsp = _____ minims or ♏ _____

Chapter readiness review answers (not given in order):

40. 125 minims or ♏ *mcxxv* 39. 2 tbsp 38. 1 glassful
37. 3 tbsp 36. 4 drams or ℨ *iv* 35. ½ dram or ℨ *ss*
34. ℥ *ss* 33. gr *ii* 32. eight pounds
31. three grains 30. 30 minims or ♏ *xxx* 29. ½ gallon or gal *ss*
28. f℥ *xvii* 27. pt *iii* or O *iii* 26. four quarts
25. one hundred minims 24. *xixss* 23. *cxl*
22. *liv* 21. *xix* 20. *ixss*
19. *cd* 18. *xl* 17. *ix*
16. *iiiss* 15. *dlxxxvi* 14. *xxiii*
13. *xx* 12. *xiss* 11. *mci*
10. *lv* 9. *xi* 8. 10
7. 500 6. 50 5. 100
4. *c* 3. *m* 2. *v*
1. *ss*

Chapter 7 Summary

Define each item in your own words; then compare your definitions with the text.

Key Words

Roman numerals (p. 164) dram (p. 173)
apothecaries' system (p. 167) ounce (p. 173)
minim (p. 168) pound (p. 173)
fluidram (p. 168) drachm (p. 174)
fluidounce (p. 168) household system (p. 179)
pint (p. 168) drop (p. 180)
quart (p. 168) teaspoonful (p. 180)
gallon (p. 168) tablespoonful (p. 180)
apothecaries' graduates (p. 168) glassful (p. 180)
grain (p. 173)

Chapter 7 Review Problems

Write each of the following as a Roman numeral (answers to all the problems are at the back of the book):

1. ½   *ss*
2. 5   *v*
3. 17 *xvii*
4. 49 *xlix*
5. 224 *ccxxiv*
6. 901 *cmi*
7. 1046 *mxlvi*
8. 2000 *mm*
9. 3008 *mmmviii*
10. 3899 *mmmdcccxcix*

Chapter 7 Apothecaries' and Household Systems

Write each of the following as a whole number:

11. xii 12
12. xxxvii 37
13. lss 50 ½
14. liv 54
15. ciii 103
16. cdxxxviii 438
17. dvi 506
18. mxlvi 1046
19. mcdi 1401
20. mmmdcccxc 3890

Solve the following word problems:

21. The *National Formulary* (*N.F.*), a common reference book on drugs, uses Roman numerals to number each new edition. What is the number of the following edition: *N.F. xiv*? 14

22. The *United States Pharmacopeia* (*U.S.P.*), another common reference book on drugs, also uses Roman numerals to number each new edition. What is the number of the following edition: *U.S.P. xix*? 19

23. Harriet Catania, pharmacist, fills a prescription for a medication for a patient to take home after he is discharged from the hospital. The prescription is for Empirin No. *xxxvi*. How many tablets will the patient take home? 36

24. Physician A wants Patient B to receive an iron supplement. The order reads Fergon® No. *xlv*. How many tablets does the order call for? 45

25. Baby Louisa needs a rectal lubricant. The pharmacist recommends Infant Glycerin Suppositories by Squibb. Each small jar contains *xii* suppositories. How many suppositories are in each jar? 12

Write the following measurements as if read aloud:

O = pints
C = gallons

26. ɱ xc 90 minims
27. fʒ ix 9 fluidram
28. pt xvi 16 pints
29. O vss 5 ½ pints
30. gal ss ½ gallon

Write the following measures in the apothecaries' abbreviated form:

31. 480 minims ɱ cdlxxx
32. 2½ fluidounces fʒ iiss
33. 26 fluidounces fʒ xxvi
34. 11 quarts qt xi
35. 7 quarts qt vii

Solve the following problems:

36. 360 minims = __¾__ fluidounces or __¾__ fʒ   480 ɱ = 1 fluid oz
37. 960 minims = __2__ fluidounces or fʒ _____
38. 256 fluidrams = __32__ fluidounces or fʒ _____
39. 8 fluidrams = __1__ fluidounces or fʒ _____
40. 24 fluidounces = __1½__ pints or pt __iss__
41. 48 fluidounces = __3__ pints or O _____
42. 5 fluidounces = __2400__ minims or ɱ _____
43. ¼ fluidounce = __120__ minims or ɱ _____
44. 3 fluidrams = __180__ minims or ɱ _____
45. ½ gallon = __512__ fluidrams or fʒ __dxii__

Chapter 7 Review Problems 193

46. ¾ gallon = __768__ fluidrams or f℥ _____

47. ¼ fluidounce = __2__ fluidrams or f℥ _____

48. 15 fluidounces = __120__ fluidrams or f℥ _____

Write the following measurements as if read aloud:

49. ℥ xii = __12 dram__   50. ℥ xiv = __14 dram__

51. ℔ xxii = __22 lbs__   52. ℔ xxviii = __28 lbs__

Write the following weights in the apothecaries' abbreviated form:

53. 1 grain = __gr i__   54. 3½ grains = __gr iii ss__

55. 11 drams = __ʒ xi__   56. 26 drams = __ʒ xxvi__

Solve the following problems

57. 60 grains = __1__ drams or ʒ __i__

58. 120 grains = __2__ drams or ʒ __ii__

59. 12 drams = __1½__ ounces or ℥ _____

60. 32 drams = __4__ ounces or ℥ _____

61. ⅓ pound = __1920__ grains or gr _____

62. ⅙ pound = __960__ grains or gr _____

63. 2½ pounds = __30__ ounces or ℥ __xxx__

64. ⅓ pound = __4__ ounces or ℥ _____

65. 13 gtt = __.78__ mℓ

66. 10 gtt = __.6__ mℓ

67. 2 tsp = __10__ mℓ

68. 1½ tsp = __7½__ mℓ

69. 5½ tbsp = __82.5__ mℓ

70. 2 glassfuls = __480__ mℓ

71. ½ glassful = __120__ mℓ

72. 2 mℓ = __30__ gtt

73. 9 mℓ = __135__ gtt

74. 12.5 mℓ = __2½__ tsp

75. 7.5 mℓ = __½__ tbsp

76. 45 mℓ = __3__ tbsp

77. 120 mℓ = __8__ tbsp

78. 17 gtt = __17__ minims or ♏ _____

79. 8 gtt = __8__ minims or ♏ _____

80. 6 tsp = __8__ fluidrams or f℥ _____

81. 12 tsp = __16__ fluidrams or f℥ __xvi__

Chapter 7 Apothecaries' and Household Systems

82. 1 tbsp = __½__ fluidounces or f℥ _____
83. ½ tbsp = __¼__ fluidounces or _____ f℥
84. 6 tsp = __1__ fluidounces or f℥ _____
85. 9 tsp = __1½__ fluidounces or f℥ _____
86. ½ glassful = __4__ fluidounces or f℥ _____
87. 3 glassfuls = __24__ fluidounces or f℥ _____
88. ♏ iii = __3__ gtt
89. ♏ v = __5__ gtt
90. 1½ tsp = __112½__ gtt
91. 4 tsp = __300__ gtt
92. f℥ iv = __3__ tsp
93. 7⅔ f℥ = __534__ tsp
94. f℥ i = __2__ tbsp
95. f℥ ii ss = __5__ tbsp
96. f℥ ss = __3__ tsp
97. f℥ iii = __18__ tsp
98. f℥ xxiv = __3__ glassfuls
99. f℥ lxiv = __1__ glassfuls
100. ♏ cdlxxx = __2⅖__ tbsp

8 Conversions and Medical Abbreviations

OBJECTIVES After studying this chapter, you should be able to:
1. Convert from one system of measurement to another.
2. Understand the Latin abbreviations commonly used in nursing practice.
3. Understand the medical abbreviations commonly used in nursing practice.

Up to now, you have been exposed to each of the commonly used systems of measurement. With the exception of Section 7.4, you have been concentrating on converting one type of unit to another *within* the same measurement system. This chapter will provide you with the necessary knowledge to switch from *one system to another*.

> **Definition:** Changing a quantity from one system to an equivalent quantity in another system is called *conversion*.

Table 8-1 shows a *conversion table* of practical approximate equivalents. Note that some equivalents are worked out to two decimal places. The table shows these equivalents because they are in widespread use. In practice the answers are rounded off according to the accuracy of the original numbers. However, as a matter of convenience the answers in this chapter should be rounded off to the nearest tenth, when applicable.

Table 8-1. Conversion Table—Equivalents

	Metric		*Apothecaries'*	*English*
Weight	0.065 grams (g) or			
	65 milligrams (mg)	≈	1 grain (gr)*	
	1 gram (g)	≈	15 grains (gr)**	
	28.35 grams (g)	≈		1 ounce (oz)
	1 kilogram (kg)	≈		2.2 pounds (lb)
Length	1 meter (m)	≈		39.37 inches (in.)
	2.54 centimeters (cm)	≈		1 inch (in.)
Volume	1 milliliter (mℓ)	≈	16 minims***	
	30 milliliters (mℓ)	≈	1 fluidounce (f℥)	
	480 milliliters (mℓ)	≈	1 pint (pt)	

*In practice, some situations will arise in which a less exact conversion, 1 gr = 60 mg, is used. In this text, however, unless otherwise noted, use 1 gr = 65 mg.
**Occasionally, you may read that 1 g = 15.432 gr, which is more precise. However, this text will use 1 g = 15 gr.
***1 mℓ = 15 minims is sometimes used in practice.

The procedures of dimensional analysis and ratio and proportion, which you learned in earlier chapters, will allow you to successfully convert from one measurement system to another as you continue your nursing studies.

Section 8.1 Conversion of Weight

> **Rule: Converting Units of Weight from One System to Another**
> **Method A: Dimensional Analysis**
> *Step 1A:* Write down the given weight.
> *Step 2A:* Using dimensional analysis and known conversion factors, express 1 in fractional form(s) in order to obtain the desired unit and eliminate the given unit.
> *Step 3A:* Cancel units and multiply the given weight by the expression(s) for 1.

Section 8.1 Conversion of Weight 197

Example 1: Use method A to convert 5.6 g to grains.
Solution:
Step 1A: 5.6 g
Step 2A: The fact in Table 8-1 that relates grams to grains is 1 g = 15 gr.

$$\frac{15 \text{ gr}}{1 \text{ g}}$$

Step 3A: $\frac{5.6 \text{ g}}{1} \times \frac{15 \text{ gr}}{1 \text{ g}} = 84 \text{ gr}$

Therefore, 5.6 grams = 84 grains, or 5.6 g = 84 gr.

Rule: Converting Units of Weight from One System to Another
Method B: Ratio and Proportion
Step 1B: Write the conversion as a ratio in fractional form, using x on the bottom to represent the number of units you are converting to.
Step 2B: Write the fact(s) from Table 8-1, Weight Equivalents, that deal with both the units you started with and the units you want to convert to. Combine fact(s) when necessary. Write the fact(s) in fractional form.
Step 3B: Write the ratios from Steps 1 and 2 as a proportion and cross multiply. Be sure that the units on each side of the equation correspond to one another.
Step 4B: To solve for the unknown quantity, divide the numbers on each side of the equals sign by the number on the same side as the x.

Now use method B to solve the same problem (convert 5.6 g to grains).

Solution:
Step 1B: $\frac{5.6 \text{ g}}{x \text{ gr}}$

Step 2B: Because 1 gram = 15 grains, we write

$$\frac{1 \text{ g}}{15 \text{ gr}}$$

Step 3B: $\frac{5.6 \text{ g}}{x \text{ gr}} = \frac{1 \text{ g}}{15 \text{ gr}}$
$x \cdot 1 = 5.6 \cdot 15$
$x \cdot 1 = 84$

Step 4B: $\frac{x \cdot \cancel{1}}{\cancel{1}} = \frac{84}{1}$
$x = 84 \text{ gr}$

Therefore, 5.6 grams = 84 grains, or 5.6 g = 84 gr.

Example 2: Use method A to convert 2.5 oz to grams.
Solution:
Step 1A: 2.5 oz

Step 2A: The fact in Table 8-1 that relates ounces to grams is 1 oz = 28.35 g.

$$\frac{28.35 \text{ g}}{1 \text{ oz}} = 1$$

Step 3A: Now cancel units and cross multiply.

$$\frac{2.5 \text{ \cancel{oz}}}{1} \times \frac{28.35 \text{ g}}{1 \text{ \cancel{oz}}} = 70.88 \text{ g} = 70.9 \text{ g}$$

Therefore, 2.5 ounces = 70.9 grams, or 2.5 oz = 70.9 g.

Example 3: Use method B to convert 150 lb to kilograms.
Solution:

Step 1B:
$$\frac{150 \text{ lb}}{x \text{ kg}}$$

Step 2B: Because 2.2 lb = 1 kg, we write

$$\frac{2.2 \text{ lb}}{1 \text{ kg}}$$

Step 3B:
$$\frac{150 \text{ lb}}{x \text{ kg}} = \frac{2.2 \text{ lb}}{1 \text{ kg}}$$
$$x \cdot 2.2 = 150 \cdot 1$$
$$x \cdot 2.2 = 150$$

Step 4B: Now divide and round the answer off to one decimal place.

$$\frac{x \cdot \cancel{2.2}}{\cancel{2.2}} = \frac{150}{2.2}$$
$$x = 68.2 \text{ kg}$$

Therefore, 150 pounds = 68.2 kilograms, or 150 lb = 68.2 kg.

Example 4: Use method A to solve the following word problem: Dr. Drake has prescribed Retin-A® Cream 0.05% for her teenage patient with acne. The patient has been instructed to apply the cream once daily, sparingly. The tube contains 20 g of cream. How many ounces of cream does it contain?

Solution:
Step 1A: 20 g
Step 2A: The fact in Table 8-1 that relates grams to ounces is 28.35 g = 1 oz.

$$\frac{1 \text{ oz}}{28.35 \text{ g}} = 1$$

Step 3A: Cancel units, cross multiply.

$$\frac{20 \text{ \cancel{g}}}{1} \times \frac{1 \text{ oz}}{28.35 \text{ \cancel{g}}} = 0.71 \text{ oz} = 0.7 \text{ oz}$$

Therefore, 20 grams = 0.7 ounces of Retin-A® Cream 0.05%.

Now use method B to solve the same word problem.

Solution:

Step 1B: $\dfrac{20 \text{ g}}{x \text{ oz}}$

Step 2B: Because 28.35 g = 1 oz, we write

$$\dfrac{28.35 \text{ g}}{1 \text{ oz}}$$

Step 3B:
$$\dfrac{20 \text{ g}}{x \text{ oz}} = \dfrac{28.35 \text{ g}}{1 \text{ oz}}$$
$$x \cdot 28.35 = 20 \cdot 1$$
$$x \cdot 28.35 = 20$$

Step 4B:
$$\dfrac{x \cdot \cancel{28.35}}{\cancel{28.35}} = \dfrac{20}{28.35}$$
$$x = 0.71 \text{ oz}$$
$$x = 0.7 \text{ oz}$$

Therefore, 20 grams = 0.7 ounces of Retin-A® Cream 0.05%.

Section 8.1 Readiness Review

1. Convert 2 g to grains _____30_____ (see Examples 1 & 2)
2. Convert gr ¼ to milligrams _____16.4_____ (see Examples 1 & 2)
3. Convert ⅛ lb to grams _____56.7_____ (see Examples 1 & 2)
4. Convert 173 lb to kilograms _____78.6_____ (see Example 3)
5. Convert 7.775 g to grains _____116.6_____ (see Example 3)
6. Convert gr *dxxv* to ounces _____1.2_____ (see Example 3)

Section readiness review answers (not given in order):

4. 78.6 kg	3. 56.8 g	1. 30 gr
2. 16.3 mg	6. 1.2 oz	5. 116.6 gr

Section 8.1 Review Problems

(Answers to the odd-numbered problems are at the back of the book.)

 45   1. Convert 3 g to grains   2. Convert 1.5 g to grains
154 3.8 3. Convert 70 kg to pounds   4. Convert 17 kg to pounds
 44 9.1 5. Convert 20 kg to pounds   6. Convert gr *x* to milligrams
     130 7. Convert gr *ii* to milligrams 8. Convert 3 oz to grams
     42.5 9. Convert 1.5 oz to grams   10. Convert 2½ oz to grams
3189.4 212.6 11. Convert 7.5 oz to grains 12. Convert 0.8 oz to grains

13. Convert gr *xii* to grams
14. Convert gr *xv* to grams
15. Convert gr *lx* to grams
16. Convert 180 lb to kilograms
17. Convert 100 lb to kilograms
18. Convert 130 mg to grains
19. Convert 500 mg to grains
20. Convert 97.5 mg to grains
21. Convert 100 g to ounces
22. Convert 141.75 g to ounces
23. Convert 50 g to ounces

Solve the following word problems:

24. A patient of Dr. Kuhn complains that the lower lid of his right eye is red and irritated. Upon examining him, Dr. Kuhn determines the eye to be infected and prescribes Neosporin® Ophthalmic Ointment ⅛ oz to be applied as directed. How many grams of Neosporin® Ophthalmic Ointment does the tube contain?

25. A Vanceril® Inhaler has been ordered for a patient with severe asthma. By providing the patient with a predetermined amount of drug to inhale, Vanceril® helps reduce inflammation. The Vanceril® Inhaler is only available as a 17-gram, 200-metered-dose inhalation package. How many grains does the package contain?

Section 8.2 Conversion of Length and Volume

This section explores the conversion of units of length and volume. Refer to Table 8-1 in Section 8-1 for help in solving length and volume conversion problems.

> **Rule: Converting Units of Length or Volume from One System to Another**
> **Method A: Dimensional Analysis**
> *Step 1A:* Write down the given length or volume.
> *Step 2A:* Using dimensional analysis and known conversion factors, express 1 in fractional form(s) in order to obtain the desired unit and eliminate the given unit.
> *Step 3A:* Cancel units and multiply the given length or volume by the expression(s) for 1.

Example 1: Use method A to convert 3 m to inches.
Solution:
Step 1A: 3 m
Step 2A: The fact in Table 8-1 that relates meters to inches is 1 m = 39.37 in., or

$$\frac{39.37 \text{ in.}}{1 \text{ m}} = 1$$

Step 3A: $\frac{3 \text{ m}}{1} \times \frac{39.37 \text{ in.}}{1 \text{ m}} = 118.1 \text{ in.}$

Therefore, 3 meters = 118.1 inches, or 3 m = 118.1 in.

> **Rule: Converting Units of Length or Volume from One System to Another**
> **Method B: Ratio and Proportion**
> *Step 1B:* Write the conversion as a ratio in fractional form, using x on the bottom to represent the number of units you are converting to.
> *Step 2B:* Write the fact(s) from Table 8-1, Equivalents, that deal with both the units you started with and the units you want to convert to. Combine fact(s) when necessary. Write the fact(s) in fractional form.
> *Step 3B:* Write the ratios from Steps 1 and 2 as a proportion and cross multiply. Be sure that the units on each side of the equation correspond to one another.
> *Step 4B:* To solve for the unknown quantity, divide the numbers on each side of the equals sign by the number on the same side as the x.

Now use method B to solve the same problem (convert 3 m to inches).

Solution:

Step 1B: $\dfrac{3 \text{ m}}{x \text{ in.}}$

Step 2B: Because 1 meter = 39.37 inches, we write

$$\dfrac{1 \text{ m}}{39.37 \text{ in.}}$$

Step 3B: $\dfrac{3 \text{ m}}{x \text{ in.}} = \dfrac{1 \text{ m}}{39.37 \text{ in.}}$

$x \cdot 1 = 3 \cdot 39.37$

$x \cdot 1 = 118.1$

Step 4B: $\dfrac{x \cdot \cancel{1}}{\cancel{1}} = \dfrac{118.1}{1}$

$x = 118.1$ in.

Therefore, 3 meters = 118.1 inches, or 3 m = 118.1 in.

Example 2: Use method B to convert 2.75 mℓ to minims.
Solution:

Step 1B: $\dfrac{2.75 \text{ m}\ell}{x \text{ m}}$

Step 2B: Because 1 mℓ = 16 ♏, we write

$$\dfrac{1 \text{ m}\ell}{16 \text{ ♏}}$$

Step 3B: $\dfrac{2.75 \text{ m}\ell}{x \text{ ♏}} = \dfrac{1 \text{ m}\ell}{16 \text{ ♏}}$

$x \cdot 1 = 2.75 \cdot 16$

$x \cdot 1 = 44$

Step 4B: $\dfrac{x \cdot \cancel{1}}{\cancel{1}} = \dfrac{44}{1}$

$x = 44$

202   Chapter 8 Conversions and Medical Abbreviations

Therefore, 2.75 milliliters = 44 minims, or 2.75 mℓ = 44 ♏︎.

Example 3: Use method A to convert 2400 mℓ into pints.
Solution:
Step 1A: 2400 mℓ
Step 2A: The fact from Table 8-1 that relates milliliters to pints is 480 mℓ = 1 pt.

$$\frac{1 \text{ pt}}{480 \text{ m}\ell} = 1$$

Step 3A: $$\frac{\overset{5}{\cancel{2400 \text{ m}\ell}}}{1} \cdot \frac{1 \text{ pt}}{\underset{1}{\cancel{480 \text{ m}\ell}}} = 5 \text{ pt}$$

Therefore, 2400 milliliters = 5 pints, or 2400 mℓ = 5 pt.

Example 4: Use method A to solve the following word problem:
Laura Hubbard would like to try a new sunscreen. She has selected Pre Sun®, a 5% para-amino benzoic acid lotion, which comes in a 4-fluidounce (f℥) size. How many milliliters of Pre Sun® does this bottle contain?

Solution:
Step 1A: 4 f℥
Step 2A: The fact in Table 8-1 that relates fluidounces to milliliters is 1 f℥ = 30 mℓ.

$$\frac{30 \text{ m}\ell}{1 \text{ f}℥} = 1$$

Step 3A: $$\frac{4 \cancel{\text{f}℥}}{1} \times \frac{30 \text{ m}\ell}{1 \cancel{\text{f}℥}} = 120 \text{ m}\ell$$

Therefore, 4 f℥ = 120 mℓ of Pre Sun®.

Now use Method B to solve the same word problem.

Solution:
Step 1B: $$\frac{4 \text{ f}℥}{x \text{ m}\ell}$$

Step 2B: Because 1 f℥ = 30 mℓ, we write

$$\frac{1 \text{ f}℥}{30 \text{ m}\ell}$$

Step 3B: $$\frac{4 \text{ f}℥}{x \text{ m}\ell} = \frac{1 \text{ f}℥}{30 \text{ m}\ell}$$
$$x \cdot 1 = 4 \cdot 30$$
$$x \cdot 1 = 120$$

Step 4B: $$\frac{x \cdot \cancel{1}}{\cancel{1}} = \frac{120}{1}$$
$$x = 120 \text{ m}\ell$$

Therefore, 4 f℥ = 120 mℓ of Pre Sun®.

Section 8.2 Conversion of Length and Volume

Section 8.2 Readiness Review

1. Convert 2 m to inches _78.7_ (see Example 1)
2. Convert f℥ vi to milliliters _180_ (see Example 2)
3. Convert ½ pt to milliliters _240_ (see Example 2)
4. Convert 30.48 cm to inches _12_ (see Example 3)
5. Convert 405.75 mℓ to minims _6492_ (see Example 2)
6. Convert 1892 mℓ to pints _3.9_ (see Example 3)

Section readiness review answers (not given in order):

2. 180 mℓ   4. 12 in.   6. 3.9 pt
1. 78.7 in.   5. 6492 ♏   3. 240 mℓ

Section 8.2 Review Problems

(Answers to the odd-numbered problems are at the back of the book.)

1. Convert 4 m to inches — 157.5
2. Convert 10 m to inches — 393.7
3. Convert 3.25 m to inches — 128
4. Convert 8 m to inches — 314.96
5. Convert 1½ in. to centimeters — 3.8
6. Convert 4 1/16 in. to centimeters
7. Convert 6 in. to centimeters — 15.2
8. Convert 8 in. to centimeters — 20.32
9. Convert 10 in. to centimeters — 25.4
10. Convert 2 mℓ to minims — 32
11. Convert 3.2 mℓ to minims — 51.2
12. Convert 6.4 mℓ to minims — 102.4
13. Convert 8.5 mℓ to minims — 136
14. Convert 10 mℓ to minims — 160
15. Convert f℥ ss to milliliters — 15
16. Convert f℥ v to milliliters — 150
17. Convert f℥ ix to milliliters — 270
18. Convert f℥ xxx to milliliters — 900
19. Convert 1 pt to milliliters — 480
20. Convert 2 pt to milliliters — 960
21. Convert 8 pt to milliliters — 3840
22. Convert 10 pt to milliliters — 4800
23. Convert ¾ pt to milliliters — 360
24. Convert 150 in. to meters — 3.8
25. Convert 24 in. to meters — .6
26. Convert 180 in. to centimeters — 457.2
27. Convert 155 in. to centimeters — 393.7
28. Convert 5.08 cm to inches — 2
29. Convert 12.7 cm to inches — 5
30. Convert 50.8 cm to inches — 20
31. Convert 76.2 cm to inches — 30
32. Convert 64.92 mℓ to minims
33. Convert 97.38 mℓ to minims — 1558.1
34. Convert 129.84 mℓ to minims — 2077.4
35. Convert 210.99 mℓ to minims — 3375.8
36. Convert 59.14 mℓ to fluidounces
37. Convert 118.28 mℓ to fluidounces — 3.9
38. Convert 325.27 mℓ to fluidounces

204 Chapter 8 Conversions and Medical Abbreviations

39. Convert 443.55 mℓ to fluidounces
40. Convert 502.69 mℓ to fluidounces
41. Convert 118.25 mℓ to pints
42. Convert 591.25 mℓ to pints
43. Convert 1419 mℓ to pints

Solve the following word problems:

44. A person who fainted was revived by a "smelling salt," namely an Aromatic Ammonia Vaporole®. This crushable capsule filled with ammonia contains 0.33 cc of drug. How many minims of drug does the capsule contain?

45. A patient has been diagnosed as having diabetes insipidus. (This disease is not to be confused with diabetes mellitus—sugar diabetes.) The patient excretes large quantities of dilute urine. The physician has prescribed an 8-mℓ Diapid® Nasal Spray, to be used one or two sprays in each nostril four times daily. The drug is absorbed from the nasal tissues and causes the kidneys to reabsorb water instead of allowing it to pass out in the urine. How many minims does the container of Diapid® spray contain?

Section 8.3 Latin Abbreviations

The last two sections of this chapter will cover abbreviations used in medicine and related areas. Section 8.3 will deal with Latin abbreviations, whereas Section 8.4 will cover common medical abbreviations. It is essential for nursing students to know these abbreviations, because they are frequently used in daily practice. This text includes the most widely used abbreviations; students are referred to basic-nursing or nursing-pharmacology texts for more complete listings.

Traditionally, physicians have used Latin words and abbreviations to write prescription orders. Latin is considered the language of medicine; however, the strict use of Latin on prescription orders has been on the decline since the beginning of the 20th century, while Latin abbreviations have been more and more commonly used.

Table 8-2 lists the Latin abbreviations plus the Latin words they come from and their exact meanings. (The Latin words are included as a reference source, because there are still a few physicians who use the words instead of the abbreviations.) To learn these abbreviations and their meanings: (1) familiarize yourself with Table 8-2, Latin Abbreviations; (2) carefully read through the examples to understand how the abbreviations are used; (3) work the problems in the Section Readiness Review (look at the answers only after you have tried the problems); (4) work the problems at the end of this section.

Table 8-2. Latin Abbreviations

Latin Abbreviation	Latin Words	Meaning
ā, or a.	ante	before
a.c.	ante cibum	before meals
a.d.	aurio dextra	right ear
ad	ad	to, up to
ad lib.	ad libitum	as desired, at pleasure, freely
a.l.	aurio laeva	left ear
alt. h.	alternis horis	every other hour
aq.	aqua	water

Table 8-2. (continued)

a.u.	aures utrae	each ear
b.i.d.	bis in die	twice a day
c̄	cum	with
cap. or caps.	capsula	capsule
cibus	cibus	food
dil.	dilue, dilutus	dilute, diluted
gr	granum, grana	grain, grains
gtt	gutta, guttae	a drop, drops
h. or hr	hora	an hour
h.s. or hor. som.	hora somni	at bedtime
l. or L.	laevo	left
M.	misce	mix
ℳ or min.	minimum	a minim
Non rep.	Non repetatur	do not repeat
nox, noctis	nox, noctis	night
O	octarius	a pint
O.D.	oculus dexter	right eye
O.L.	oculus laevus	left eye
O.S.	oculus sinister	left eye
O.U., O$_2$	oculo utro	each eye
os, oris	os, oris	mouth
p.c.	post cibum	after meals
p.o.	per ora	by mouth
p.r.n.	pro re nata	as needed, when necessary
q.	quaque, quisque	each, every
q.d.	quaque die	every day, daily
q.i.d.	quater in die	four times a day
q.s. or Q.S.	quantum sufficit	as much as is sufficient
℞	recipe	take
Rept.	repetatur	let it be repeated
Sig.	Signa, Signetur	write, let be written
ss or s̄s̄	semi, semis	one-half
stat.	statim	immediately
suppos.	suppositorium	suppository
suppos. rect.	suppositorium rectalium	rectal suppository
tab.	tabella, tabletta	tablet
t.i.d.	ter in die	three times a day
ut dict. or u.d.	ut dictum	as directed

Example 1: Fill in the following blanks:

	Latin Abbreviation	*Meaning*
a.	ā	_____
b.	b.i.d.	_____
c.	_____	grain, grains
d.	_____	a pint

Solution:

a. before
b. twice a day
c. gr
d. O

When using abbreviations in the apothecaries' system, remember to write the amount of drug ordered in Roman numerals *following* the abbreviation (see Chapter 7, Apothecaries' and Household Systems). For example, eight drams is written as ʒ *viii*. If the abbreviation is not used and the unit is spelled out, then the amount is written as a whole number (mixed number or fraction, when necessary) *before* the unit. In this case, then, eight drams is written as 8 drams.

Example 2: Rewrite the following directions, using the meanings of the abbreviations: Give Robitussin DM® fʒ ii q. 4 h. p.r.n. for cough.

Solution: Give Robitussin DM® 2 fluidrams every four hours as needed for cough.

Example 3: Rewrite the following directions, using the meanings of the abbreviations: Keflex® 500 mg caps.: Take 1 cap. q.i.d. a.c. and h.s. for 10 days.

Solution: Keflex® 500 mg caps.: Take 1 capsule four times daily, before meals and at bedtime, for ten days.

Example 4: Rewrite the following directions, using the meanings of the abbreviations: Cortisporin Otic Solution®: Instill gtt ii a.d. q. 6 h.

Solution: Cortisporin Otic Solution®: Instill 2 drops in the right ear every six hours.

Example 5: Rewrite the following directions, using the meanings of the abbreviations: Insert 1 suppos. rect. h.s.

Solution: Insert one rectal suppository at bedtime.

Section 8.3 Readiness Review

Fill in the following blanks (see Example 1):

Latin Abbreviation	Meaning
1. _____	tablet
2. a.u.	_____
3. _____	food
4. u.d.	_____

Rewrite the following directions, using the meanings of the abbreviations (see Examples 2-5):

5. PenVee K® 250 mg/tsp: Give 1 tsp q.i.d. until all taken. _____
6. Librium® 10 mg caps.: Take 1 or 2 caps. b.i.d. p.r.n. for nerves. _____

Section readiness review answers (not given in order):

1. tab.   3. cibus   2. each ear   4. as directed
6. Librium® 10 mg caps.: Take 1 or 2 capsules twice daily as needed for nerves.
5. PenVee K® 250 mg/tsp: Give 1 teaspoonful four times daily until all taken.

Section 8.3 Review Problems

(Answers to the odd-numbered problems are at the back of the book.)

Latin Abbreviation	Meaning
1. suppos.	_____
2. _____	dilute, diluted

3. _____ write, let it be written
4. _____ before meals
5. O.L. _____
6. h. or hr _____
7. _____ left ear
8. _____ noctis
9. stat. _____
10. _____ as directed
11. p.o. _____
12. O.S. _____
13. Non rep. _____
14. _____ at bedtime
15. _____ take
16. q.d. _____

Rewrite the following directions, using the meanings of the abbreviations:

17. Propoxyphene HCl 32 mg: Take 2 caps. q. 6–8 hr p.r.n. for pain.

18. Inderal® 40-mg tablets: Take ī tab. p.o. b.i.d. Do not go off medication abruptly.

19. Amoxil® 250-mg caps.: Give ī cap. p.o. t.i.d. a.c. until gone.

20. Maalox® and Gelusil®: Give f℥ ī alt. h. around the clock.

Section 8.4 Common Medical Abbreviations

Medical abbreviations, like Latin abbreviations, are the "shorthand" of the medical community. Table 8-3 shows the common medical abbreviations; however, it is not possible to list all the medical abbreviations used in North America. Therefore, check with the hospital where you work for additional or differently used abbreviations. Use the same steps mentioned in the previous section on Latin abbreviations to learn the common medical abbreviations. Good luck.

Table 8-3. Common Medical Abbreviations

Abbreviation	For
ASHD	arteriosclerotic heart disease
AV	atrioventricular
BC	blood culture
BE	barium enema
BP	blood pressure
BRP	bathroom privileges
BUN	blood urea nitrogen
BS	blood sugar
CA	carcinoma, cancer
CBC	complete blood count
CHF	congestive heart failure
COPD	chronic obstructive pulmonary disease
C and S	culture and sensitivity
CSF	cerebral spinal fluid
CVA	cerebrovascular accident

Table 8-3. (continued)

Abbreviation	For
D/C	discontinue
D and C	dilation and curettage
D and E	dilation and evacuation
DM	diabetes mellitus
D5W	dextrose 5% in water
Dx	diagnosis
ECG	electrocardiogram
EEG	electroencephalogram
EKG	electrocardiogram
FBS	fasting blood sugar
GI	gastrointestinal
GU	genitourinary
H	hypodermic
HA	headache
Hct	hematocrit
H_2O	water
IM	intramuscular
I and O	intake and output
IPPB	intermittent positive pressure breathing
I.V.	intravenous
LYTES	electrolytes
mEq	milliequivalents
MI	myocardial infarction
MS	mitral stenosis; morphine sulfate
NG	nasogastric
NPO	nothing by mouth
NS	normal saline
N and V	nausea and vomiting
P	pulse
PID	pelvic inflammatory disease
PPD	purified protein derivative
PR	per rectum
pt.	patient
PT	physical therapy; prothrombin time
RBC	red blood cell
Rh	Rhesus factor
R/O	rule out
SC	subcutaneous
SGOT	serum glutamic oxaloacetic transaminase
SGPT	serum glutamic pyruvic transaminase
SLE	systemic lupus erythematosus
SOB	shortness of breath
SQ	subcutaneous
T	temperature
T and A	tonsillectomy and adenoidectomy
TB	tuberculosis
TKO	to keep open
UA	urinalysis
UGI	upper gastrointestinal series
URI	upper respiratory infection
UTI	urinary tract infection
VS	vital signs
WBC	white blood cell; white blood count

Section 8.4 Common Medical Abbreviations   209

Example 1:   Fill in the following blanks:

Abbreviation   *For*
a. LYTES   _____
b. _____   congestive heart failure
c. PR   _____
d. UA   _____

Solution:

a. electrolytes
b. CHF
c. per rectum
d. urinalysis

Example 2:   The following orders have been written by the physician. Rewrite these orders, using the meanings instead of the abbreviations:

a. Dx: possible UTI
b. Laboratory tests: CBC
c. UA-Stat. (C and S is requested)
d. Start Pyridium® 200 mg t.i.d. and Septra DS® ī tab. b.i.d. after UA.

Solution:

a. Diagnosis: possible urinary tract infection
b. Laboratory tests: complete blood count
c. Urinalysis-immediately (culture and sensitivity is requested)
d. Start Pyridium® 200 milligrams three times daily and Septra DS® one tablet two times daily after urinalysis.

Section 8.4 Readiness Review

Fill in the following blanks:

Abbreviation   *For*
1. _____   tuberculosis
2. NS   _____
3. VS   _____
4. UGI   _____

The following orders have been written by the physician. Rewrite these orders, using the meanings instead of the abbreviations.

5. Dx: possible CHF (right failure)

6. Laboratory tests: CBC, LYTES, BUN

7. EKG-Stat.

8. 2 g sodium diet

9. I.V. fluids: D5W TKO

10. Digoxin 0.5 mg p.o. now, then in morning start 0.25 mg q.d.

11. Lasix 20 mg I.V. now, then in morning start 20 mg p.o. q.d.

12. I and O

Section readiness review answers (not given in order):

1. TB
2. normal saline
3. vital signs
4. upper gastrointestinal series
12. intake and output
10. Digoxin 0.5 milligram by mouth now, then in the morning start 0.25 milligram every day.
5. Diagnosis: possible congestive heart failure (right failure)
7. electrocardiogram immediately
6. Laboratory tests: complete blood count, electrolytes, blood urea nitrogen
9. Intravenous fluids: dextrose 5% in water to keep open
11. Lasix 20 milligrams intravenously now, then in morning start 20 milligrams by mouth daily.
8. 2-gram sodium diet

Section 8.4 Review Problems

Fill in the following blanks (answers to the odd-numbered problems are at the back of the book):

	Abbreviation	*For*
1.	_____	serum glutamic pyruvic transaminase
2.	_____	temperature
3.	BRP	_____
4.	pt.	_____
5.	_____	physical therapy
6.	SC	_____
7.	SQ	_____
8.	_____	systemic lupus erythematosus
9.	_____	intermittent positive pressure breathing
10.	I.V.	_____
11.	_____	purified protein derivative
12.	RBC	_____
13.	_____	cerebral spinal fluid
14.	_____	discontinue
15.	DM	_____
16.	GU	_____
17.	IM	_____
18.	NPO	_____
19.	ECG	_____
20.	_____	hypodermic
21.	EKG	_____

The following orders have been written by the physician. Rewrite these orders, using the meanings instead of the abbreviations.

22. Dx: PID
23. Laboratory tests: CBC
24. Be sure to get C and S of vaginal discharge
25. Start ampicillin 500 mg p.o. q.i.d. for seven days.
26. BRP

Chapter 8 Readiness Review

1. Convert 100 g to grains _____
2. Convert gr iii to milligrams _____
3. Convert 156 lb to kilograms _____
4. Convert 3 m to inches _____
5. Convert fʒ vii to milliliters _____
6. Convert 120 in. to meters _____
7. Convert 568.05 mℓ to minims _____
8. Convert 2160 mℓ to pints _____

Fill in the following blanks:

	Latin Abbreviation	Meaning
9.	_____	to, up to
10.	c̄	_____
11.	_____	three times a day
12.	q.s. or Q.S.	_____
13.	℞	_____

Rewrite the following directions, using the meanings instead of the abbreviations:

14. Dimetapp Extentabs®: Give i b.i.d. p.r.n. for congestion.
    _____

15. Saturated solution of potassium iodide: gtt x dil. in fʒ viii aq. t.i.d. or q.i.d.
    _____

Fill in the following blanks:

	Abbreviation	For
16.	_____	Arteriosclerotic heart disease
17.	BUN	_____

18. CBC   _____

19. _____   blood urea nitrogen

20. R/O   _____

21. _____   serum glutamic oxaloacetic transaminase

The following orders have been written by the physician. Rewrite these orders, using the meanings instead of the abbreviations.

22. Dx: DM

23. Laboratory tests: CBC, FBS, UA

24. Give: regular U-100 insulin—rainbow coverage
    10 units for 2% urine sugar
    5 units for 1% urine sugar
    2 units for ¾% urine sugar
    0 units for ½% or less urine sugar

25. Test urine with Diastix® a.c. and h.s.

Chapter readiness review answers (not given in order):

25. Test urine with Diastix® before meals and at bedtime.
24. This problem is o.k. as written.
23. Laboratory tests: complete blood count, fasting blood sugar, urinalysis
22. Diagnosis: diabetes mellitus   21. SGOT
20. rule out   19. BUN   18. complete blood count
17. blood urea nitrogen   16. ASHD
15. Saturated solution of potassium iodide: 10 drops diluted in 8 fluidounces of water three times daily or four times daily
14. Dimetapp Extentabs: Give 1 tablet two times daily as needed for congestion.
13. take   12. as much as is sufficient   11. t.i.d.
10. with   9. ad   8. 4.5 pt
7. 9088.8 ℳ   6. 3.0 m   5. 210 mℓ
4. 118.1 in.   3. 70.9 kg   2. 195 mg
1. 1500 gr

Chapter 8 Summary

Define each item in your own words, then compare your definitions with the text.

Key Words

conversion   (p. 196)
Latin abbreviations   (p. 204)
medical abbreviations   (p. 207)

Chapter 8 Review Problems

(Answers to all the problems are at the back of the book.)

1. Convert 4.6 g to grains
2. Convert 7.5 g to grains
3. Convert 75 kg to pounds
4. Convert 16 kg to pounds
5. Convert gr *v* to milligrams
6. Convert gr *vii ss* to milligrams
7. Convert 4 oz to grams
8. Convert 1.75 oz to grams
9. Convert 1.75 oz to grains
10. Convert 6 oz to grains
11. Convert gr *x* to grams
12. Convert gr *xvi* to grams
13. Convert 150 lb to kilograms
14. Convert 125 lb to kilograms
15. Convert 750 mg to grains
16. Convert 113.75 mg to grains
17. Convert 75 g to ounces
18. Convert 99.225 g to ounces
19. Convert 56.7 g to ounces
20. Convert 5 m to inches
21. Convert 12 m to inches
22. Convert 1 3/16 in. to centimeters
23. Convert 3 1/8 in. to centimeters
24. Convert 3 mℓ to minims
25. Convert 7.2 mℓ to minims
26. Convert 8.1 mℓ to minims
27. Convert 20 mℓ to minims
28. Convert f℥ *ss* to milliliters
29. Convert f℥ *x* to milliliters
30. Convert f℥ *xv* to milliliters
31. Convert 1/4 pt to milliliters
32. Convert 12 pt to milliliters
33. Convert 1 1/2 pt to milliliters
34. Convert 100 in. to meters
35. Convert 1608 in. to meters
36. Convert 7.62 cm to inches
37. Convert 38.1 cm to inches
38. Convert 81.15 mℓ to minims
39. Convert 162.3 mℓ to minims
40. Convert 227.22 mℓ to minims
41. Convert 88.71 mℓ to fluidounces
42. Convert 236.56 mℓ to fluidounces
43. Convert 354.84 mℓ to fluidounces
44. Convert 473.12 mℓ to fluidounces
45. Convert 520.3 mℓ to pints
46. Convert 709.5 mℓ to pints
47. Convert 3311 mℓ to pints
48. Convert 5203 mℓ to pints

Fill in the following blanks:

Latin Abbreviation	Meaning
49. _____	a minim
50. suppos. rect.	_____
51. q.i.d.	_____
52. p.r.n.	_____
53. caps.	_____

214   Chapter 8 Conversions and Medical Abbreviations

54. _____ right ear

55. _____ as desired, at pleasure, freely

56. _____ water

57. alt. h. _____

58. _____ after meals

59. O.U. or O₂ _____

60. gtt _____

61. *ss* or $\overline{ss}$ _____

62. _____ left eye

63. q. _____

64. _____ let it be repeated

Rewrite the following directions, using the meanings instead of the abbreviations:

65. Lorelco® 250-mg tablets: Take ii tabs. b.i.d. c̄ the morning and evening meal.

66. Aldomet® 250-mg tablets: Take i tab. b.i.d. for 48 h., then increase to ii b.i.d.

67. Norpace® 150-mg capsules: Take ii caps. stat. then i q. 6 h.

68. Sorbitrate® 2.5 mg, sublingual tablets: Place i–ii under tongue p.r.n. for angina pain.

69. Dyazide® capsules: Take i b.i.d. p.c.

70. Prednisone 5-mg tablets: Take iv b.i.d. for two days, iii b.i.d. for two days, ii b.i.d. for two days, i b.i.d. for two days, i q.d. until gone.

71. K-Lyte® tabs: Dissolve 1 tab. in f℥ vi aq. b.i.d.

Fill in the following blanks:

Abbreviation	For
72. _____	shortness of breath
73. _____	tonsillectomy and adenoidectomy
74. Rh	_____
75. TKO	_____
76. _____	nasogastric
77. LYTES	_____
78. mEq	_____
79. _____	diagnosis
80. FBS	_____

81. HA   _____

82. GI   _____

83. _____ electroencephalogram

84. UTI   _____

85. _____ white blood count

86. _____ water

87. COPD   _____

88. _____ pulse

89. PR   _____

90. _____ blood culture

Rewrite the following orders, using the meanings instead of the abbreviations:

91. Dx: N and V

92. Laboratory tests: CBC, LYTES

93. NPO

94. I.V.: NS 500 mℓ TKO c̄ potassium chloride 30 mEq now.

95. Place NG tube, low suction

96. Arrange for BE tomorrow morning.

Solve the following word problems:

97. Vena Dyer has developed a slight headache (probably due to the pressures of working at a publishing company). Her wise pharmacist/author has recommended 325 mg of aspirin. How many grains of aspirin is this?

98. A patient has been placed on a daily dose of 130 mg of thyroid USP (desiccated thyroid), after removal of his thyroid gland. How many grains of thyroid USP is this patient taking?

99. A tube of Nitrol® 2% ointment (long-acting nitroglycerin ointment) has been prescribed for Rabbi Weinburg. The patient has been instructed to apply a 2-inch squeeze of ointment to his arm every four hours for angina. How many centimeters of ointment will he apply per dose?

100. Ms. Monroe has been diagnosed as having a vaginal infection. The organism causing the problem is *Haemophilus*. The doctor has prescribed Sultrin®, a triple sulfa cream with an applicator. The Sultrin® comes in a 78-gram tube. How many ounces does the tube contain?

Three
Oral Dosages

9
Tablets, Capsules, and Oral Solutions

OBJECTIVES After studying this chapter, you should be able to:

1. Determine the whole number of capsules or tablets (or fractions of tablets) to be given when the drug required may not be available in the exact strength ordered by the doctor.
2. Determine the correct volume of oral solution to be given when the dose required may not be available in the exact strength ordered by the doctor.

Section 9.1 Tablets and Capsules

Tablets are the most frequently prescribed oral-dosage form, followed by capsules and oral solutions.

> A *tablet* may be defined as a molded or compressed preparation of an active ingredient. (A diluent may be present to give substance to the tablet.)

Some tablets are scored to allow easy breaking. Other tablets are easily crushed and dissolved. There are tablets that are covered with enteric coating, which delays the disintegration of the tablet until it reaches the intestine. Figure 9-1 shows the two basic forms of tablets, scored and unscored.

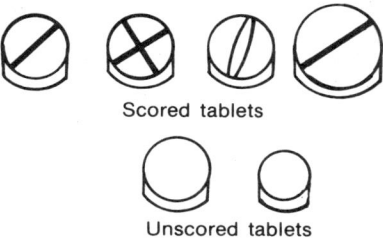

Figure 9-1.

Tablets come in many different strengths. The strength of a tablet may be expressed using the metric system or the apothecaries' system, either in units or in milliequivalents. For example,

Aldomet® 250-mg tablets (metric)
Thyroid U.S.P. gr $\bar{iss}$ tablets (apothecaries')
V Cillin K® 400,000-unit tablets (units)[1]
Slow K® 8.3-mEq tablets (milliequivalents)

> A *capsule* is a solid-dosage form with a covering of gelatin, which encases the active ingredient and, in some cases, the diluent as well. The soluble capsule may be hard or soft.

A capsule *cannot* be divided accurately. Therefore, don't pull it apart and try to measure the contents. Figure 9-2 shows a capsule.

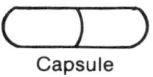

Figure 9-2.

---

[1] In problems dealing with penicillin, the strength of the drug may be expressed in units or milligrams. One milligram is equal to 1600 units; therefore, 250 mg are equal to 400,000 units, and so on.

> **Rule: Solving Tablet- and Capsule-Dosage Problems**
> **Method A: Dimensional Analysis**
> *Step 1A:* Write down the dosage strength desired by the physician.
> *Step 2A:* Using dimensional analysis and known conversion factors, express 1 in fractional form(s) in order to obtain the desired unit and eliminate the given unit.
> *Step 3A:* Cancel units and multiply the given strength by the expression(s) for 1. (Only scored tablets can be used for a fraction-of-tablet answer.)
> *Step 4A:* Using the information from Step 3A, rewrite the orders. Do not use abbreviations.

The following example shows a prescription ordered by the physician along with the available form of the prescribed item. Use method A to determine the appropriate number of tablets per dose.

Example 1: *Ordered:* Keflex® 500 mg q.i.d.
On hand: Keflex® 1-g scored tablets

Solution:

Step 1A: Write down the dosage strength desired by the physician.

500 mg

Step 2A: The on-hand information tells us that

$$\frac{1 \text{ tablet}}{1 \text{ g}} = 1$$

and we know from metric equivalents that

$$\frac{1 \text{ g}}{1000 \text{ mg}} = 1$$

Step 3A: Cancel units and multiply the given strength by the expressions for 1.

$$\overset{1}{\cancel{500 \text{ mg}}} \times \frac{1 \text{ tablet}}{\cancel{1 \text{ g}}} \times \frac{\cancel{1 \text{ g}}}{\underset{2}{\cancel{1000 \text{ mg}}}} = \frac{1}{2} \text{ tablet}$$

Step 4A: Give ½ Keflex® 1-g tablet four times daily.

> **Rule: Solving Tablet- and Capsule-Dosage Problems**
> **Method B: Ratio and Proportion**
> *Step 1B:* Write the conversion as a ratio in fractional form, using x on the bottom to represent the dosage strength desired by the physician.
> *Step 2B:* Write down the amount of the on-hand item first in the units you are starting with and then in the units you want to convert to. Use the fraction form and combine fact(s) when necessary.
> *Step 3B:* Write the ratios from Steps 1B and 2B as a proportion and cross multiply. Be sure that the units on each side of the equation correspond to one another.
> *Step 4B:* To solve for the unknown quantity, divide the numbers on each side of the equals sign by the number on the same side as the x. (Only scored tablets can be used for a fraction-of-tablet answer.)
> *Step 5B:* Using the information from Step 4B, rewrite the orders. Do not use abbreviations.

Section 9.1 Tablets and Capsules 221

Now use method B to determine the appropriate number of tablets per dose in the same example.

>*Ordered:* Keflex® 500 mg q.i.d.
>*On hand:* Keflex® 1-g scored tablets

Solution:

Step 1B: Write the conversion as a ratio in fractional form, using x to represent the unknown quantity.

$$\frac{500}{x}$$

Step 2B: From the on-hand information, we write

1 g = 1 tablet

or

1000 mg = 1 tablet

or

$$\frac{1000}{1}$$

Step 3B: Write the two ratios as a proportion and cross multiply.

$$\frac{500}{x} = \frac{1000}{1}$$
$$x \cdot 1000 = 500 \cdot 1$$
$$x \cdot 1000 = 500$$

Step 4B: To solve for the unknown, divide the numbers on each side of the equals sign by the number on the same side as the x.

$$\frac{x \cdot \cancel{1000}}{\cancel{1000}} = \frac{\cancel{500}}{\cancel{1000}}$$
$$x = \frac{1}{2} \text{ tablet}$$

Step 5B: Give ½ Keflex® 1-g tablet four times daily.

The next example also shows another prescription ordered by the physician along with the available form of the prescribed item. Use method B to determine the appropriate number of capsules per dose.

Example 2: *Ordered:* Colace® 250 mg p.o. q.d.
On hand: Colace® 50-mg capsules

Solution:

Step 1B: Write the conversion as a ratio in fractional form, using x to represent the unknown quantity.

$$\frac{250}{x}$$

Step 2B: From the on-hand information, we write

$$50 \text{ mg} = 1 \text{ capsule}$$

or

$$\frac{50}{1}$$

Step 3B: Write the two ratios as a proportion and cross multiply.

$$\frac{250}{x} = \frac{50}{1}$$
$$x \cdot 50 = 250 \cdot 1$$
$$x \cdot 50 = 250$$

Step 4B: To solve for the unknown, divide the numbers on each side of the equals sign by the number on the same side as the *x*.

$$\frac{x \cdot \cancel{50}}{\cancel{50}} = \frac{\cancel{250}^{5}}{\cancel{50}}$$
$$x = 5 \text{ capsules}$$

Step 5B: Give 5 Colace® 50-mg capsules by mouth daily.

Section 9.1 Readiness Review

Each of the following problems shows the prescription ordered by the physician along with the available form of the prescribed item. In each case, determine the appropriate number of tablets or capsules to be given from the on-hand supply.

1. *Ordered:*   Thyroid U.S.P. gr i $\overline{ss}$ p.o. q.d.
   On hand: Thyroid U.S.P. gr $\overline{ss}$ tablets

2. *Ordered:*   KCl 15 mEq² p.o. q.d.
   On hand: KCl 5-mEq tablets

3. *Ordered:*   Tegopen® 1000 mg p.o. q. 6 h. for severe infection
   On hand: Tegopen® 250-mg capsules

4. *Ordered:*   Penicillin G 400,000 units q. 6 h.
   On hand: Penicillin G 800,000-unit tablets

Section readiness review answers (not given in order):

3. 4 capsules. Give 4 Tegopen® 250-mg capsules by mouth every six hours for severe infection.
2. 3 tablets. Give 3 KCl 5-mEq tablets by mouth each day.
4. ½ tablet. Give ½ penicillin G 800,000-unit tablet every six hours.
1. 3 tablets. Give 3 thyroid U.S.P. gr $\overline{ss}$ tablets by mouth each day.

---

²The topic of milliequivalents is covered in Application II.

Section 9.1 Review Problems

Each of the following problems shows a prescription ordered by the physician along with the available form of the prescribed item. Determine the appropriate number of tablets or capsules per dose (answers to the odd-numbered problems are at the back of the book).

1. *Ordered:* Furadantin® 100 mg q.i.d.
   On hand: Furadantin® 50-mg tablets

2. *Ordered:* Urecholine® 20 mg p.o. t.i.d.
   On hand: Urecholine® 5-mg tablets

3. *Ordered:* Chloral hydrate gr *xv* p.o. h.s.
   On hand: Chloral hydrate gr *viiss* capsules

4. *Ordered:* Dalmane® 30 mg p.o. h.s. p.r.n.
   On hand: Dalmane® 15-mg capsules

5. *Ordered:* Cleocin® 150 mg p.o. q. 6 h.
   On hand: Cleocin® 75-mg capsules

6. *Ordered:* V Cillin K® 500 mg q. 6 h.
   On hand: V Cillin K® 125-mg tablets

7. *Ordered:* Vibramycin® 100 mg p.o. q.d.
   On hand: Vibramycin® 50-mg capsules

8. *Ordered:* Keflex® 1 g p.o. q.i.d.
   On hand: Keflex® 250-mg capsules

9. *Ordered:* Periactin® 2 mg p.o. q.i.d. p.r.n. itching
   On hand: Periactin® 4-mg scored tablet

10. *Ordered:* Dramamine® 100 mg p.o. q.i.d. p.r.n. nausea
    On hand: Dramamine® 50-mg scored tablet

11. *Ordered:* Benadryl® 50 mg p.o. h.s.
    On hand: Benadryl® 25-mg capsules

12. *Ordered:* Fulvicin U/F® 500 mg p.o. b.i.d. for fungal infection
    On hand: Fulvicin U/F® 250-mg tablets

13. *Ordered:* Isoniazid (INH) 300 mg q.d. in A.M.
    On hand: Isoniazid 100-mg tablets

14. *Ordered:* Azulfidine® 1.5 g p.o. q.i.d.
    On hand: Azulfidine® 500-mg tablets

15. *Ordered:* Ferrous sulfate 600 mg p.o. t.i.d. p.c.
    On hand: Ferrous sulfate 300-mg tablets

Section 9.2 Oral Solutions

Not all patients can swallow tablets and capsules. You may recall your own difficulty as a child trying to swallow a large tablet or capsule. To remedy this situation, some drugs are available in a liquid form (oral solution) or in the form of a powder to be put into solution (reconstituted).

> An *oral solution* usually contains a drug that has been manufactured in a liquid form.

> *Reconstitution* involves the addition of a solvent (usually distilled water) to a drug that is in the form of a powder.

Why are some drugs made to be reconstituted? The main reason is stability. Once the water (solvent) is added to the powdered drug, the resulting solution will usually last only 10 to 14 days. The powdered drug alone, however, may have an expiration date of 2 to 5 years. Quite a difference, wouldn't you say? Figure 9-3 shows a drug before and after reconstitution.

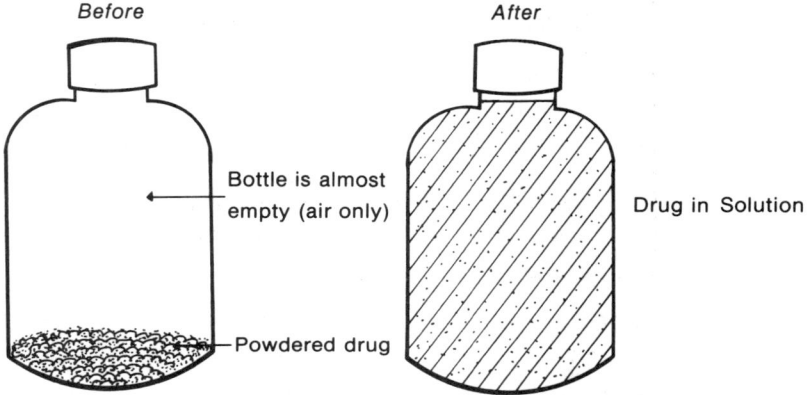

Figure 9-3. Reconstitution

The containers of drugs that need to be reconstituted will be labeled with directions; for example,

Keflex® 125 mg/5mℓ   100-mℓ size
for oral suspension[3]

Each 100-mℓ-size package consists of a bottle containing cephalexin monohydrate equivalent to 2.5 g cephalexin in a dry, pleasantly flavored mixture. At the time of dispensing, add 63 mℓ of water to the dry mixture in the bottle in two portions. Shake well after each addition. When mixed as directed, each 5 mℓ (approximately 1 teaspoonful) will contain cephalexin monohydrate equivalent to 125 mg cephalexin. After mixing, store in a refrigerator. The mixture may be kept for fourteen days without significant loss of potency.

Most oral solutions (whether to be reconstituted or not) will be labeled according to the amount of drug per so many milliliters or drams, and so on.

---

[3] Directions for oral suspension of Keflex® courtesy of Eli Lilly & Company.

Section 9.2 Oral Solutions   225

To solve oral-solution problems, you must read these labels carefully. A typical label will contain the following information:

> Name of medication (trade name or generic or both)
> Strength per volume
> Container size
> Directions for reconstitution (when applicable)

> **Rule: Solving Oral-Solution Dosage Problems**
> **Method A: Dimensional Analysis**
> *Step 1A:* Write down the dosage strength desired by the physician.
> *Step 2A:* If the drug is to be reconstituted, follow the manufacturer's printed instructions; the final concentration (after reconstitution) is printed on the label. Use this information in your calculation.
> *Step 3A:* Using dimensional analysis and known conversion factors, express 1 in fractional form(s) in order to obtain the desired unit and eliminate the given unit.
> *Step 4A:* Cancel units and multiply the given volume by the expression(s) for 1.
> *Step 5A:* Using the information from Step 4A, rewrite the orders. Do not use abbreviations. (For our purposes, 5 mℓ = 1 teaspoonful.)

The following example shows a prescription ordered by the physician along with the available form of the prescribed item. Use method A to determine the appropriate volume per dose.

Example 1:   *Ordered:* Tylenol® 60 mg q. 4–6 hr p.r.n.
On hand: Tylenol® elixir
   120 mg per 5 mℓ
   4 oz

Solution:
Step 1A: Write down the strength desired by the physician.

   60 mg

Step 2A: The concentration of the on-hand drug per volume is 120 mg per 5 mℓ.

Step 3A: $\dfrac{5 \text{ m}\ell}{120 \text{ mg}} = 1$ and $\dfrac{1 \text{ tsp}}{5 \text{ m}\ell} = 1$

Step 4A: Cancel units and multiply the given strength times the expression for 1.

$$\overset{1}{\cancel{60 \text{ mg}}} \times \frac{\cancel{5 \text{ m}\ell}}{\underset{2}{\cancel{120 \text{ mg}}}} \times \frac{1 \text{ tsp}}{\cancel{5 \text{ m}\ell}} = \frac{1}{2} \text{ tsp}$$

Step 5A: Give ½ teaspoonful Tylenol® 120 mg/5 mℓ elixir every four to six hours as needed.

> **Rule: Solving Oral-Solution Dosage Problems**
> **Method B: Ratio and Proportion**
> *Step 1B:* Write the conversion as a ratio in fractional form, using x to represent the number of units you are converting to.
> *Step 2B:* If the drug is to be reconstituted, follow the manufacturer's printed instructions. The final concentration (after reconstitution) is printed on label. Use this information in your calculation.
> *Step 3B:* Write down the amount of the on-hand item first in the units you are starting with and then in the units you want to convert to. Use the fractional form and combine fact(s) when necessary.
> *Step 4B:* Write the ratios from Steps 1 and 2 as a proportion and cross multiply. Be sure that the units on each side of the equation correspond to one another.
> *Step 5B:* To solve for the unknown quantity, divide the numbers on each side of the equals sign by the number on the same side as the x.
> *Step 6B:* Using the information from Step 5B, rewrite the orders. Do not use abbreviations. (For our purposes, 5 mℓ = 1 teaspoonful.)

Now use method B to determine the appropriate volume per dose in the same example.

Ordered: Tylenol® 60 mg q. 4–6 hr p.r.n.
On hand: Tylenol® elixir
120 mg per 5 mℓ
4 oz

Solution:

Step 1B: Write the conversion as a ratio in fractional form, using x to represent the unknown quantity.

$$\frac{60}{x}$$

Step 2B: The concentration of the drug per volume is 120 mg per 5 mℓ.

Step 3B: $\frac{120}{5}$

Step 4B: Write the two ratios as a proportion and cross multiply.

$$\frac{60}{x} = \frac{120}{5}$$
$$x \cdot 120 = 60 \cdot 5$$
$$x \cdot 120 = 300$$

Step 5B: To solve for the unknown, divide the numbers on each side of the equals sign by the number on the same side as the x.

$$\frac{x \cdot \cancel{120}^{1}}{\cancel{120}_{1}} = \frac{\cancel{300}^{5}}{\cancel{120}_{2}}$$

$$x = 2.5 \text{ mℓ (or } \frac{1}{2} \text{ teaspoonful)}$$

Step 6B: Give ½ teaspoonful Tylenol® 120 mg/5 mℓ elixir every four to six hours as needed.

Section 9.2 Oral Solutions

The following example again shows a prescription ordered by the physician along with the available form of the prescribed item. Use method A to determine the appropriate volume per dose.

Example 2: *Ordered:* Chloral hydrate 500 mg h.s.
On hand: Chloral hydrate syrup
250 mg per 5 mℓ
1 pt

Solution:

Step 1A: Write down the dosage strength ordered by the physician.

500 mg

Step 2A: The concentration of the on-hand drug per volume is 250 mg per 5 mℓ.

Step 3A: $\dfrac{5 \text{ mℓ}}{250 \text{ mg}} = 1$

Step 4A: Cancel units and multiply the ordered strength times the expression for 1.

$$\overset{2}{\cancel{500}} \text{ mg} \times \dfrac{5 \text{ mℓ}}{\underset{1}{\cancel{250}} \text{ mg}} = 10 \text{ mℓ (or 2 teaspoonsful)}$$

Step 5A: Take 2 teaspoonsful chloral hydrate syrup 250 mg/5 mℓ at bedtime.

Section 9.2 Readiness Review

Each of the following examples shows a prescription ordered by the physician along with the available form of the prescribed item. Determine the appropriate volume per dose.

1. *Ordered:* V Cillin K® 125 mg q. 6 h.
   On hand: V Cillin K® powder for oral solution
   250 mg per 5 mℓ
   100 mℓ
   Reconstitute as directed by manufacturer

2. *Ordered:* Aquasol A® 15,000 I.U.[4] q.d.
   On hand: Aquasol A® drops
   5000 I.U. per 0.1 mℓ
   30 mℓ with dropper

3. *Ordered:* Klorvess® 30 mEq dil. in juice q.d.
   On hand: Klorvess® 10% liquid
   20 mEq per 15 mℓ
   1 pt

4. *Ordered:* Lanoxin® 150 mcg q.d.
   On hand: Lanoxin® pediatric elixir
   0.05 mg per 1 mℓ
   2 oz with dropper

---

[4]I.U. stands for International Units.

228   Chapter 9 Tablets, Capsules, and Oral Solutions

Section readiness review answers (not given in order):

1. Give ½ teaspoonful V Cillin K® 250 mg/5 mℓ every six hours.
3. Take 1½ tablespoonsful Klorvess® 10% liquid diluted in juice every day.
4. Give 3 mℓ Lanoxin® pediatric elixir 0.05 mg per 1 mℓ every day.
2. Take 0.3 mℓ Aquasol A® drops 5000 I.U. per 0.1 mℓ every day.

Section 9.2 Review Problems

Each of the following problems shows a prescription ordered by the physician along with the available form of the prescribed item. Determine the appropriate volume per dose (answers to the odd-numbered problems are at the back of the book).

1. *Ordered:* Lasix® 40 mg q.d.
   On hand: Lasix® oral solution
   10 mg per 1 mℓ
   120 mℓ

2. *Ordered:* Mellaril® 60 mg t.i.d.
   On hand: Mellaril® concentrate
   30 mg per 1 mℓ
   4 oz

3. *Ordered:* Haldol 0.5 mg t.i.d.
   On hand: Haldol concentrate
   2 mg per 1 mℓ
   15 mℓ

4. *Ordered:* Choledyl® 200 mg q.i.d.
   On hand: Choledyl® elixir
   100 mg per 5 mℓ
   1 pt

5. *Ordered:* Benadryl® 25 mg h.s. p.r.n. sleep
   On hand: Benadryl® elixir
   12.5 mg per 5 mℓ
   1 pt

6. *Ordered:* Periactin® 6 mg t.i.d. p.r.n. itching
   On hand: Periactin® syrup
   2 mg per 5 mℓ
   473 mℓ

7. *Ordered:* Phenergan® 25 mg stat. for nausea
   On hand: Phenergan® syrup
   6.25 mg per 5 mℓ
   1 pt

8. *Ordered:* Tegopen® 250 mg q. 6 h.
   On hand: Tegopen® powder for oral solution
   125 mg per 5 mℓ
   200 mℓ
   Reconstitute as directed by manufacturer

9. *Ordered:* Ampicillin 500 mg q. 6 h.
   On hand: Ampicillin powder for oral suspension
   125 mg per 5 mℓ
   200 mℓ
   Reconstitute as directed by manufacturer

10. *Ordered:* Amoxil® 500 mg q. 6 h.
    On hand: Amoxil® for oral suspension
    250 mg per 5 mℓ
    150 mℓ
    Reconstitute as directed by manufacturer

11. *Ordered:* Thorazine® 60 mg t.i.d. and p.r.n.
    On hand: Thorazine® concentrate
    30 mg per 1 mℓ
    4 oz

12. *Ordered:* Compazine® 2.5 mg t.i.d. p.r.n. nausea
    On hand: Compazine® syrup
    5 mg per 5 mℓ
    4 oz

13. *Ordered:* Dilantin® 100 mg t.i.d.
    On hand: Dilantin®-125 pediatric elixir
    125 mg per 5 mℓ
    8 oz

14. *Ordered:* Potassium chloride 10 mEq dil. in f℥ *vi* water or juice q.d.
    On hand: Potassium chloride 10% liquid
    20 mEq per 15 mℓ
    480 mℓ

Chapter 9 Readiness Review

Each of the following examples shows a prescription ordered by the physician along with the available form of the prescribed item. Determine the appropriate number of tablets, capsules, or volume per dose.

1. *Ordered:* Mycostatin® p.o. 1,000,000 units t.i.d.
   On hand: Mycostatin® 500,000-U tabs

2. *Ordered:* E-Mycin® p.o. 500 mg q. 6 h. until gone
   On hand: E-Mycin® 250-mg tabs

3. *Ordered:* Ampicillin p.o. 500 mg q. 6 h.
   On hand: Ampicillin 250-mg capsules

4. *Ordered:* Geocillin® 764 mg p.o. q.i.d.
   On hand: Geocillin® 382-mg capsules

5. *Ordered:* Phenobarbital 30 mg b.i.d.
   On hand: Phenobarbital elixir
   20 mg per 5 mℓ
   1 pt

6. *Ordered:* Mysoline® 125 mg b.i.d.
   On hand: Mysoline® suspension
   250 mg per 5 mg
   8 oz

7. *Ordered:* Lomotil® 5 mg q.i.d.
   On hand: Lomotil® liquid
   2.5 mg/5 mℓ
   2 oz

8. *Ordered:* SK-Penicillin G® 400,000 U a.c. and h.s.
   On hand: SK-Penicillin G® powder for solution
   200,000 U per 5 mℓ
   200 mℓ
   Reconstitute as directed by manufacturer

Chapter readiness review answers (not given in order):

8. Take 2 teaspoonsful SK-Penicillin G® 200,000 U per 5 mℓ before meals and at bedtime.
7. Take 2 teaspoonsful Lomotil® 2.5 mg per 5 mℓ four times daily.
6. Give ½ teaspoonful Mysoline® 250 mg per 5 mℓ suspension two times daily.
5. Give 1½ teaspoonsful phenobarbitol 20 mg per 5 mℓ elixir two times daily.
4. 2 capsules. Give 2 Geocillin® 382-mg capsules by mouth four times daily.
3. 2 capsules. Give 2 Ampicillin® 250-mg capsules every six hours.
2. 2 tablets. Give 2 E-Mycin® 250-mg tablets by mouth every six hours until gone.
1. 2 tablets. Give 2 Mycostatin® 500,000-U tablets by mouth three times daily.

Chapter 9 Summary

Define each item in your own words, then compare your definitions with the text.

Key Words

oral dosages   (p. 219)
tablets   (p. 219)
capsules   (p. 219)
diluent   (p. 219)

scored   (p. 219)
oral solution   (p. 224)
reconstitution   (p. 224)

Chapter 9 Review Problems

Each of the following problems shows a prescription ordered by the physician along with the available form of the prescribed item. Determine the appropriate number of tablets, capsules, or volume per dose (answers to all the problems are at the back of the book).

1. *Ordered:* Aldomet® 500 mg p.o. t.i.d. for BP
   On hand: Aldomet® 125-mg tablets

2. *Ordered:* Reserpine 0.125 mg p.o. q.d.
   On hand: Reserpine 0.25-mg scored tablets

3. *Ordered:* Apresoline® 50 mg p.o. q.i.d.
   On hand: Apresoline® 10-mg scored tablets

4. *Ordered:* Sudafed® 30 mg p.o. q.i.d. for inner-ear pressure
   On hand: Sudafed® 60-mg scored tablets

5. *Ordered:* Dilantin® 300 mg p.o. q.d.
   On hand: Dilantin® 100-mg capsules

6. *Ordered:* Inderal® 10 mg p.o. q.i.d.
   On hand: Inderal® 20-mg scored tablets

7. *Ordered:* Quinidine sulfate 200 mg q.i.d.
   On hand: Quinidine sulfate 100-mg tablets

8. *Ordered:* Sudafed® 60 mg p.o. t.i.d. for congestion
   On hand: Sudafed® 30-mg tablets

9. *Ordered:* Penicillin G 250 mg q. 6 h.
   On hand: Penicillin G 125-mg tablets

10. *Ordered:* V Cillin K® 800,000 units q.i.d., a.c. and h.s.
    On hand: V Cillin K® 200,000-unit tablets

11. *Ordered:* Coumadin® 5 mg p.o. q.d.
    On hand: Coumadin® 2.5-mg tablets

12. *Ordered:* Lanoxin® 0.25 mg p.o. q.d.
    On hand: Lanoxin® 0.125-mg tablets

13. *Ordered:* Keflex® 1 g p.o. q.i.d.
    On hand: Keflex® 500-mg capsules

14. *Ordered:* Inderal® 20 mg p.o. b.i.d.
    On hand: Inderal® 10-mg scored tablets

15. *Ordered:* Depakene® 125 mg b.i.d.
    On hand: Depakene® syrup
    250 mg per 5 mℓ
    1 pt

16. *Ordered:* Symmetrel® 100 mg b.i.d.
    On hand: Symmetrel® syrup
    50 mg per 5 mℓ
    1 pt

17. *Ordered:* Mylicon® 80 mg p.c. and h.s.
    On hand: Mylicon drops
    40 mg per 0.6 mℓ
    30 mℓ

18. *Ordered:* Velosef® 250 mg q.i.d.
    On hand: Velosef® oral suspension
    125 mg per 5 mℓ
    100 mℓ
    Reconstitute as directed by manufacturer

19. *Ordered:* Ilosone® 250 mg q.i.d.
    On hand: Ilosone® liquid 125
    125 mg per 5 mℓ
    1 pt

20. *Ordered:* NegGram® 1 g q.i.d.
    On hand: NegGram® suspension
    250 mg per 5 mℓ
    1 pt

21. *Ordered:* Sudafed® 60 mg q. 4 h. p.r.n.
    On hand: Sudafed® liquid
    30 mg per 5 mℓ
    4 oz

22. *Ordered:* Elixophyllin® 200 mg q. 6 h. $\bar{c}$ food
    On hand: Elixophyllin® elixir
    80 mg per 15 mℓ
    1 pt

23. *Ordered:* Kaon-CL® 50 mEq dil. in fl℥ x aq. q.d.
    On hand: Kaon-CL® 20% liquid
    40 mEq per 15 mℓ
    1 pt

24. *Ordered:* Feosol® 330 mg t.i.d. p.c.
    On hand: Feosol® elixir
    220 mg per 5 mℓ
    12 oz

25. *Ordered:* Decadron® 1.5 mg b.i.d. p.c.
    On hand: Decadron® elixir
    0.5 mg per 5 mℓ
    237 mℓ

Four
Parenteral Dosages

10
Injectable Drugs in Solution

OBJECTIVES After studying this chapter, you should be able to:

1. Solve problems that involve injectable drugs already in solution.
2. Solve problems that involve hypodermic tablets to be dissolved for injection.
3. Reconstitute dry drugs correctly and determine the proper volume of solution to be given to the patient.

Section 10.1 Injectable Drug Calculations — Overview

Drugs that are given by the *parenteral* route are injected under or through layers of the skin and thus are also called *injections*. The skin acts as a protective barrier, however, and so a needle breaking the skin can do more harm than good by allowing harmful microorganisms to penetrate the barrier. It is for this reason that nurses and other professionals must observe good aseptic technique when they give injections.

> **Definition:** *Parenteral* refers to the administration of drugs via injection.

> **Definition:** *Aseptic* means free of microorganisms—for example, bacteria. *Aseptic techniques*, therefore, prevent contamination of equipment and people by microorganisms.

The most common types of injections are: (1) subcutaneous—the drug is injected into tissue just under the skin; (2) intramuscular—the drug is injected into a muscle; and (3) intravenous—the drug is injected into a vein (see Figure 10-1).

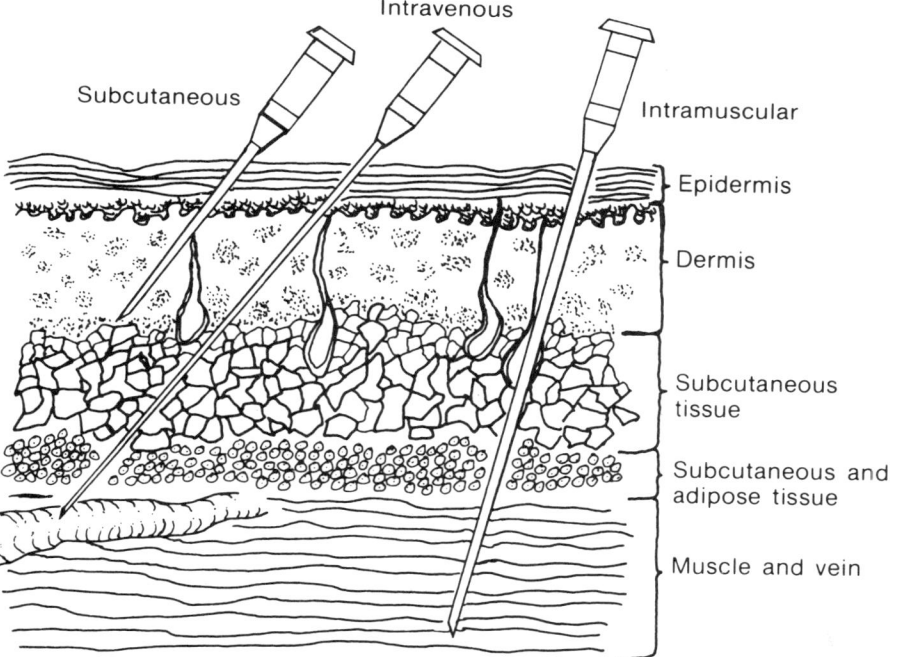

Figure 10-1. Routes of parenteral administration

You may recall from Chapter 9 that oral solutions are available either in liquid form or in the form of a powder to be put into solution. The same is true of injectable drugs. However, because injectable drugs are delivered to the body via a needle, they have to be sterile.

Chapter 10 Injectable Drugs in Solution

Injectable drugs are packaged in vials, ampules, or as prefilled syringes. Table 10-1 shows the basic characteristics of each.

Table 10-1. Packaging of injectable drugs

Vial	Ampule	Prefilled Syringe
multiple dose or single dose	single dose only	single dose only
made of glass	made of glass	made of glass or plastic
rubber stopper	glass neck designed to snap open	ready to use

Figure 10-2 shows several ampules, single-dose and multiple-dose vials, and prefilled syringes.

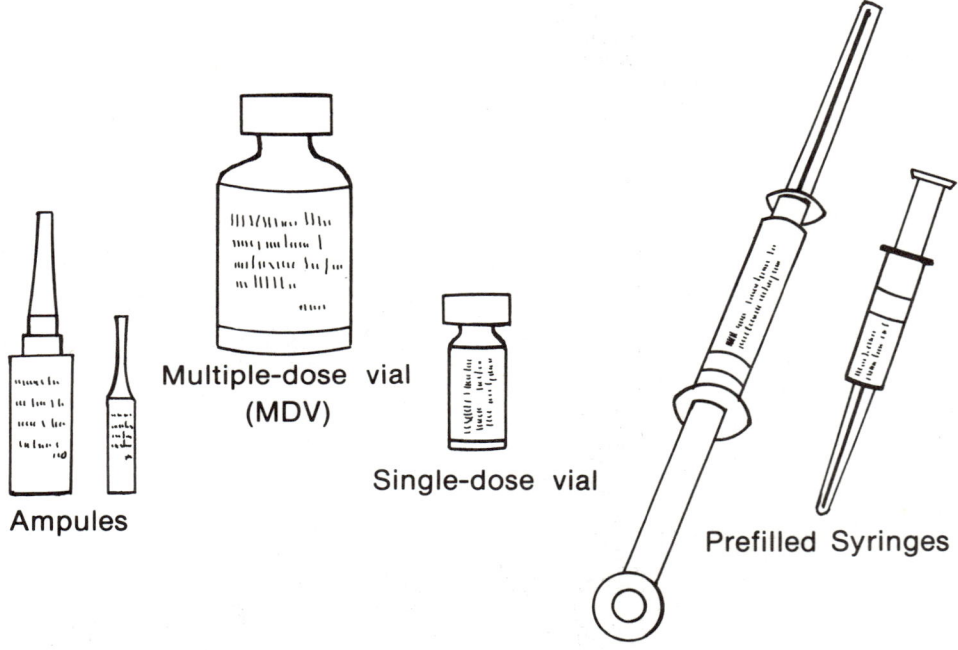

Figure 10-2. Parenteral containers

A general technique for withdrawing injectable medication from either an ampule or a vial is shown in Figures 10-3 and 10-4. (Although the primary objective of this text is to teach nursing mathematics, this technique is important enough to review here.) Because a prefilled syringe is nothing more than a syringe with medication already inside, you are probably already familiar with it.

Section 10.1 Injectable Drug Calculations — Overview 237

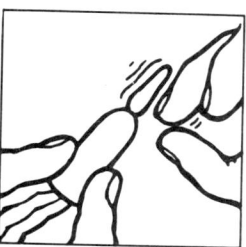
(1) Tap (flicking motion) top of ampule to displace fluid from neck.

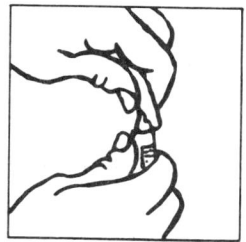
(2) Wipe neck of ampule with alcohol swab. Using the swab as protection, snap off (away from you) neck of ampule.

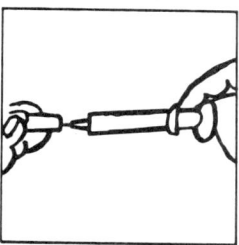
(3) Tilt ampule and insert needle into solution. Withdraw solution.

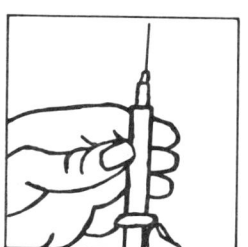
(4) Hold syringe upright and remove air bubbles.

Figure 10-3. Withdrawing medication from an ampule

(1) If vial has plastic protective cap, flick it off.

(2) Remove metal cap.

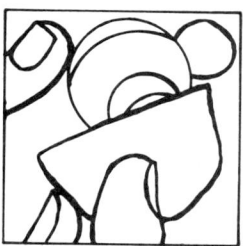
(3) Wipe off exposed rubber stopper with alcohol swab.

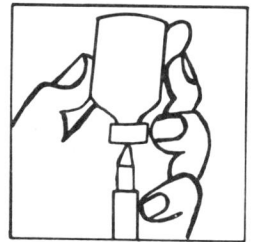
(4) Invert the vial. Inject amount of air equal to the volume desired into the vial. Withdraw the amount of solution desired.

Figure 10-4. Withdrawing medication from a vial

Chapter 10 Injectable Drugs in Solution

The first section of this chapter emphasizes calculations involving injectable drugs that do *not* require reconstitution. The problems in this section deal with I.M., SQ, and, occasionally, I.V. routes of administration. (The balance of the I.V. calculations are in Chapter 11.)

Most injectable drugs (whether they are to be reconstituted or not) will be labeled according to the amount of drug per a certain volume of solvent. Be sure to read these labels carefully when you are solving parenteral dosage problems. A typical label will show the following information:

> Name of medication (trade name, generic name, or both)
> Strength per volume
> Container type and size
> Route of administration
> Directions for reconstitution (where applicable)

For example,

> Imferon® injection
> 50 mg/1 mℓ
> 10-mℓ vial
> I.M.

> **Rule: Solving Problems Involving Injectable Drugs That Do Not Require Reconstitution**
> **Method A: Dimensional Analysis**
> *Step 1A:* Read the label of the on-hand drug carefully, and identify the concentration of drug per volume and the route of administration.
> *Step 2A:* Write down the dosage strength ordered by the physician.
> *Step 3A:* Write the concentration of the on-hand drug as a fraction. Be sure, when using dimensional analysis, that the desired unit will remain and the given unit will be eliminated.
> *Step 4A:* Cancel units and multiply the ordered strength by the fraction from Step 3A.
> *Step 5A:* Using the information from Step 4A, rewrite the orders. Do not use abbreviations.
> *Step 6A:* Shade in the correct dosage on the syringe when required.

Example 1: Use method A to determine the correct amount of injectable medication that satisfies the following order using the drug available.

Ordered: Dramamine® 35 mg I.M. p.r.n.
On hand: Dramamine® injection
50 mg/mℓ
1-mℓ ampule
I.M.

Shade in the correct dosage on the syringe.

Section 10.1 Injectable Drug Calculations — Overview 239

Solution:

Step 1A: The concentration of the on-hand drug per volume is 50 mg/mℓ, and it is to be given by I.M. injection.

Step 2A: The dosage strength ordered is 35 mg.

Step 3A: Write the concentration of the on-hand drug as a fraction.

$$\frac{1 \text{ m}\ell}{50 \text{ mg}}$$

Step 4A: Now cancel units and multiply.

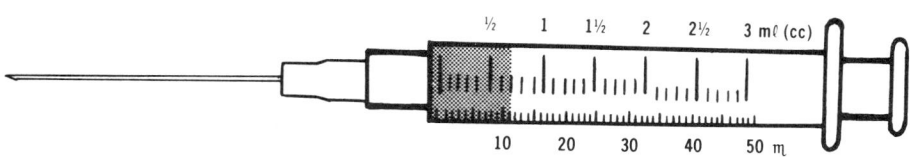

Step 5A: Give 0.7 mℓ Dramamine® 50 mg/mℓ intramuscularly as needed.

Step 6A: Shade in the correct dosage on the syringe.

> **Rule: Solving Problems Involving Injectable Drugs That Do Not Require Reconstitution**
> **Method B: Ratio and Proportion**
>
> *Step 1B:* Read the label of the on-hand drug carefully, and identify the concentration of drug per volume and the route of administration.
>
> *Step 2B:* Write the strength ordered as a ratio in fractional form, using x to represent the unknown volume of drug product.
>
> *Step 3B:* Write the concentration of the on-hand drug as a ratio in fractional form.
>
> *Step 4B:* Write the ratios from Steps 2B and 3B as a proportion and cross multiply. Be sure that the units on each side of the equation correspond to one another.
>
> *Step 5B:* To solve for the unknown quantity, divide the numbers on each side of the equals sign by the number on the same side as the x.
>
> *Step 6B:* Using the information from Step 5B, rewrite the orders. Do not use abbreviations.
>
> *Step 7B:* Shade in the correct dosage on the syringe when required.

Now use method B in the same example to determine the correct amount of injectable medication that satisfies the order using the drug available.

Ordered: Dramamine® 35 mg I.M. p.r.n.

On hand: Dramamine® injection
50 mg/mℓ
1-mℓ ampule
I.M.

240   Chapter 10 Injectable Drugs in Solution

Solution:

Step 1B: The concentration of on-hand drug per volume is 50 mg/mℓ, and it is to be given by I.M. injection.

Step 2B: $\dfrac{35}{x}$

Step 3B: On the basis of the information on the on-hand drug label, we write

$$\dfrac{50}{1}$$

Step 4B: Write the ratios as a proportion and cross multiply.

$$\dfrac{35}{x} = \dfrac{50}{1}$$
$$x \cdot 50 = 35 \cdot 1$$
$$x \cdot 50 = 35$$

Step 5B: Now solve for the unknown quantity.

$$\dfrac{x \cdot \overset{1}{\cancel{50}}}{\underset{1}{\cancel{50}}} = \dfrac{\overset{7}{\cancel{35}}}{\underset{10}{\cancel{50}}}$$
$$x = 0.7 \text{ mℓ}$$

Step 6B: Give 0.7 mℓ Dramamine® 50 mg/mℓ intramuscularly as needed.
Step 7B: The syringe diagram is the same as Example 1—Step 6A.

Example 2: Use method A to determine the correct amount of injectable medication that satisfies the following order using the drug available.

Ordered: Urecholine® 2.5 mg SQ t.i.d.
On hand: Urecholine® injection
5 mg/mℓ
1-mℓ vial
SQ

Shade in the correct dosage on the syringe.

How many milliliters of air must be injected into the vial to withdraw the correct amount?

Solution:

Step 1A: The concentration of the on-hand drug per volume is 5 mg/mℓ, and it is to be given SQ.
Step 2A: The dosage strength ordered is 2.5 mg.
Step 3A: The concentration of on-hand drug is written as

$$\dfrac{1 \text{ mℓ}}{5 \text{ mg}}$$

Step 4A: $\overset{1}{\cancel{2.5 \text{ mg}}} \times \dfrac{1 \text{ mℓ}}{\underset{2}{\cancel{5 \text{ mg}}}} = 0.5 \text{ mℓ}$

Section 10.1 Injectable Drug Calculations — Overview   241

Step 5A: Give 0.5 mℓ Urecholine® 5 mg/mℓ subcutaneously three times daily.
Step 6A: Shade in the correct dosage on the syringe.

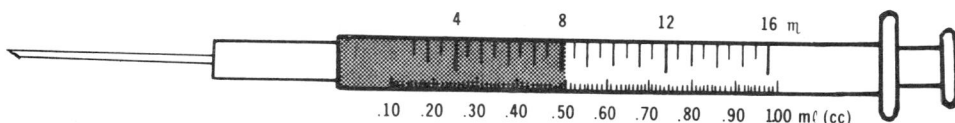

Figure 10-4 showed that the amount of air to be injected into the vial is equal to the volume of drug to be injected. In this case, you want 0.5 mℓ of drug; therefore, you must inject 0.5 mℓ of air into the vial before withdrawing the drug.

Section 10.1 Readiness Review

Determine the correct amount of injectable medication that satisfies each of the following orders using the drug available.

1. *Ordered:* Valium® 4 mg I.M. q. 3–4 h. p.r.n. muscle spasm
   On hand: Valium® injection
   5 mg/mℓ
   2-mℓ ampule
   I.M. or I.V.

   Shade in the correct dosage on the syringe.

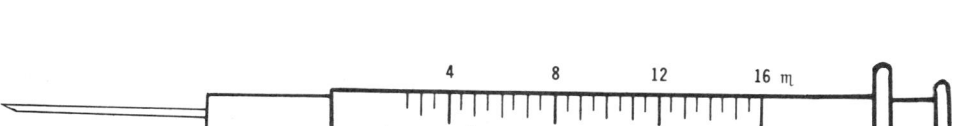

2. *Ordered:* Robinul® 0.1 mg I.V. q. 4 h.
   On hand: Robinul® injection
   0.2 mg/mℓ
   1-mℓ vial
   I.M. or I.V.

   Shade in the correct dosage on the syringe.

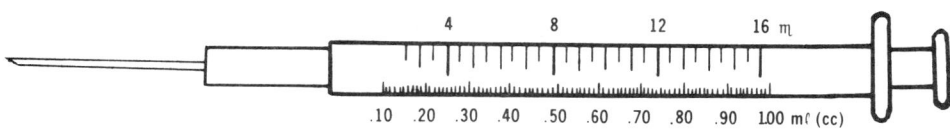

How many milliliters of air must be injected into the vial to withdraw the correct amount of drug?

3. *Ordered:* Marezine® 40 mg I.M. q. 4–6 h. p.r.n. for N & V
   On hand: Marezine® (as lactate) injection
              50 mg/ml
              1-ml ampules
              I.M.

Shade in the correct dosage on the syringe.

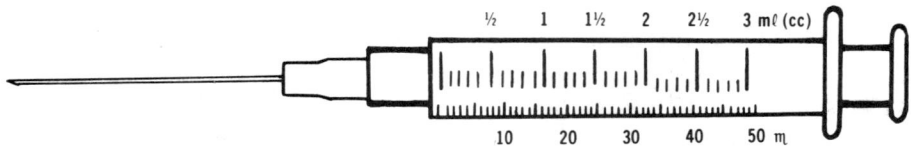

4. *Ordered:* Nebcin® 50 mg I.M. q. 8 h.
   On hand: Nebcin® injection
              40 mg/ml
              2-ml vials
              I.M. or I.V.

Section readiness review answers (not given in order):

1. Give 0.8 ml Valium® 5 mg/ml intramuscularly, every three to four hours as needed for muscle spasm.

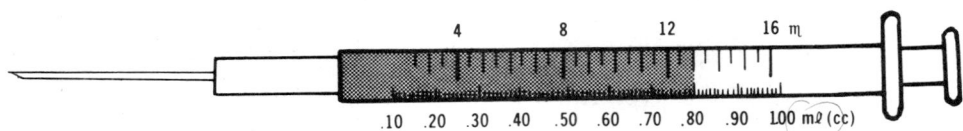

4. Give 1.25 ml Nebcin® 40 mg/ml intramuscularly, every eight hours.
3. Give 0.8 ml Marezine® 50 mg/ml intramuscularly every four to six hours as needed for nausea and vomiting.

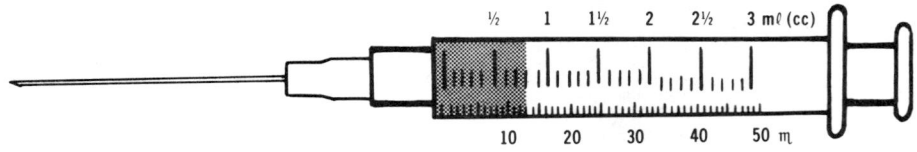

2. Give 0.5 ml Robinul® intravenously every four hours.

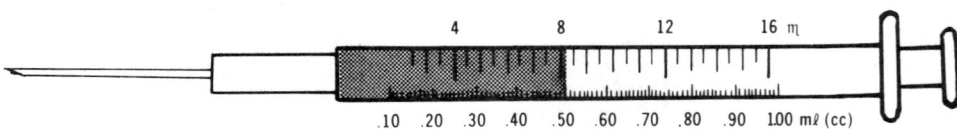

0.5 ml of air.

Section 10.1 Review Problems

Determine the correct amount of injectable medication that satisfies each of the following orders using the drug available (answers to the odd-numbered problems are at the back of the book).

1. *Ordered:* Lanoxin® 0.5 mg I.V. q. 6 h. for two doses
   On hand: Lanoxin® injection
   0.25 mg/mℓ
   2-mℓ ampule
   I.V.

   Shade in the correct dosage on the syringe.

   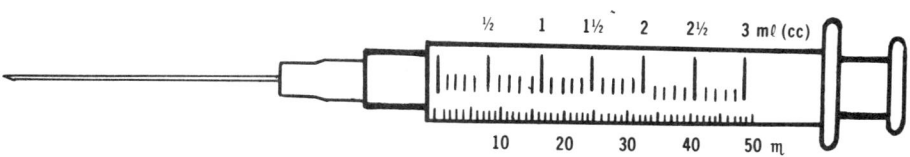

2. *Ordered:* Quinidine 400 mg I.M. stat.
   On hand: Quinidine Gluconate® injection
   80 mg/mℓ
   10-mℓ vial
   I.M. or I.V.

3. *Ordered:* Gynergen® 0.75 mg SQ at onset of attack
   On hand: Gynergen® injection
   0.5 mg/mℓ
   1-mℓ ampule
   I.M. or SQ

4. *Ordered:* Myochrysine® 10 mg I.M. intragluteally (first week). Give c̄ patient lying down.
   On hand: Myochrysine® injection
   25 mg/mℓ
   1-mℓ ampule
   I.M.

   Shade in the correct dosage on the syringe.

   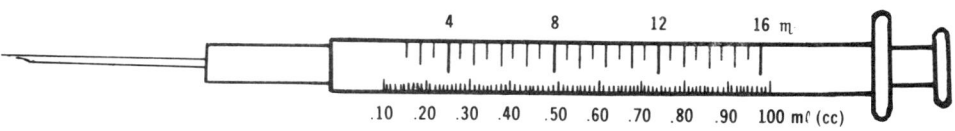

5. *Ordered:* Kantrex® 400 mg I.M. q. 8 h.
   On hand: Kantrex® injection
   1 g/3 mℓ
   3-mℓ vial
   I.M. or I.V.

Shade in the correct dosage on the syringe.

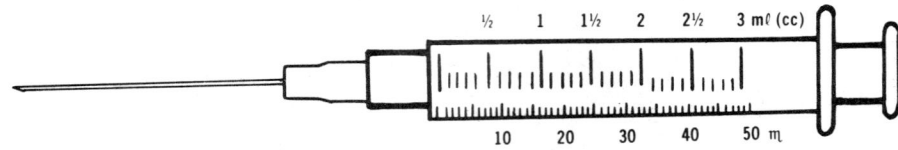

How many milliliters of air must be injected into the vial to withdraw the correct amount?

6. *Ordered:* Garamycin® 60 mg I.M. q. 8 h.
   On hand: Garamycin® injection
   40 mg/mℓ
   2-mℓ vial
   I.M. or I.V.

7. *Ordered:* ACTH 20 units SQ q.i.d.
   On hand: ACTH injection (corticotropin)
   80 units/mℓ
   5-mℓ vial
   I.M., SQ, or I.V. (as diagnostic)

Shade in the correct dosage on the syringe.

How many milliliters of air must be injected into the vial to withdraw the correct amount?

8. *Ordered:* M.S. 5 mg SQ q. 4 h. p.r.n. severe pain
   On hand: Morphine sulfate injection
   15 mg/mℓ
   20-mℓ vial
   SQ, I.V. (used rarely)

9. *Ordered:* Demerol® 75 mg I.M. q. 3–4 h. p.r.n. pain
   On hand: Demerol® injection
   50 mg/mℓ
   2-mℓ ampule
   I.M., SQ, I.V.

Shade in the correct dosage on the syringe.

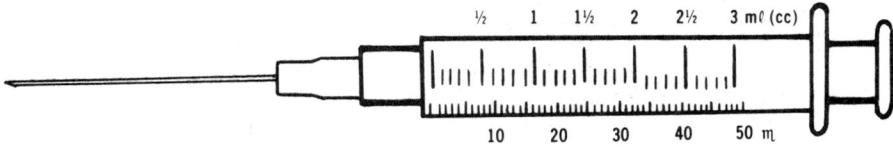

Section 10.1 Injectable Drug Calculations — Overview

10. *Ordered:* Narcan® 0.2 mg I.M. stat.
    On hand: Narcan® injection
    0.4 mg/mℓ
    1-mℓ ampule
    I.V., I.M., or SQ

11. *Ordered:* Robaxin® 300 mg I.M. q. 8 h. p.r.n. (inject in gluteal region)
    On hand: Robaxin® injection
    100 mg/mℓ
    10-mℓ vial
    I.M. or I.V.

    Shade in the correct dosage on the syringe.

    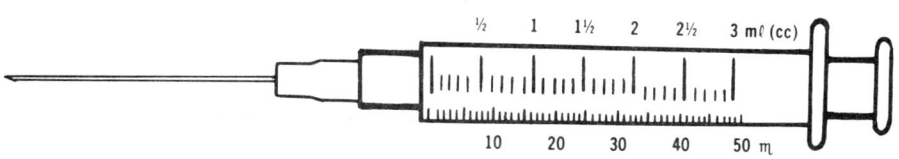

    How many milliliters of air must be injected into the vial to withdraw the correct amount?

12. *Ordered:* Norflex® 60 mg I.M. q. 12 h. p.r.n.
    On hand: Norflex® injection
    30 mg/mℓ
    2-mℓ vial
    I.V. or I.M.

13. *Ordered:* Methergine® 0.2 mg I.M. after delivery of placenta
    On hand: Methergine® injection
    0.2 mg/mℓ
    1-mℓ ampule
    I.M. (preferred) or I.V.

    Shade in the correct dosage on the syringe.

    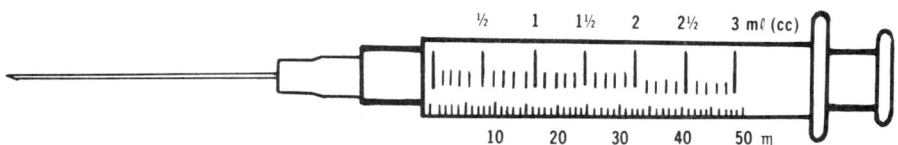

14. *Ordered:* Lasix® 40 mg I.V. stat. (given slowly over one to two minutes)
    On hand: Lasix® injection
    10 mg/mℓ
    10-mℓ ampule
    I.M. or I.V.

15. *Ordered:* Terramycin I.M.® 150 mg q. 24 h.
    On hand: Terramycin I.M.® injection
    50 mg/mℓ
    10-mℓ vial
    I.M.

Shade in the correct dosage on the syringe.

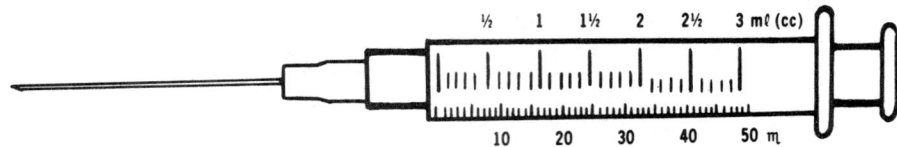

How many milliliters of air must be injected into the vial to withdraw the correct amount?

16. *Ordered:* Erythromycin 100 mg I.M. q. 8 h.
    On hand: Erythrocin Ethyl Succinate I.M.®
    50 mg/mℓ
    10-mℓ vial
    I.M.

17. *Ordered:* Tigan® 200 mg I.M. t.i.d. (deep injection into upper outer quadrant of the gluteal region)
    On hand: Tigan® injection
    100 mg/mℓ
    20-mℓ vial
    I.M. only

18. *Ordered:* Emete-Con® 40 mg I.M. q. 3–4 h. p.r.n.
    On hand: Emete-Con® injection
    50 mg/2 mℓ
    I.M. or I.V.

Shade in the correct dosage on the syringe.

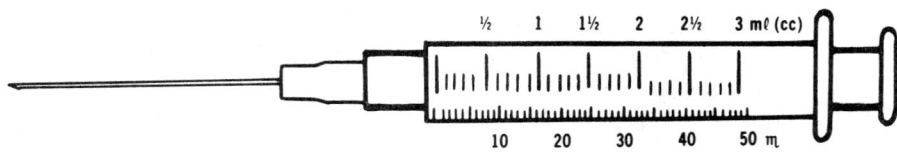

How many milliliters of air must be injected into the vial to withdraw the correct amount?

19. *Ordered:* Ephedrine 35 mg SQ stat.
    On hand: Ephedrine sulfate injection
    50 mg/mℓ
    1-mℓ ampule
    SQ, I.M., or I.V.

20. *Ordered:* Serpasil® 0.5 mg I.M. initially. Monitor BP
    On hand: Serpasil® injection
    2.5 mg/mℓ
    2-mℓ ampule
    I.M.

Shade in the correct dosage on the syringe.

Section 10.2 Hypodermic Tablets

> A *hypodermic tablet* is an extremely soft tablet that dissolves rapidly and completely when placed in solution.

The hypodermic tablet has traditionally served as a source of injectable medication. Today, however, hypodermic tablets are rarely, if ever, used for injection. But it is conceivable that, in the case of a shortage of certain injectable drugs or an extreme emergency, nurses might be asked to inject solutions made from hypodermic tablets. It is for this reason that Section 10.2 is included.

Because hypodermic tablets are rather soft, they can't be broken. Therefore, when a dosage of injectable drug is prescribed that requires only a fraction of a tablet, a solution is made using the whole tablet, but only a portion of the solution is administered. Problems that involve hypodermic tablets consist of two parts: (1) determination of the quantity of tablets needed to make a solution, and (2) determination of the amount of solution to be injected.

Rule: Solving Hypodermic-Tablet Problems
Method A: Dimensional Analysis

Step 1A: Write down the dosage strength ordered by the physician.

Step 2A: Write the concentration of the on-hand drug as a fraction. Be sure, when using dimensional analysis, that the desired unit will remain and the given unit will be eliminated.

Step 3A: Cancel units and multiply the ordered strength by the fraction from Step 2A. This will give you the *actual* number of hypodermic tablets required. If this number is a fraction or a mixed number, round off to the next whole number to get the number of whole tablets to be dissolved.

Step 4A: Express the number of whole tablets to be dissolved and the amount of diluent as a fraction. Multiply this fraction by the *actual* number of tablets required. This will give you the volume of solution to be administered to the patient.

Step 5A: Using the information from Step 4A, rewrite the orders. Do not use abbreviations.

The following example shows a prescription ordered by the physician along with the available form of the prescribed item. Use method A to determine the number of hypodermic tablets needed and the volume of solution to be administered.

248 Chapter 10 Injectable Drugs in Solution

Example 1: *Ordered:* Codeine sulfate 15 mg SQ q.i.d. p.r.n. pain
On hand: Codeine sulfate 30-mg hypodermic tablets (diluent: sterile water for injection: 0.5 mℓ)

Number of tablets used to make solution? _____
How many milliliters of solution are injected? _____

Solution:
Step 1A: 15 mg

Step 2A: We have 30-mg-strength tablets on hand; therefore, we write

$$\frac{1 \text{ tablet}}{30 \text{ mg}}$$

Step 3A: $15 \text{ mg} \times \dfrac{1 \text{ tablet}}{30 \text{ mg}} = \dfrac{1}{2} \text{ tablet}$

Because we can't break hypodermic tablets, we'll have to dissolve 1 tablet in 0.5 mℓ of diluent.

Step 4A: $\dfrac{1}{2} \text{ tablet} \times \dfrac{0.5 \text{ m}\ell}{1 \text{ tablet}} = 0.25 \text{ m}\ell$

Step 5A: Dissolve one 30-mg codeine sulfate tablet in 0.5 mℓ of sterile water for injection. Give 0.25 mℓ of this solution, which contains 15 mg of the drug, subcutaneously four times daily as needed for pain.

Number of tablets used to make solution? __1__
How many milliliters of solution are given? __0.25__

Rule: Solving Hypodermic-Tablet Problems
   Method B: Ratio and Proportion

Step 1B: Write the strength ordered as a ratio in fractional form, using x to represent the unknown volume of drug product.

Step 2B: Write the concentration of the on-hand drug as a ratio in fractional form.

Step 3B: Write the ratios from Steps 1B and 2B as a proportion and cross multiply. Be sure that the units on each side of the equation correspond to one another.

Step 4B: To solve for the unknown quantity, divide the numbers on each side of the equals sign by the number on the same side as the x. This will give you the *actual* number of hypodermic tablets required. If the number is a fraction or a mixed number, round off to the next whole number to get the number of whole tablets to be dissolved.

Step 5B: Now write the *actual* number of tablets from Step 4B as a ratio in fractional form, using y to represent the volume of solution to be administered.

Step 6B: Express the number of whole tablets to be dissolved and the given volume of diluent as a ratio in fractional form.

Step 7B: Write the two ratios as a proportion and solve for the unknown quantity, y. (y is equal to the volume of solution to be administered to the patient.)

$$\frac{\text{actual no. of tablets}}{\text{volume needed } (y)} = \frac{\text{whole no. of tablets}}{\text{given volume}}$$

Step 8B: Using the information from Step 7B, rewrite the orders. Do not use abbreviations.

Section 10.2 Hypodermic Tablets

Now use method B in the same example to determine the correct number of hypodermic tablets and the volume of solution to be administered.

Ordered: Codeine sulfate 15 mg SQ q.i.d. p.r.n. pain
On hand: Codeine sulfate 30-mg hypodermic tablets (diluent: sterile water for injection: 0.5 mℓ)

Number of tablets used to make solution? _____
How many milliliters of solution are given? _____

Solution:

Step 1B: $\dfrac{15}{x}$

Step 2B: $\dfrac{30}{1}$

Step 3B:
$$\dfrac{15}{x} = \dfrac{30}{1}$$
$$x \cdot 30 = 15 \cdot 1$$
$$x \cdot 30 = 15$$

Step 4B:
$$\dfrac{x \cdot \cancel{30}^{1}}{\cancel{30}_{1}} = \dfrac{\cancel{15}^{1}}{\cancel{30}_{2}}$$
$$x = \dfrac{1}{2} \text{ tablet}$$

Because the hypodermic tablet can't be broken, dissolve one tablet in 0.5 mℓ of diluent.

Step 5B: $\dfrac{\frac{1}{2}}{y}$

Step 6B: $\dfrac{1}{0.5}$

Step 7B:
$$\dfrac{\frac{1}{2}}{y} = \dfrac{1}{0.5}$$
$$y \cdot 1 = \dfrac{1}{2} \cdot 0.5$$
$$y \cdot 1 = 0.25$$
$$y = 0.25 \text{ mℓ}$$

Step 8B: Dissolve one 30-mg codeine sulfate tablet in 0.5 mℓ of sterile water for injection. Give 0.25 mℓ of this solution, which contains 15 mg of the drug, subcutaneously four times daily as needed for pain.

Number of tablets used to make solution? ___1___
How many milliliters of solution are given? ___0.25___

The next example shows a different prescription ordered by the physician along with the available form of the prescribed item. Use method A to determine the correct number of hypodermic tablets and the volume of solution to be administered.

250   Chapter 10 Injectable Drugs in Solution

Example 2:   *Ordered:* Scopolamine HBr $1/100$ gr I.M. q. 6–8 h. p.r.n.
On hand: Scopolamine HBr $1/200$-gr hypodermic tablets (diluent: sterile water for injection: ♏ *vi*)

Number of tablets used to make solution? _____
How many minims of solution are given? _____

Solution:

Step 1A: $1/100$ gr

Step 2A: The tablets we have on hand are $1/200$ gr; therefore, we write

$$\frac{1 \text{ tablet}}{\frac{1}{200} \text{ gr}} = 1$$

Step 3A:

$$\frac{1}{100} \text{ gr} \times \frac{1 \text{ tablet}}{\frac{1}{200} \text{ gr}} = \frac{\frac{1}{100}}{\frac{1}{200}}$$

$$= \frac{1}{100} \times \frac{200}{1} = 2 \text{ tablets}$$

Step 4A: $2 \text{ tablets} \times \frac{6 \text{ ♏}}{2 \text{ tablets}} = 6$ ♏ or ♏ *vi*

Step 5A: Dissolve two $1/200$-gr scopolamine HBr tablets in 6 minims of sterile water for injection. Give 6 minims of this solution, which contains $1/100$ gr of the drug, intramuscularly every six to eight hours as needed.

Number of tablets used to make solution? __2__
How many minims of solution are given? __6__

Section 10.2 Readiness Review

Each of the following problems shows a prescription ordered by the physician along with the available form of the prescribed item. In each case, determine the correct number of hypodermic tablets and the volume of solution to be administered.

1. *Ordered:* Codeine sulfate $3/4$ gr SQ q. 4 h. around the clock
   On hand: Codeine sulfate gr i hypodermic tablets (diluent: sterile water for injection: ♏ *viii*)

   Number of tablets used to make solution? _____
   How many minims of solution are given? _____

2. *Ordered:* Codeine phosphate 20 mg SQ q. 12 h. p.r.n.
   On hand: Codeine phosphate 30-mg hypodermic tablets (diluent: sterile water for injection: 0.5 mℓ)

   Number of tablets used to make solution? _____
   How many milliliters of solution are given? _____

Section readiness review answers (not given in order):

2. Dissolve one 30-mg codeine phosphate tablet in 0.5 mℓ of sterile water for injection. Give 0.33 mℓ of this solution, which contains 20 mg of the drug, subcutaneously every twelve hours as needed.

Number of tablets used to make solution? ____1____
How many milliliters of solution are given? ___0.33___

1. Dissolve one gr ī codeine sulfate tablet in 8 minims of sterile water for injection. Give 6 minims of this solution, which contains ¾ gr of the drug, subcutaneously every four hours around the clock.

Number of tablets used to make solution? ____1____
How many minims of solution are given? ____6____

Section 10.2 Review Problems

Each of the following problems shows a prescription ordered by the physician along with the available form of the prescribed item. Determine the correct number of hypodermic tablets and the volume of solution to be administered (answers to the odd-numbered problems are at the back of the book).

1. *Ordered:* Codeine sulfate 45 mg SQ t.i.d. p.r.n.
   On hand: Codeine sulfate 60-mg hypodermic tablets (diluent: sterile water for injection: 1.0 mℓ)

   Number of tablets used to make solution? _____
   How many milliliters of solution are given? _____

2. *Ordered:* Atropine sulfate ¹⁄₂₀₀ gr I.M. p.r.n.
   On hand: Atropine sulfate ¹⁄₁₀₀-gr hypodermic tablets (diluent: sterile water for injection: ♏ *xvi*)

   Number of tablets used to make solution? _____
   How many minims of solution are given? _____

3. *Ordered:* Scopolamine HBr SQ 0.32 mg q. 8 h.
   On hand: Scopolamine HBr 0.4-mg hypodermic tablets (diluent: sterile water for injection: 0.5 mℓ)

   Number of tablets used to make solution? _____
   How many milliliters of solution are given? _____

4. *Ordered:* Codeine sulfate ¼ gr SQ q. 3–4 h. p.r.n.
   On hand: Codeine sulfate gr s̄s̄ hypodermic tablets (diluent: sterile water for injection: ♏ *xii*)

   Number of tablets used to make solution? _____
   How many minims of solution are given? _____

5. *Ordered:* Codeine phosphate ⅛ gr SQ q. 10 h. p.r.n.
   On hand: Codeine phosphate gr ī hypodermic tablets (diluent: sterile water for injection: ♏ *xvi*)

   Number of tablets used to make solution? _____
   How many minims of solution are given? _____

6. *Ordered:* Scopolamine HBr 0.7 mg I.M. now
   On hand: Scopolamine HBr 0.3-mg hypodermic tablets (diluent: sterile water for injection: 1 mℓ)

Number of tablets used to make solution? _____
How many milliliters of solution are given? _____

7. *Ordered:* Atropine sulfate 0.2 mg I.M. pre-op
   On hand: Atropine sulfate 0.6-mg hypodermic tablets (diluent: sterile water for injection: 1 ml)

Number of tablets used to make solution? _____
How many milliliters of solution are given? _____

8. *Ordered:* Codeine phosphate 10 mg SQ q. 2 h. p.r.n. severe pain
   On hand: Codeine phosphate 15-mg hypodermic tablets (diluent: sterile water for injection: 0.8 ml)

Number of tablets used to make solution? _____
How many milliliters of solution are given? _____

9. *Ordered:* Atropine sulfate 1/100 gr I.M. q. 11 h.
   On hand: Atropine sulfate 1/200-gr hypodermic tablets (diluent: sterile water for injection: ℞ v)

Number of tablets used to make solution? _____
How many minims of solution are given? _____

10. *Ordered:* Codeine phosphate 55 mg SQ q. 4–6 h. p.r.n.
    On hand: Codeine phosphate 60-mg hypodermic tablets (diluent: sterile water for injection: 0.9 ml)

Number of tablets used to make solution? _____
How many milliliters of solution are given? _____

Section 10.3 Injectable Drugs Requiring Reconstitution

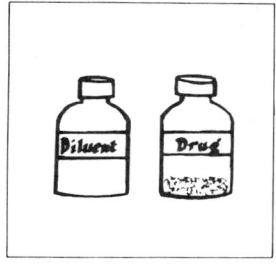

(1) If vials have plastic protective caps, flick them off.

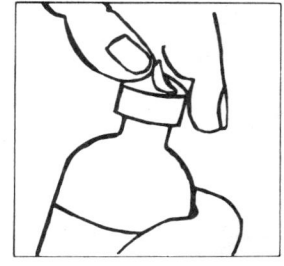

(2) Remove metal cap.

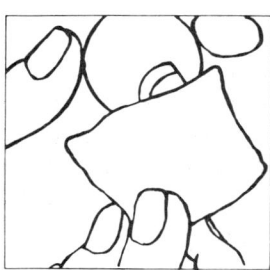

(3) Wipe off exposed rubber stopper of diluent with alcohol swab.

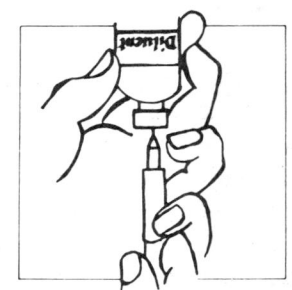

(4) Invert the vial. Inject amount of air equal to the volume of diluent desired into the vial. Now withdraw the amount of diluent required.

Figure 10-5. Reconstituting an injectable drug

Section 10.3 Injectable Drugs Requiring Reconstitution   253

Figure 10-5. (continued)

The concept of reconstitution was introduced in Chapter 9. You'll recall that reconstitution involves the addition of a solvent to a drug that is in the form of a dry powder. The most important thing to remember when reconstituting injectable drugs is that *both the drug and the diluent must be sterile.* You must use good aseptic technique (see Figure 10-5).

The label on the vial containing the sterile powdered drug will usually tell you the correct diluent to use. The most common diluents are sterile water for injection, sodium chloride 0.9% injection, and bacteriostatic water for injection. There are a few drugs that require special diluents; these are supplied with the drugs as a convenience.

The original drug container will usually be labeled with directions on how to reconstitute the drug; for example,

Penicillin G Potassium for Injection, USP (Buffered)[1]

For intramuscular or intravenous administration, penicillin G potassium is provided in rubber-stoppered vials as a dry powder and is intended for solution in sterile, pyrogen-free isotonic sodium chloride.

For intravenous use, it may be preferable under appropriate circumstances to dissolve the powder in sterile, pyrogen-free distilled water or 5 per cent dextrose solution. The antibiotic may be diluted initially as suggested below and then added to an intravenous solution being administered intermittently every four to six hours or on a continuous basis. Prompt, high peak serum levels are obtained with intravenous administration, but the levels are of short duration since the penicillin is rapidly excreted by the kidneys.

The dry salt is stable and does not need to be refrigerated. The aqueous solutions should be refrigerated.

The following dilution table may prove helpful. If desired, a further dilution may be made after the solution has been withdrawn from the vial. The solvent should be added with aseptic precautions. After preparation, the sterile solution may be kept in the refrigerator for one week without significant loss of potency.

Dilution Table

Ampoule (vial) No.	Amount of diluent	Units of penicillin per mℓ
522	4 mℓ	50,000
	2 mℓ	100,000
	1 mℓ	200,000
525	5 mℓ	100,000
	4 mℓ	125,000
	2 mℓ	250,000
526	9.6 mℓ	100,000
	4.6 mℓ	200,000
	1.6 mℓ	500,000
544	18 mℓ	250,000
	8 mℓ	500,000
	3 mℓ	1,000,000

How Supplied

Ampoules (vials)
    200,000 units, 5-mℓ size (No. 522) 1 and Traypacks* of 100
    500,000 units, 5-mℓ size (No. 525) 1
    1,000,000 units, 10-mℓ size (No. 526) 1 and Traypacks of 100
    5,000,000 units, 20-mℓ size (No. 544) 1 and Traypacks of 100

*Traypacks (multivial cartons, Lilly).

---

[1] Directions for using penicillin G potassium courtesy of Eli Lilly & Company.

Section 10.3 Injectable Drugs Requiring Reconstitution 255

> **Rule: Solving Problems That Involve Injectable Drugs Needing Reconstitution**
> **Method A: Dimensional Analysis**
> *Step 1A:* Read the label of the on-hand drug carefully. Identify the recommended diluent, volume of diluent required, concentration of solution after reconstitution, and route of administration.
> *Step 2A:* Write down the dosage strength ordered by the physician.
> *Step 3A:* Write the concentration of the on-hand drug as a fraction. Be sure, when using dimensional analysis, that the desired unit will remain and the given unit will be eliminated.
> *Step 4A:* Cancel units and multiply the ordered strength by the fraction from Step 3A.
> *Step 5A:* Using the information from Step 4A, rewrite the orders. Do not use abbreviations.
> *Step 6A:* Shade in the correct dosage on the syringe.

Example 1: Use method A to determine the correct amount of injectable medication that satisfies the following order using the drug available. Then shade in the correct dosage on the syringe.

Ordered: Ampicillin 300 mg I.M. q. 6 h.
On hand: Ampicillin sodium injection
2 g/vial
I.M. or I.V.
Reconstitution: Add 6.8 mℓ of sterile water for injection to contents of vial. Resultant solution has a concentration of 250 mg/mℓ.

Solution:
Step 1A: The recommended diluent is sterile water for injection, and 6.8 mℓ is needed. The concentration is 250 mg/mℓ. The route of administration ordered by the doctor is I.M.

Step 2A: The dosage strength ordered is

300 mg

Step 3A: The concentration of drug is written as

$$\frac{1 \text{ m}\ell}{250 \text{ mg}} = 1$$

Step 4A: $\overset{6}{\cancel{300}} \text{ mg} \times \frac{1 \text{ m}\ell}{\underset{5}{\cancel{250}} \text{ mg}} = 1.2 \text{ m}\ell$

Step 5A: Give 1.2 mℓ ampicillin 250 mg/mℓ intramuscularly every six hours.

Step 6A: Shade in the correct dosage on the syringe.

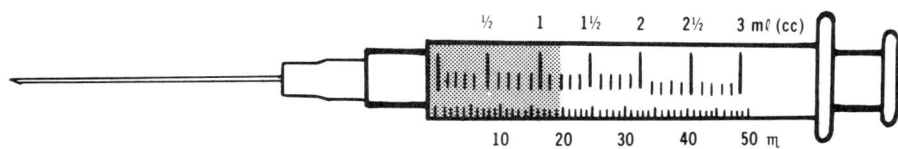

> **Rule: Solving Problems That Involve Injectable Drugs Needing Reconstitution**
> **Method B: Ratio and Proportion**
> *Step 1B:* Read the on-hand drug label carefully. Identify the recommended diluent, volume of diluent required, concentration of solution after reconstitution, and route of administration.
> *Step 2B:* Write the strength ordered as a ratio in fractional form, using x to represent the unknown volume of drug product.
> *Step 3B:* Write the concentration of the on-hand drug as a ratio in fractional form.
> *Step 4B:* Write the ratios from Steps 2B and 3B as a proportion and cross multiply. Be sure that the units on each side of the equation correspond to one another.
> *Step 5B:* To solve for the unknown quantity, divide the numbers on each side of the equals sign by the number on the same side as the x.
> *Step 6B:* Using the information from Step 5B, rewrite the orders. Do not use abbreviations.
> *Step 7B:* Shade in the correct dosage on the syringe.

Now use method B in the same example to determine the correct amount of injectable medication that satisfies the order using the drug available. Then shade in the correct dosage on the syringe.

Ordered: Ampicillin 300 mg I.M. q. 6 h.
On hand: Ampicillin sodium injection
2 g/vial
I.M. or I.V.
Reconstitution: Add 6.8 mℓ of sterile water for injection to contents of vial. Resultant solution has a concentration of 250 mg/mℓ.

Solution:

Step 1B: The recommended diluent is sterile water for injection, and 6.8 mℓ is needed. The concentration is 250 mg/mℓ. The route of administration ordered by the doctor is I.M.

Step 2B: $\dfrac{300}{x}$

Step 3B: Based on the information on the on-hand drug label, we write

$\dfrac{250}{1}$

Step 4B: $\dfrac{300}{x} = \dfrac{250}{1}$
$x \cdot 250 = 300 \cdot 1$
$x \cdot 250 = 300$

Step 5B: $\dfrac{x \cdot \cancel{250}^{1}}{\cancel{250}_{1}} = \dfrac{\cancel{300}^{6}}{\cancel{250}_{5}}$

$x = 1.2 \text{ mℓ}$

Step 6B: Give 1.2 mℓ ampicillin 250 mg/mℓ intramuscularly every six hours.
Step 7B: The syringe diagram is the same as Example 1—Step 6A.

Example 2: Use method B to determine the correct amount of injectable medication that satisfies the following order using the drug available.

Ordered: Staphcillin® 1.45 g I.M. q. 6 h.
On hand: Staphcillin® injection
6 g/vial
I.M. or I.V.
Reconstitution: Add 8.6 mℓ of NS to contents of vial. Resultant solution has a concentration of 500 mg/mℓ.

Solution:
Step 1B: The recommended diluent is NS (normal saline) and 8.6 mℓ is needed. The concentration is 500 mg/mℓ. The route of administration ordered by the doctor is I.M.

Step 2B:
$$\frac{1.45 \text{ g}}{x \text{ mℓ}} \quad \text{or} \quad \frac{1450 \text{ mg}}{x \text{ mℓ}}$$

Therefore,

$$\frac{1450}{x}$$

Step 3B: Based on the information on the on-hand drug label, we write

$$\frac{500}{1}$$

Step 4B:
$$\frac{1450}{x} = \frac{500}{1}$$
$$x \cdot 500 = 1450 \cdot 1$$
$$x \cdot 500 = 1450$$

Step 5B:
$$\frac{x \cdot \cancel{500}^{1}}{\cancel{500}_{1}} = \frac{\cancel{1450}^{29}}{\cancel{500}_{10}}$$
$$x = 2.9 \text{ mℓ}$$

Step 6B: Give 2.9 mℓ Staphcillin® 500 mg/mℓ intramuscularly every six hours.

Section 10.3 Readiness Review

Determine the correct amounts of injectable medications that satisfy the following orders using the drugs available.

1. *Ordered:* Aqueous Pen G 500,000 U I.M. q. 6 h.
   On hand: Penicillin G potassium injection
   1,000,000 units/vial
   I.M. or I.V.
   Reconstitution: Add 4.6 mℓ of NS to contents of vial. Resultant solution has a concentration of 200,000 units/mℓ.

Shade in the correct dosage on the syringe.

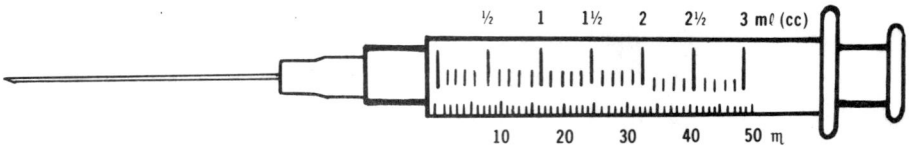

2. *Ordered:* Loridine® 250 mg I.M. q.i.d.
   On hand: Loridine® injection
   500 mg/vial
   I.M. or I.V.
   Reconstitution: Add 2.4 mℓ of NS for injection. Resultant solution has a concentration of 200 mg/mℓ.

Section readiness review answers:

1. Give 2.5 mℓ Aqueous Pen G 200,000 units/mℓ intramuscularly every six hours.

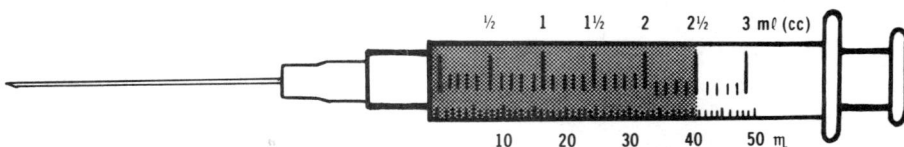

2. Give 1.25 mℓ Loridine® 200 mg/mℓ intramuscularly four times daily.

Section 10.3 Review Problems

Determine the correct amounts of injectable medications that satisfy the following orders using the drugs available (answers to the odd-numbered problems are at the back of the book).

1. *Ordered:* Penicillin G potassium 1,000,000 U I.M. q. 4 h.
   On hand: Penicillin G potassium injection
   10,000,000 units/vial
   I.M. or I.V.
   Reconstitution: Add 15.5 mℓ of sterile water for injection to contents of vial. Resultant solution has a concentration of 500,000 units/mℓ.

   Shade in the correct dosage on the syringe.

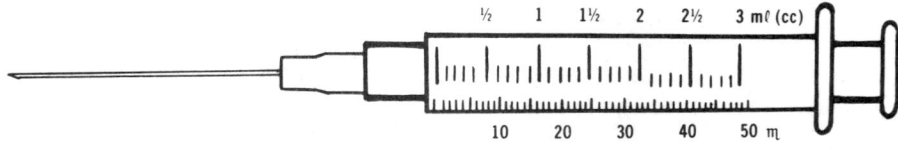

Section 10.3 Injectable Drugs Requiring Reconstitution   259

2. *Ordered:* Cefazolin 500 mg I.M. q. 8 h.
   On hand: Cefazolin sodium injection
   1 g/vial
   I.M. or I.V.
   Reconstitution: Add 2.5 mℓ of sterile water for injection. Resultant solution has a concentration of 330 mg/mℓ.

3. *Ordered:* Unipen® 250 mg I.M. q. 4 h.
   On hand: Unipen® injection
   1 g/vial
   I.M. or I.V.
   Reconstitution: Add 3.4 mℓ of sterile water for injection to contents of vial. Resultant solution has a concentration of 250 mg/mℓ.

4. *Ordered:* Oxacillin 350 mg I.M. q. 4 h.
   On hand: Oxacillin sodium injection
   500 mg/vial
   I.M. or I.V.
   Reconstitution: Add 2.7 mℓ of sterile water for injection to contents of vial. Resultant solution has a concentration of 250 mg/1.5 mℓ.

   Shade in the correct dosage on the syringe.

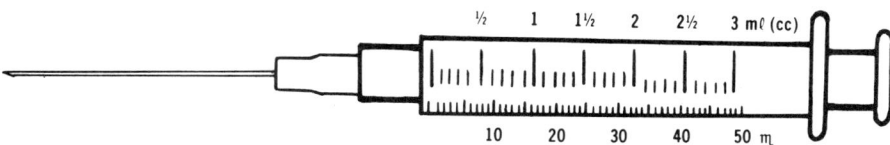

5. *Ordered:* Ampicillin 250 mg I.M. q. 6 h.
   On hand: Ampicillin sodium injection
   125 mg/vial
   I.M. or I.V.
   Reconstitution: Add 1.2 mℓ of sterile water for injection to contents of vial. Resultant solution has a concentration of 125 mg/mℓ.

   Shade in the correct dosage on the syringe.

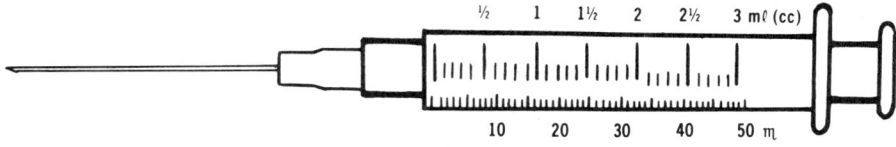

6. *Ordered:* Librium® 75 mg I.M. stat.
   On hand: Librium® injection
   5-mℓ ampule containing 100 mg of powdered drug supplied with 2-mℓ ampule of I.M. diluent
   I.M. or I.V.
   Reconstitution: For I.M. injection, add 2 mℓ of special diluent to contents of 5-mℓ ampule of powder. The resultant solution has a concentration of 50 mg/mℓ.

7. *Ordered:* Penicillin G potassium 1,500,000 units I.M. q. 2 h.
   On hand: Penicillin G potassium injection
   5,000,000 units/vial
   I.M. or I.V.
   Reconstitution: Add 3 mℓ of D5W to contents of vial. Resultant solution has a concentration of 1,000,000 units/mℓ.

8. *Ordered:* Keflin® 400 mg I.M. q. 6 h.
   On hand: Keflin® neutral injection
   1 g/vial
   I.M. or I.V.
   Reconstitution: Add 5 mℓ of sterile water for injection. Resultant solution has a concentration of 500 mg/2.7 mℓ.

   Shade in the correct dosage on the syringe.

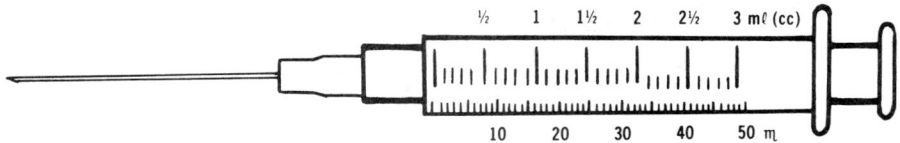

9. *Ordered:* Carbenicillin 1.2 g I.M. q. 6 h.
   On hand: Carbenicillin disodium injection
   2 g/vial
   I.M. or I.V.
   Reconstitution: Add 4 mℓ of sterile water for injection to contents of vial. Resultant solution has a concentration of 1 g/2.5 mℓ.

   Shade in the correct dosage on the syringe.

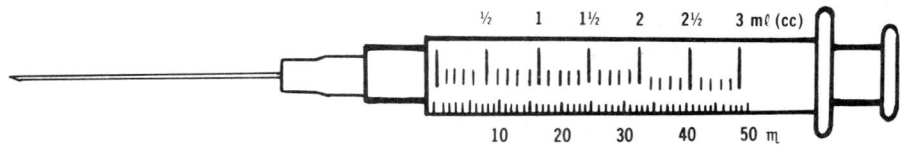

10. *Ordered:* Ampicillin 500 mg I.M. q. 6 h.
    On hand: Ampicillin sodium injection
    1 g/vial
    I.M. or I.V.
    Reconstitution: Add 3.5 mℓ of sterile water for injection to contents of vial. Resultant solution has a concentration of 250 mg/mℓ.

Chapter 10 Readiness Review

Determine the correct amounts of injectable medications that satisfy the following orders using the drugs available.

1. *Ordered:* Pituitrin® 8 units I.M. stat.
   On hand: Pituitrin® injection
   10 units/mℓ
   1-mℓ ampule
   SQ or I.M.

   Shade in the correct dosage on the syringe.

   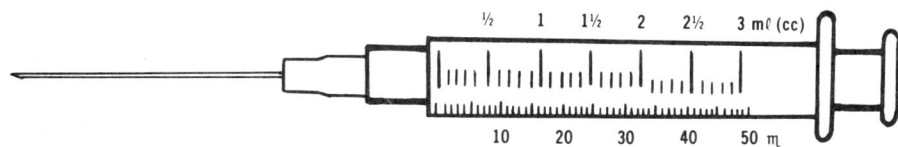

2. *Ordered:* Amikin® 450 mg I.M. q. 8 h.
   On hand: Amikin® injection
   500 mg/2 mℓ
   2-mℓ vial
   I.M. or I.V.

3. *Ordered:* Atropine 0.1 mg now
   On hand: Atropine sulfate injection
   0.4 mg/mℓ
   10-mℓ vial
   SQ, I.M., or I.V.

4. *Ordered:* Prolixin Enanthate® 12.5 mg SQ to initiate therapy.
   On hand: Prolixin Enanthate® injection
   25 mg/mℓ
   5-mℓ vial
   I.M. or SQ

   Shade in the correct dosage on the syringe.

   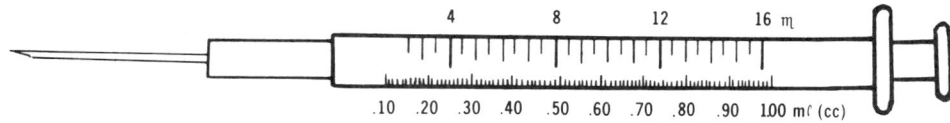

   How many milliliters of air must be injected into the vial to withdraw this amount?

5. *Ordered:* Codeine sulfate 30 mg SQ q. 6 h.
   On hand: Codeine sulfate 60-mg hypodermic tablets (diluent: sterile water for injection 1.0 mℓ)

   Number of tablets used to make solution? _____
   How many milliliters of solution are given? _____

6. *Ordered:* Scopolamine HBr 1/200 gr I.M. q. 8 h. p.r.n.
   On hand: Scopolamine HBr 1/150-gr hypodermic tablets (diluent: sterile water for injection: ℳ x)

   Number of tablets used to make solution? _____
   How many minims of solution are given? _____

262   Chapter 10 Injectable Drugs in Solution

7. *Ordered:* Keflin® 350 mg I.M. q. 4 h.
On hand: Keflin® neutral injection
1 g/vial
I.M. or I.V.
Reconstitution: Add 5 mℓ of sterile water for injection. Resultant solution has a concentration of 500 mg/2.7 mℓ.

Shade in the correct dosage on the syringe.

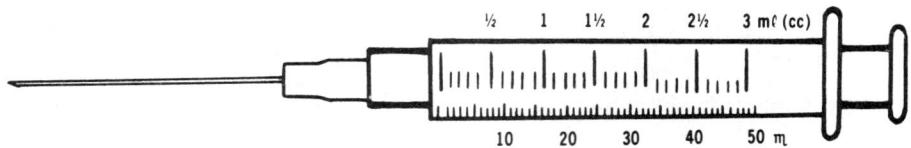

8. *Ordered:* Prostaphlin® 450 mg I.M. q. 6 h.
On hand: Prostaphlin® injection
1 g/vial
I.M. or I.V.
Reconstitution: Add 5.7 mℓ of sterile water for injection to contents of vial. Resultant solution has a concentration of 250 mg/1.5 mℓ.

Chapter readiness review answers (not given in order):

8. Give 2.7 mℓ Prostaphlin® 250 mg/1.5 mℓ intramuscularly every six hours.
7. Give 1.9 mℓ (1.89 mℓ = 350 mg) Keflin® 500 mg/2.7 mℓ intramuscularly every four hours.

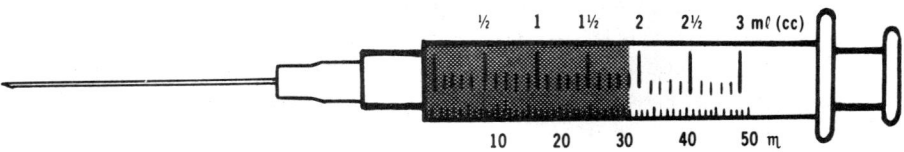

6. Dissolve one $1/150$-gr scopolamine HBr tablet in 10 minims of sterile water for injection. Give 7.5 minims of this solution, which contains $1/200$ gr of the drug, intramuscularly every eight hours as needed.

Number of tablets used to make solution? ___1___
How many minims of solution are given? ___7.5___

5. Dissolve one 60-mg codeine sulfate tablet in 1 mℓ of sterile water for injection. Give 0.5 mℓ of this solution, which contains 30 mg of the drug, subcutaneously every six hours as needed.

Number of tablets used to make solution? ___1___
How many milliliters of solution are given? ___0.5___

4. Give 0.5 mℓ Prolixin Enanthate® 25 mg/mℓ subcutaneously to initiate therapy.

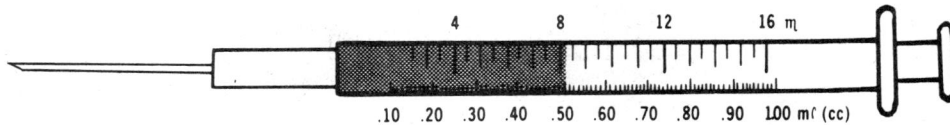

3. Give 0.25 mℓ atropine sulfate 0.4 mg/mℓ now.
2. Give 1.8 mℓ Amikin® 500 mg/2 mℓ intramuscularly every eight hours.
1. Give 0.8 mℓ Pituitrin® 10 units/mℓ intramuscularly immediately.

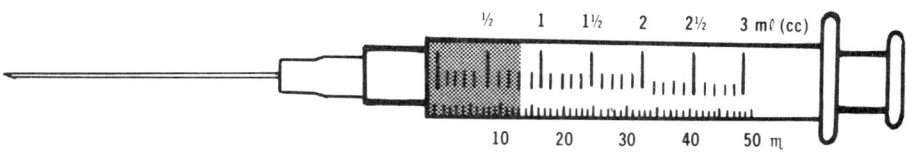

Chapter 10 Summary

Define each item in your own words, then compare your definitions with the text.

Key Words

parenteral   (p. 235)
aseptic technique   (p. 235)
subcutaneous   (p. 235)
intramuscular   (p. 235)
intravenous   (p. 235)
contamination   (p. 235)

prefilled syringes   (p. 236)
multiple-dose vial   (p. 236)
single-dose vial   (p. 236)
ampule   (p. 236)
hypodermic tablets   (p. 247)
reconstitution   (p. 252)

Chapter 10 Review Problems

Determine the correct amounts of injectable medications that satisfy the following orders using the drugs available (answers to all the problems are at the back of the book).

1. *Ordered:* Vasodilan® 10 mg I.M. b.i.d.
   On hand: Vasodilan® injection
   5 mg/mℓ
   2-mℓ ampule
   I.M.

2. *Ordered:* Phenergan® 25 mg I.M. q. 4–6 h. p.r.n. N & V
   On hand: Phenergan® injection
   50 mg/mℓ
   1-mℓ ampule
   I.M.

   Shade in the correct dosage on the syringe.

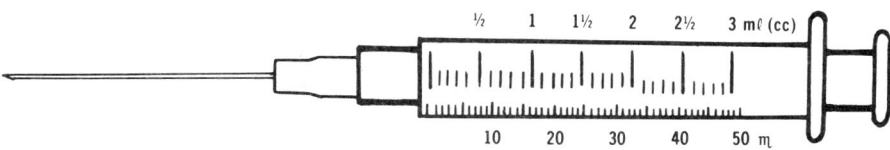

3. *Ordered:* Haldol® 3 mg I.M. q. 4–8 h. p.r.n. acute agitation.
   On hand: Haldol® (as lactate) injection
   5 mg/mℓ
   10-mℓ vial
   I.M.

   Shade in the correct dosage on the syringe.

   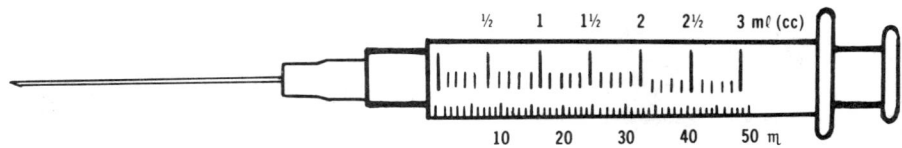

   How many milliliters of air must be injected into the vial to withdraw this amount?

4. *Ordered:* Nembutal® 150 mg I.M. h.s.
   On hand: Nembutal® injection
   50 mg/mℓ
   30-mℓ vial
   I.M. or I.V.

5. *Ordered:* Isuprel® 0.2 mℓ SQ stat.
   On hand: Isuprel® injection
   1 : 5000 solution
   5-mℓ ampule
   I.M., SQ, I.V.

   Shade in the correct dosage on the syringe.

   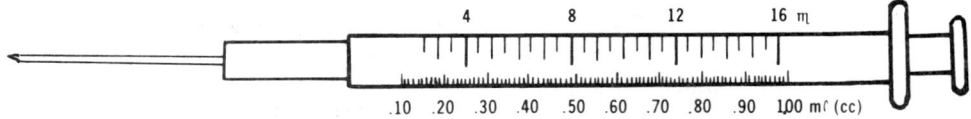

6. *Ordered:* Aramine® 3 mg SQ now
   On hand: Aramine® injection
   10 mg/mℓ
   1-mℓ vial
   I.M., SQ, or I.V.

7. *Ordered:* Compazine® 10 mg I.M. now (deeply into the upper outer quadrant of the buttock)
   On hand: Compazine® injection
   5 mg/mℓ
   10-mℓ vial
   I.M. or I.V.

8. *Ordered:* Vistaril® 50 mg I.M. now (injected into the gluteus maximus)
   On hand: Vistaril® injection
   25 mg/mℓ
   10-mℓ vial
   I.M. only

Shade in the correct dosage on the syringe.

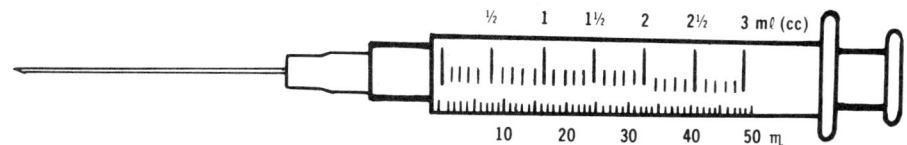

How many milliliters of air must be injected into the vial to withdraw this amount?

9. *Ordered:* Apresoline® 15 mg I.M. now, rept. p.r.n.
   On hand: Apresoline® injection
   20 mg/mℓ
   1-mℓ ampule
   I.M. or I.V.

   Shade in the correct dosage on the syringe.

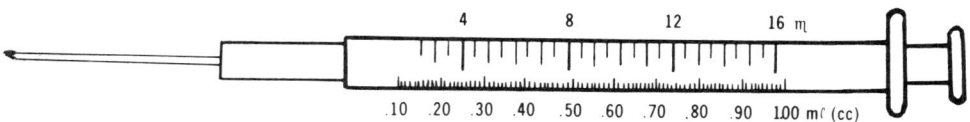

10. *Ordered:* Brethine® 0.25 mg SQ into lateral deltoid area
    On hand: Brethine® injection
    1 mg/mℓ
    1-mℓ solution in a 2-mℓ ampule
    SQ

    Shade in the correct dosage on the syringe.

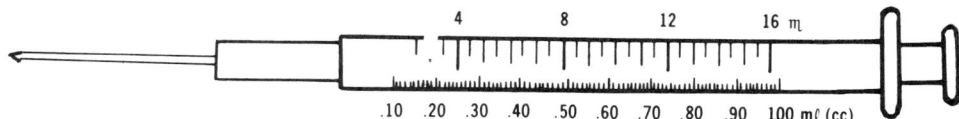

11. *Ordered:* Theelin® aqueous 0.5 mg I.M. today, rept. in three days
    On hand: Theelin® aqueous injection
    2 mg/mℓ
    10-mℓ vial
    I.M.

12. *Ordered:* Pronestyl® 600 mg I.M. q. 6 h.
    On hand: Pronestyl® injection
    500 mg/mℓ
    2-mℓ vial
    I.M. or I.V.

Shade in the correct dosage on the syringe.

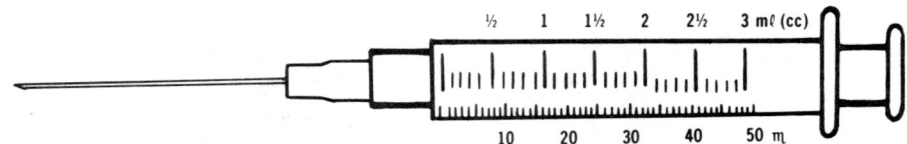

How many milliliters of air must be injected into the vial to withdraw this amount?

13. *Ordered:* Dilaudid® 1 mg SQ q. 4–6 h. p.r.n. severe pain
    On hand: Dilaudid® sulfate injection
    2 mg/mℓ
    10-mℓ vial
    SQ, I.M., or I.V.

Shade in the correct dosage on the syringe.

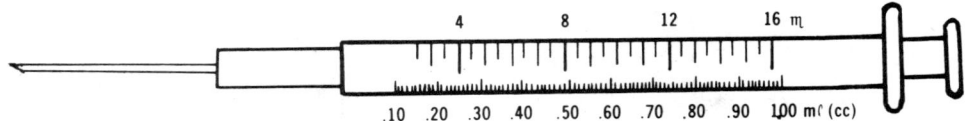

How many milliliters of air must be injected into the vial to withdraw this amount?

14. *Ordered:* Talwin® 20 mg SQ q. 3–4 h. p.r.n.
    On hand: Talwin® lactate injection
    30 mg/mℓ
    10-mℓ vial
    I.M., SQ, or I.V.

15. *Ordered:* Wyamine® 30 mg I.M. stat.
    On hand: Wyamine® injection
    15 mg/mℓ
    10-mℓ vial
    I.M., I.V.

16. *Ordered:* Benadryl® 20 mg I.M. (deeply) now
    On hand: Benadryl® injection
    10 mg/mℓ
    10-mℓ vial
    I.V. or I.M. (deep); avoid SQ

17. *Ordered:* Pitressin® 5 units I.M. t.i.d. p.r.n.
    On hand: Pitressin® injection
    20 units/mℓ
    0.5-mℓ ampule
    I.M. or SQ, *not* I.V.

Shade in the correct dosage on the syringe.

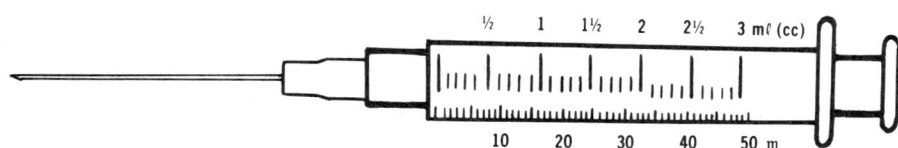

Each of the following problems shows a prescription ordered by the physician along with the available form of the prescribed item. In each case, determine the correct number of hypodermic tablets and the volume of solution to be administered.

18. *Ordered:* Atropine sulfate 0.1 mg I.M. q. 8 h. p.r.n.
 On hand: Atropine sulfate 0.4-mg hypodermic tablets (diluent: sterile water for injection 4 mℓ)

 Number of tablets used to make solution? _____
 How many milliliters of solution are given? _____

19. *Ordered:* Codeine phosphate gr i SQ stat.
 On hand: Codeine phosphate ¼-gr hypodermic tablets (diluent: sterile water for injection ♏ viii)

 Number of tablets used to make solution? _____
 How many minims of solution are given? _____

20. *Ordered:* Atropine sulfate ¹⁄₁₅₀ gr SQ q. 3–4 h.
 On hand: Atropine sulfate ¹⁄₁₀₀-gr hypodermic tablets (diluent: sterile water for injection ♏ x)

 Number of tablets used to make solution? _____
 How many minims of solution are given? _____

21. *Ordered:* Scopolamine HBr 0.5 mg SQ q. 8–12 h. p.r.n.
 On hand: Scopolamine 0.4 mg hypodermic tablets (diluent: sterile water for injection 0.8 mℓ)

 Number of tablets used to make solution? _____
 How many milliliters of solution are given? _____

Determine the correct amount of injectable medication that satisfies each of the following orders using the drug available.

22. *Ordered:* Ampicillin 450 mg I.M. q. 6 h.
 On hand: Ampicillin sodium injection
  500 mg/vial
  I.M. or I.V.
  Reconstitution: Add 1.8 mℓ of sterile water for injection to contents of vial. Resultant solution has a concentration of 250 mg/mℓ.

Shade in the correct dosage on the syringe.

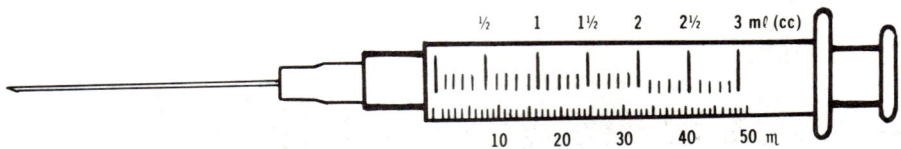

23. *Ordered:* Geopen® 800 mg I.M. q. 6 h.
    On hand: Geopen® injection
    1 g/vial
    I.M. or I.V.
    Reconstitution: Add 2.5 mℓ of sterile water for injection to contents of vial. Resultant solution has a concentration of 1 g/3 mℓ.

Shade in the correct dosage on the syringe.

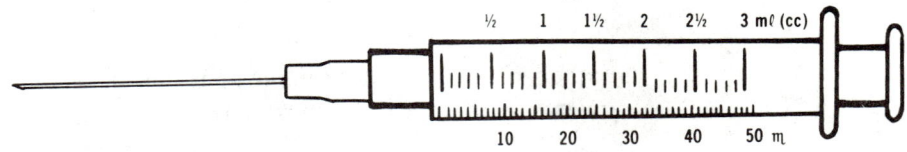

24. *Ordered:* Kefzol® 250 mg I.M. q. 6 h.
    On hand: Kefzol® injection
    500 mg/vial
    I.M. or I.V.
    Reconstitution: Add 2 mℓ of NS. Resultant solution has a concentration of 225 mg/mℓ.

25. *Ordered:* Loridine® 500 mg I.M. q. 12 h.
    On hand: Loridine® injection
    1 g/vial
    I.M. or I.V.
    Reconstitution: Add 2.5 mℓ of sterile water for injection. Resultant solution has a concentration of 300 mg/mℓ.

11 Intravenous Fluids and Medication

OBJECTIVES   After studying this chapter, you should be able to:

1. Understand the general concepts of I.V.-fluid and medication administration.
2. Solve problems involving drops per minute.
3. Determine the length of time it takes for an I.V. fluid to run to completion.
4. Solve problems involving intravenous admixtures.

Section 11.1 General Concepts and Drops Per Minute

Intravenous fluids and medication must be carefully administered. Remember, giving drugs intravenously means that the drug will be carried throughout the body in a matter of minutes. Once the drug has been given, it is extremely difficult to reverse its effects (see Figure 11-1).

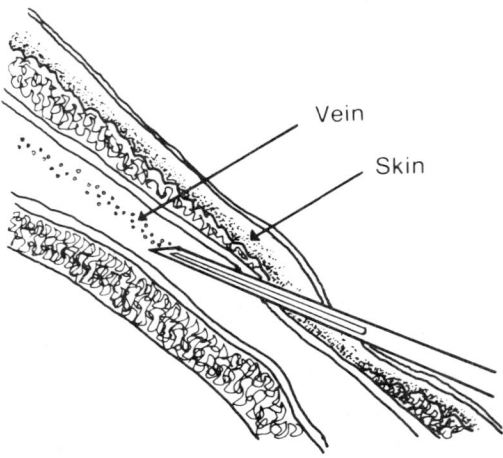

Figure 11-1. Intravenous injection

The use of intravenous fluids administered as large-volume parenterals (L.V.P.s) has increased dramatically.

> A *large-volume parenteral (L.V.P.)* is a sterile solution of 100 mℓ or more that is administered by injection.

The most common large-volume parenteral solutions are: dextrose 5% in water, sodium chloride 0.9%, dextrose (various strengths) in water, sodium chloride 0.45%, Ringer's, lactated Ringer's, and so on.

Many L.V.P.s are used to replace fluid and other nutrients (for example, glucose, electrolytes, vitamins, and so on). In addition, it is becoming increasingly common to add drugs to L.V.P.s and administer them in combination. This practice may help to make drugs less irritating to the vein because they are less concentrated than when they are used alone. It also allows for more continuous drug therapy over specified time intervals. Incompatibility, however, can be a major problem when one or more medications are added to a large-volume parenteral.

> *Incompatibility* occurs when one or more drugs are added to an L.V.P., and the resulting mixture is not suitable for administration to the patient.

The topic of incompatibility is actually quite complex and beyond the scope of this text. The following information is provided as a short guide to signs of possible incompatibility. Consult your hospital pharmacist if you have questions about specific drugs in combination with other drugs or L.V.P.s.

Obvious Signs of Possible Incompatibility*
- Crystals in the solution
- Precipitates (particles, flakes, clumping) formed in the solution
- Cloudiness or change in color of solution
- Formation of gas bubbles
- Separation of drugs into layers

*Don't administer L.V.P. mixture; check with pharmacist.

How are I.V. fluids and drugs actually administered to the patient? The most common way is with an I.V. administration set.

Definition: An *I.V. administration set* consists of plastic tubing that connects a needle (or catheter) in the vein to a hanging I.V. bottle (or I.V. bag) (see Figure 11-2).

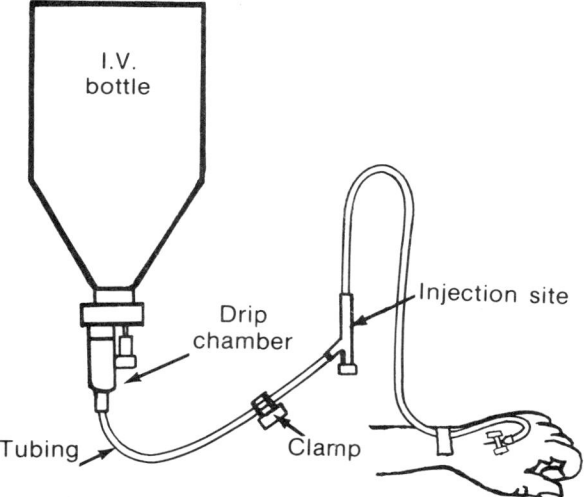

Figure 11-2. I.V. administration set

When a physician writes an order for I.V. fluids, he or she will commonly request a certain volume of fluid to be given over a fixed period of time. The following order is an example.

1000 mℓ D5W I.V. over eight hr

Imagine that you, as the nurse, must now administer this large-volume parenteral as ordered. To do this, you must convert the order to a rate of flow. The I.V.-bottle labels will usually state the rate of flow for I.V. fluids in *drops per minute*. The I.V. administration set will be labeled according to number of drops per milliliter. The rates vary from manufacturer to manufacturer—that is, 10 drops per milliliter, 20 drops per milliliter, and so on, because the administration sets differ according to the size of the drops they produce. For instance, there are I.V. administration sets that deliver 60 microdrops per milliliter and are usually intended for pediatric use. (These "microdrip" sets are covered in Application I—Pediatric and Geriatric Dosages.) Use dimensional analysis to solve rate-of-flow problems in terms of I.V. fluid drops per minute.

Rule: Solving I.V.-Fluid-Drops-per-Minute Problems
Method A: Dimensional Analysis

Step 1A: Identify the parts of the problem:
  a. number of milliliters ordered (if ordered in liters, convert to milliliters, as described in Chapter 6)
  b. number of minutes I.V. is to flow (if ordered in hours, convert to minutes, as described in Chapter 4)
  c. drops per milliliter of administration set (this information is printed on the I.V. administration set)

Step 2A: Write down the drops per milliliter given on the I.V. administration set.

Step 3A: Using dimensional analysis, write the number of milliliters ordered over the number of minutes the I.V. is to flow (fractional form).

Step 4A: Cancel units and multiply the number of drops per milliliter given on the I.V. administration set by the fraction from Step 3A.

Step 5A: Round off your answer to the nearest whole number.

Example 1: Use method A to determine how many drops per minute are needed to satisfy the following order:

Ordered: 1000 mℓ D5W I.V. over eight hr
Administration set: 15 gtt/mℓ

Solution:

Step 1A:
  a. 1000 mℓ
  b. $8 \text{ hr} \times \dfrac{60 \text{ min}}{1 \text{ hr}} = 480 \text{ min}$
  c. 15 gtt/mℓ (remember, gtt = drops)

Step 2A: $\dfrac{15 \text{ gtt}}{\text{mℓ}}$

Step 3A: $\dfrac{1000 \text{ mℓ}}{480 \text{ min}}$

Step 4A: $\dfrac{15 \text{ gtt}}{\text{mℓ}} \times \dfrac{\overset{25}{\cancel{1000}} \cancel{\text{mℓ}}}{\underset{12}{\cancel{480}} \text{min}} = \dfrac{375}{12} = 31.25 \text{ gtt/min}$

Step 5A: Round off answer to nearest whole number: 31.25 = 31 gtt/min.

Example 2: How many drops per minute are needed to satisfy the following order?

Ordered: 1000 mℓ D5NS (dextrose 5% in normal saline) I.V. over 600 minutes
Administration set: 20 gtt/mℓ

Section 11.1 General Concepts and Drops Per Minute 273

Solution:

Step 1A:
    a. 1000 mℓ
    b. 600 minutes
    c. 20 gtt/mℓ

Step 2A: $\dfrac{20 \text{ gtt}}{\text{mℓ}}$

Step 3A: $\dfrac{1000 \text{ mℓ}}{600 \text{ min}}$

Step 4A: $\dfrac{20 \text{ gtt}}{\cancel{\text{mℓ}}} \times \dfrac{\cancel{1000}^{5} \cancel{\text{mℓ}}}{\cancel{600}_{3} \text{ min}} = 33.33 \text{ gtt/min}$

Step 5A: Round off answer to nearest whole number: 33.33 = 33 gtt/min.

Section 11.1 Readiness Review

Find the rate of flow for each of the following L.V.P.s (see Examples 1 & 2):

1. *Ordered:* 500 mℓ D5W I.V. over eight hr
   Administration set: 15 gtt/mℓ

2. *Ordered:* 1000 mℓ NS I.V. over six hr
   Administration set: 20 gtt/mℓ

3. *Ordered:* 1 ℓ D5NS (dextrose 5% in normal saline) I.V. over 780 min
   Administration set: 10 gtt/mℓ

4. *Ordered:* 1000 mℓ D5LR (dextrose 5% in lactated Ringers) I.V. over five hr
   Administration set: 15 gtt/mℓ

Section readiness review answers (not given in order):

3. 13 gtt/min   4. 50 gtt/min

2. 56 gtt/min   1. 16 gtt/min

Section 11.1 Review Problems

Find the rate of flow for each of the following L.V.P.s (answers to odd-numbered problems are at the back of the book):

1. *Ordered:* 100 mℓ LR (lactated Ringer's) I.V. over 480 min
   Administration set: 10 gtt/mℓ

2. *Ordered:* 1000 mℓ LR (lactated Ringer's) I.V. over 560 min
   Administration set: 15 gtt/mℓ

3. *Ordered:* 1000 mℓ D5NS (dextrose 5% in normal saline) I.V. over 14 hr
   Administration set: 15 gtt/mℓ

4. *Ordered:* 1000 mℓ D5W I.V. over eight hr
   Administration set: 10 gtt/mℓ

5. *Ordered:* 1000 mℓ D5W I.V. over six hr
   Administration set: 10 gtt/mℓ

6. *Ordered:* 1000 mℓ D5LR (dextrose 5% in lactated Ringer's) I.V. over eight hr
   Administration set: 10 gtt/mℓ

7. *Ordered:* 1ℓ D5W I.V. over four hr
   Administration set: 20 gtt/mℓ

8. *Ordered:* 1 ℓ D5LR (dextrose 5% in lactated Ringer's) I.V. over eight hr
   Administration set: 20 gtt/mℓ

9. *Ordered:* 1000 mℓ NS I.V. over 12 hr
   Administration set: 15 gtt/mℓ

Section 11.2 I.V. Running Time

Sometimes a physician will write the I.V. order in such a way that the total length of time the I.V. is to run is not given. For example,

Ordered: 750 mℓ D5W I.V. at 30 gtt/min
Administration set: 20 gtt/mℓ

Ordered: 1000 mℓ D5¼NS (dextrose 5% in 0.225% normal saline) I.V. at 100 mℓ/hr
Administration set: 10 gtt/mℓ

In such cases it is up to you to determine how long each I.V. bottle will last. Here's how to do it.

Rule: Solving I.V.-Running-Time Problems—Drops per Minute
Method A: Dimensional Analysis

Step 1A: Identify the parts of the problem:
   a. number of milliliters ordered (if ordered in liters, convert to milliliters)
   b. number of drops per minute
   c. drops per milliliter of administration set
Step 2A: Write down the number of milliliters ordered.
Step 3A: Using dimensional analysis, express the drops per milliliter as one fraction and the drops per minute as a separate fraction.
Step 4A: Cancel units and multiply the number of milliliters ordered by both fractions from Step 3A.
Step 5A: Convert minutes into hours (covered in Chapter 4). Round off to one decimal place.

Example 1: How long will the following I.V. run?

Ordered: 750 mℓ D5W I.V. at 30 gtt/min
Administration set: 20 gtt/mℓ

Solution:
Step 1A: a. 750 mℓ
            b. 30 gtt/min
            c. 20 gtt/mℓ

Section 11.2 I.V. Running Time

Step 2A: 750 mℓ

Step 3A: $\dfrac{20 \text{ gtt}}{1 \text{ mℓ}}$

and

$\dfrac{1 \text{ min}}{30 \text{ gtt}}$

Step 4A: $\overset{25}{\cancel{750}} \text{ mℓ} \times \dfrac{20 \cancel{\text{ gtt}}}{1 \cancel{\text{ mℓ}}} \times \dfrac{1 \text{ min}}{\underset{1}{\cancel{30} \cancel{\text{ gtt}}}} = 500 \text{ min}$

Step 5A: $\overset{25}{\cancel{500}} \cancel{\text{ min}} \times \dfrac{1 \text{ hr}}{\underset{3}{\cancel{60} \cancel{\text{ min}}}} = 8.3$ hours or 8 hr 18 min

Example 2: How long will the following I.V. run?

Ordered: 600 mℓ NS at 44 gtt/min
Administration set: 15 gtt/mℓ

Solution:

Step 1A:
a. 600 mℓ
b. 44 gtt/min
c. 15 gtt/mℓ

Step 2A: 600 mℓ

Step 3A: $\dfrac{15 \text{ gtt}}{1 \text{ mℓ}}$

and

$\dfrac{1 \text{ min}}{44 \text{ gtt}}$

Step 4A: $\overset{150}{\cancel{600}} \cancel{\text{ mℓ}} \times \dfrac{15 \cancel{\text{ gtt}}}{1 \cancel{\text{ mℓ}}} \times \dfrac{1 \text{ min}}{\underset{11}{\cancel{44} \cancel{\text{ gtt}}}} = 204.5 = 205 \text{ min}$

Step 5A: $\overset{41}{\cancel{205}} \cancel{\text{ min}} \times \dfrac{1 \text{ hr}}{\underset{12}{\cancel{60} \cancel{\text{ min}}}} = 3.4$ hours or 3 hr 25 min

---

Rule: Solving I.V.-Running-Time Problems—Milliliters per Hour
  Method A: Dimensional Analysis
  Step 1A: Identify the parts of the problem:
  a. number of milliliters ordered (if ordered in liters, convert to milliliters)
  b. number of milliliters per hour
  Step 2A: Write down the number of milliliters ordered.
  Step 3A: Using dimensional analysis, express the number of milliliters per hour as a fraction.
  Step 4A: Cancel units and multiply the number of milliliters ordered by the fraction from Step 3A.
  Step 5A: Round off answer to one decimal place when necessary.

Example 3: How long will the following I.V. run?

Ordered: 1000 mℓ D5¼NS (dextrose 5% in 0.225% normal saline) I.V. at 100 mℓ/hr

Administration set: 10 gtt/mℓ

Solution:

Step 1A: **a.** 1000 mℓ
             b. 100 mℓ/hr

Step 2A: 1000 mℓ

Step 3A: $\dfrac{1 \text{ hr}}{100 \text{ mℓ}}$

Step 4A: $\cancel{1000}^{10} \text{ mℓ} \times \dfrac{1 \text{ hr}}{\cancel{100}_{1} \text{ mℓ}} = 10 \text{ hours}$

Step 5A: Rounding off is not required.

Example 4: How long will the following I.V. run?

Ordered: 1000 mℓ LR (lactated Ringer's) 85 mℓ/hr

Administration set: 15 gtt/mℓ

Solution:

Step 1A: **a.** 1000 mℓ
             b. 85 mℓ/hr

Step 2A: 1000 mℓ

Step 3A: $\dfrac{1 \text{ hr}}{85 \text{ mℓ}}$

Step 4A: $1000 \text{ mℓ} \times \dfrac{1 \text{ hr}}{85 \text{ mℓ}} = 11.76 \text{ hr}$

Step 5A: 11.76 hr, rounded off to one decimal place, is 11.8 hr, or 11 hr 48 min.

Section 11.2 Readiness Review

How long will each of the following I.V.s run?

1. *Ordered:* 1000 mℓ NS at 55 gtt/min (see Examples 1 & 2)
   Administration set: 20 gtt/mℓ

2. *Ordered:* 800 mℓ D5W at 60 gtt/min (see Examples 1 & 2)
   Administration set: 15 gtt/mℓ

3. *Ordered:* 1 ℓ LR (lactated Ringer's), 80 mℓ/hr (see Examples 3 & 4)
   Administration set: 20 gtt/mℓ

4. *Ordered:* 1 ℓ D5NS (dextrose 5% in normal saline), 100 mℓ/hr (see Examples 3 & 4)
   Administration set: 10 gtt/mℓ

Section readiness review answers (not given in order):

    2. 3.3 hr   **1.** 6.1 hr   **4.** 10 hr   **3.** 12.5 hr

Section 11.2 Review Problems

How long will the following I.V.s run? (Answers to the odd-numbered problems are at the back of the book.)

1. *Ordered:* 1200 mℓ D5NS (dextrose 5% in normal saline) at 42 gtt/min
   Administration set: 10 gtt/mℓ

2. *Ordered:* 1150 mℓ LR (lactated Ringer's) at 51 gtt/min
   Administration set: 20 gtt/mℓ

3. *Ordered:* 700 mℓ D5NS (dextrose 5% in normal saline) at 24 gtt/min
   Administration set: 15 gtt/mℓ

4. *Ordered:* 650 mℓ D5NS (dextrose 5% in normal saline) at 31 gtt/min
   Administration set: 10 gtt/mℓ

5. *Ordered:* 1.5 ℓ NS at 60 gtt/min
   Administration set: 20 gtt/mℓ

6. *Ordered:* 1000 mℓ D5¼NS (dextrose 5% in 0.225% normal saline) at 90 mℓ/hr
   Administration set: 20 gtt/mℓ

7. *Ordered:* 1 ℓ D5½NS (dextrose 5% in 0.45% normal saline) 80 mℓ/hr
   Administration set: 15 gtt/mℓ

8. *Ordered:* 980 mℓ LR (lactated Ringer's), 125 mℓ/hr
   Administration set: 10 gtt/mℓ

9. *Ordered:* 1000 mℓ LR (lactated Ringer's), 100 mℓ/hr
   Administration set: 15 gtt/mℓ

10. *Ordered:* 1 ℓ D5LR (dextrose 5% in lactated Ringer's), 125 mℓ/hr
    Administration set: 20 gtt/mℓ

Section 11.3 Intravenous Admixtures

> An *intravenous admixture* is a large volume of sterile solution that contains drugs to be delivered into the vein via injection.

There are different types of I.V. admixtures available. Two of the more common varieties are I.V. infusions—that is, the drug is added directly to the L.V.P. and the drug/fluid mixture is dripped into the vein—and I.V. "piggybacks" or I.V.P.B.s—that is, the drug is contained in a minibottle to which a specific volume of diluent (D5W or NS) is added. The "piggyback" is normally plugged into the injection site of the primary I.V. (see Figure 11-3).

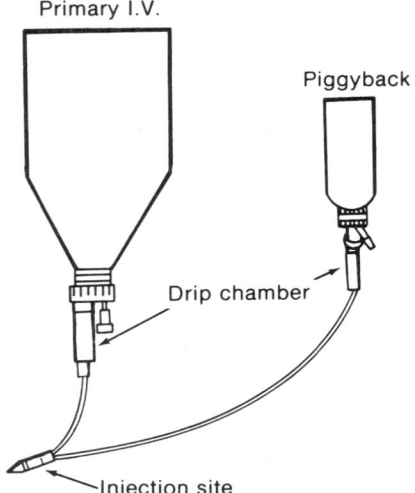

Figure 11-3. "Piggyback" connected to primary L.V.P.

The use of pumps to administer I.V. fluids and medications is rapidly gaining acceptance. The reason for this trend is that pumps increase the ability of I.V. equipment to deliver a more accurate dose of drug over a set period of time. Two popular kinds of pump differ in that one pump works on a dropper-minute basis, whereas the other operates according to volume (cc) per minute. If you would like to know more about I.V.-infusion pumps, ask your instructor to recommend some nursing-skills books. This text will not go into this topic in greater detail; rather, it emphasizes the traditional method of I.V. administration of drugs and, of course, the mathematics necessary to solve I.V.-administration problems. In addition, this section of the text will give you the opportunity to apply the skills that you have learned up to now.

The following problem presents a situation that is similar to the ones you will face in your everyday hospital work. After studying the problem, answer the questions that follow.

Example 1: Jan Olmsted is hospitalized with symptoms of chills, fever, and chest pain. After examining her, the physician makes a diagnosis of pneumococcal pneumonia.

Ordered: Penicillin G potassium 4 million units I.V.P.B. in 100 mℓ D5W q. 6 h. over one hour

On hand: Penicillin G potassium injection
5 million units/vial
I.M. or I.V.
Reconstitution: Add 8 mℓ of sterile water for injection to contents of vial. Resultant solution has a concentration of 500,000 units/mℓ.

Method of administration: I.V.P.B.
Administration set: 20 gtt/mℓ

a. How many milliliters of the reconstituted solution should be delivered to the 100-mℓ, D5W "additive bottle" to provide the ordered amount? _____
b. How many drops per minute are required to administer the drug as ordered? _____
c. Rewrite the orders. Do not use abbreviations.

Solution:

a. How many milliliters of the reconstituted solution should be delivered to the 100-mℓ, D5W "additive bottle" to provide the ordered amount? __8 mℓ__

The first question (**a**) represents the kind of problem that was covered in Section 10.3, Injectable Drugs Requiring Reconstitution. For our purposes, method A—Dimensional Analysis—is used to answer it.

First: The recommended diluent is sterile water for injection, and 8 mℓ is needed. The concentration is 500,000 units/mℓ. The route of administration is I.V.

Second: The strength desired by the physician is 4,000,000 units.

Third: Using dimensional analysis, express the concentration of drug product as a fraction.

$$\frac{1 \text{ mℓ}}{500,000 \text{ units}}$$

Fourth: The answer to this question is 8 mℓ.

$$\overset{8}{\cancel{4,000,000}} \text{ units} \times \frac{1 \text{ mℓ}}{\underset{1}{\cancel{500,000}} \text{ units}} = 8 \text{ mℓ}$$

b. How many drops per minute are required to administer the drug as ordered? __36 gtt/min__

This kind of problem (**b**) was covered in Section 11.1, General Concepts and Drops Per Minute.

First: (a) The number of milliliters of fluid ordered is 100 mℓ of diluent *plus* 8 mℓ of drug (100 mℓ + 8 mℓ = 108 mℓ).
(b) The number of minutes I.V. is to flow is one hour, or 60 min.
(c) The number of drops per milliliter of the administration set is 20 gtt/mℓ.

Second: Write down the drops per milliliter of the I.V. administration set.

20 gtt/mℓ

Third: Using dimensional analysis, express the milliliters per minute as a fraction.

$$\frac{108 \text{ mℓ}}{60 \text{ min}}$$

Fourth: $\dfrac{\overset{1}{\cancel{20}} \text{ gtt}}{1 \text{ }\cancel{\text{mℓ}}} \times \dfrac{\overset{36}{\cancel{108}} \text{ }\cancel{\text{mℓ}}}{\underset{\underset{1}{3}}{\cancel{60}} \text{ min}} = 36 \text{ gtt/min}$

Fifth: Rounding off is not required.

c. Rewrite the orders. Do not use abbreviations.

Give Penicillin G potassium, 4 million units intravenous "piggyback" diluted in 100 mℓ dextrose 5% in water, every six hours over one hour.

Section 11.3 Readiness Review

To aid you in answering the questions at the end of the following problem, re-examine the Readiness Reviews in Sections 10.1, 10.3, 11.1, and 11.2.

John Wood, a patient with a history of asthma flare-ups, is admitted to the hospital. The physician decides to start aminophylline.

Ordered: Aminophylline 400 mg qd. in 250 mℓ D5W over two hours
On hand: Aminophylline injection
    500 mg/20 mℓ
    20-mℓ ampule
    I.M. or I.V.
Method of administration: I.V. infusion
Administration set: 15 gtt/mℓ

1. How many milliliters of drug should be added to the 250 mℓ D5W to equal the amount of drug ordered? _____
2. How many drops per minute are required to administer the drug as ordered? _____
3. Rewrite the orders. Do not use abbreviations.

Section readiness review answers (not given in order):

3. Give aminophylline 400 mg daily, diluted in 250 mℓ dextrose 5% in water over two hours.
2. 33 gtt/min
1. 16 mℓ

Section 11.3 Review Problems

Calculate the answers to the questions at the end of the problems (answers to the questions in Problem 1 are at the end of the book).

1. Barbra Briggs presents to the emergency room in acute asthmatic distress.

    Ordered: Aminophylline 375 mg qd. in 50 mℓ D5W over 45 min
    On hand: Aminophylline injection
        500 mg/20 mℓ
        20-mℓ ampule
        I.M. or I.V.
    Method of administration: I.V. infusion
    Administration set: 15 gtt/mℓ

    a. How many milliliters of drug should be added to the 50 mℓ D5W to equal the amount of drug ordered? _____
    b. How many drops per minute are required to administer the drug as ordered? _____
    c. Rewrite the above orders. Do not use abbreviations.

2. After reading the results of the culture and sensitivity tests, the physician orders an I.V. antibiotic to treat Lowell Christiansen's upper respiratory infection.

Ordered: Ampicillin 450 mg I.V.P.B. q. 6 h. in 50 mℓ NS over five minutes
On hand: Ampicillin sodium injection
  2 g/vial
  I.M. or I.V.
  Reconstitution: Add 6.8 mℓ of sterile water for injection to contents of vial. Resultant solution has a concentration of 250 mg/mℓ.
Method of administration: I.V.P.B.
Administration set: 16 gtt/mℓ

a. How many milliliters of the reconstituted solution should be added to the 50-mℓ NS "additive bottle" to provide the ordered amount? _____
b. How many drops per minute are required to administer the drug as ordered? _____
c. Rewrite the above orders. Do not use abbreviations.

Chapter 11 Readiness Review

Find the rate of flow for each of the following L.V.P.s:

1. *Ordered:* 1 ℓ LR (lactated Ringer's) I.V. over 400 minutes
   Administration set: 20 gtt/mℓ

2. *Ordered:* 1000 mℓ D5W I.V. over 12 hours
   Administration set: 10 gtt/mℓ

How long will each of the following I.V.s run?

3. *Ordered:* 1000 mℓ D5W at 45 gtt/min
   Administration set: 20 gtt/mℓ

4. *Ordered:* 1 ℓ D5NS (dextrose 5% in normal saline) 100 mℓ/hr
   Administration set: 20 gtt/mℓ

Answer the questions at the end of the next problem.

5. A laboratory worker has accidentally exposed himself to the organism *Spirillum minus*. He shows the symptoms of rat-bite fever and is hospitalized.

   Ordered: Penicillin G potassium 15 million units daily, I.V. infusion in 1000 mℓ D5W over 24 hr
   On hand: Penicillin G potassium injection
     20,000,000 units/vial
     I.M. or I.V.
     Reconstitution: Add 31.6 mℓ of D5W to contents of vial. Resultant solution has a concentration of 500,000 units/mℓ.
   Method of administration: I.V. infusion
   Administration set: 16 gtt/mℓ

   a. How many milliliters of the drug should be added to the 1000 mℓ D5W to equal the amount of drug ordered? _____
   b. How many drops per minute are required to administer the drug as ordered? _____
   c. Rewrite the above orders. Do not use abbreviations.

Chapter readiness review answers:

1. 50 gtt/min
2. 14 gtt/min
3. 7.4 hours
4. 10 hours
5. a. 30 mℓ
   b. 12 gtt/min
   c. Give penicillin G potassium 15 million units daily by intravenous infusion diluted in 1000 mℓ dextrose 5% in water over 24 hours.

Chapter 11 Summary

Define each item in your own words, then compare your definitions with the text.

Key Words

intravenous fluids   (p. 270)
large-volume parenteral   p. 270)
incompatibility   (p. 270)
drops per minute   (p. 271)
precipitates   (p. 271)
I.V. administration set   (p. 271)
I.V. running time   (p. 274)
intravenous admixture   (p. 277)
I.V. infusions   (p. 277)
I.V. "piggybacks"   (p. 277)
primary L.V.P.   (p. 278)

Chapter 11 Review Problems

Find the rate of flow for each of the following L.V.P.s (answers to all the problems are at the back of the book).

1. *Ordered:* 1000 mℓ NS I.V. over four hr
   Administration set: 10 gtt/mℓ

2. *Ordered:* 1.5 ℓ D5W I.V. over eight hr
   Administration set: 20 gtt/mℓ

3. *Ordered:* 1 ℓ D5W I.V. over six hr
   Administration set: 20 gtt/mℓ

4. *Ordered:* 1000 mℓ D5W I.V. over four hr
   Administration set: 10 gtt/mℓ

5. *Ordered:* 1000 mℓ LR (lactated Ringer's) I.V. over 760 min
   Administration set: 15 gtt/mℓ

How long will each of the following I.V.s run?

6. *Ordered:* 500 mℓ D5W at 36 gtt/min
   Administration set: 20 gtt/mℓ

7. *Ordered:* 1 ℓ NS at 43 gtt/min
   Administration set: 10 gtt/mℓ

8. *Ordered:* 900 mℓ D5W at 75 gtt/min
   Administration set: 10 gtt/mℓ

9. *Ordered:* 1.5 ℓ D5 LR (dextrose 5% in lactated Ringer's) 90 mℓ/hr
   Administration set: 15 gtt/mℓ

10. A patient is to be placed on a course of I.V. antibiotics to treat a recurrent urinary tract problem. After several tests, the urologist decides to start the patient on Mandol®.

   Ordered: Mandol® 1 g in 100 mℓ D5W I.V.P.B. q. 8 hr over 20 min
   On hand: Mandol® injection
          1 g/vial (partial-filled)[1]
          I.M. or I.V.
          Reconstitution: Add 100 mℓ of D5W to the 1-g "piggyback" (100-mℓ) vial.
   Method of administration: I.V.P.B.
   Administration set: 16 gtt/mℓ

   a. How many drops per minute are required to administer the drug as ordered? _____
   b. Rewrite the above orders. Do not use abbreviations.
   c. If the physician had ordered Mandol® 1 g in 100 mℓ D5W I.V.P.B. q. 8 hr at a rate of 2.5 mℓ/min, how long would it take for the "piggyback" to run in? _____

[1] The term *partial-filled* means that the drug is in powdered form in a "piggyback" bottle supplied by the manufacturer. The drug is reconstituted according to directions. In this problem, the partial-filled bottle is filled to the 100-mℓ mark. Therefore, the total volume is 100 mℓ.

12
Insulin and Heparin

OBJECTIVES After studying this chapter, you should be able to:

1. Understand how insulin is used in the treatment of diabetes.
2. Solve insulin-dosage problems and select the correct syringe when applicable.
3. Understand how heparin helps prevent thrombi and/or emboli.
4. Solve heparin-dosage problems.

Section 12.1 Insulin

It might seem strange for a textbook on nursing mathematics to include a chapter on two specific drugs—insulin and heparin. The use of these two drugs, however, signifies major advances in modern medicine, and nurses must know how to do the special calculations needed to solve insulin- and heparin-dosage problems.

Insulin is a natural substance in the body that helps to maintain sugar balance. The pancreas releases insulin automatically when the body needs it. When the body cannot produce insulin and the sugar continues to accumulate in the bloodstream, the resulting condition is called *diabetes mellitus.*

For many years, the only treatment for diabetes was diet, which in severe cases led to starvation. However, in 1921-1922, the drug *insulin* was isolated by Banting and Best. During early 1922, the insulin extract was given to a 14-year-old diabetic boy, Leonard Thompson, in Canada. The boy made dramatic improvement. Insulin had saved his life.

Today insulin still saves many lives. However, it has changed since the original extract was made in 1922. Insulin is more pure now and is obtained primarily from cows and pigs. Most insulins are a combination of beef and pork, although beef-only and pork-only extracts also exist. There are different kinds of insulins that are classified according to how quickly they act (see Table 12-1).

Table 12-1. Insulin action speed

Action	Type of Insulin
Fast	Regular
	Semilente
Intermediate	NPH (isophane)
	Lente
Long	Protamine zinc
	Ultralente

Insulin preparations are measured in units, abbreviated U. The insulin unit was initially defined in terms of the ability of the drug to lower the sugar in the bloodstream of rabbits used in testing. Now, however, insulin units are based on the weight of the drug. The techniques used in purifying insulin have changed the standard from approximately 8 units per milligram, in the early days, to 26 units per milligram today.

What should the nursing student know about solving problems involving insulin units?

Traditionally, there have been three insulin strengths available—40, 80, and 100 units per milliliter (cc) of solution. The U-40 and U-80 strengths are gradually being phased out in favor of the U-100 strength. Regular insulin also comes in 500 units per milliliter, but its use is usually restricted to hospitals dealing with severe cases of diabetes. There are special syringes for administering the different strengths of insulin:

Insulin 40 units per milliliter[1] syringe calibrated for 40 units
Insulin 80 units per milliliter[1] syringe calibrated for 80 units
Insulin 100 units per milliliter syringe calibrated for 100 units

---

[1] There is a syringe that is calibrated for 40 units on one side and 80 units on the other.

It is extremely important for nurses to use the correct syringe when giving insulin to patients. In some cases, when an insulin syringe is not available, a tuberculin or equivalent syringe can be substituted.

Let's take a moment to study the different insulin, tuberculin, and typical 3-mℓ (cc) syringes available (see Figures 12-1 through 12-6).

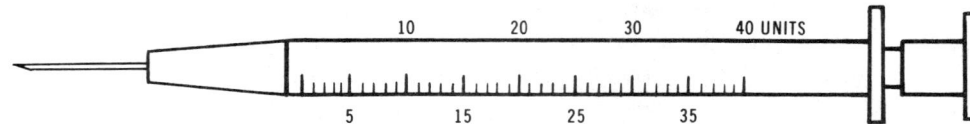

Figure 12-1. 40-unit insulin syringe—1 mℓ (cc). This syringe delivers 40 units in 1 mℓ.

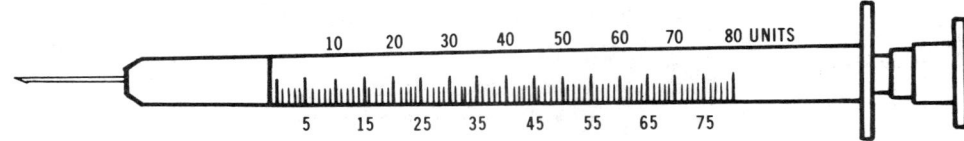

Figure 12-2. 80-unit insulin syringe—1 mℓ (cc). This syringe delivers 80 units in 1 mℓ.

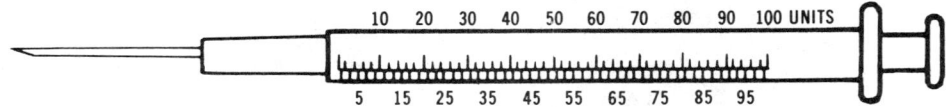

Figure 12-3. 100-unit insulin syringe—1 mℓ (cc). This syringe delivers 100 units in 1 mℓ.

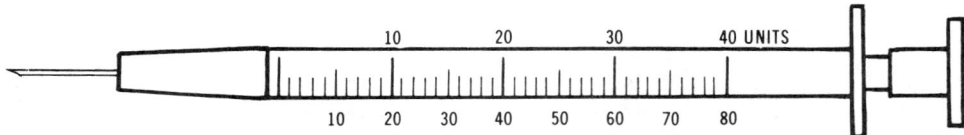

Figure 12-4. 40–80-unit insulin syringe—1 mℓ (cc) combination. A 40-unit scale is printed on one side of this syringe, and an 80-unit scale is printed on the other side. It is important to identify which scale you wish to use. This syringe delivers 40 units in 0.5 mℓ and 80 units in 1 mℓ.

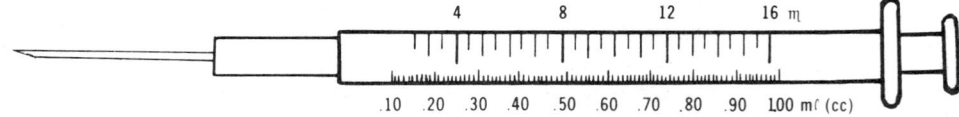

Figure 12-5. Tuberculin syringe—1 mℓ (cc) calibrated in milliliters (cc) and minims (♏). This syringe will deliver up to 1 mℓ (cc) and is broken down in tenths of a milliliter (cc).

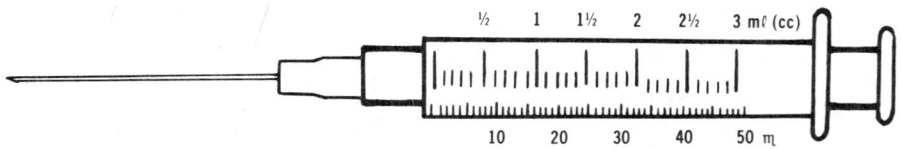

Figure 12-6. 3-mℓ (cc) syringe calibrated in milliliters (cc) and minims (♏).

Section 12.1 Insulin   287

> **Rule: Solving Insulin Problems**
> *Step 1:* Determine the kind of insulin and the amount required (both items will appear in the doctor's order).
> *Step 2:* Select the type of syringe that corresponds to the strength of insulin ordered.
> *Step 3:* Pull back the syringe to the desired amount (for our purposes, shade in the amount on the drawing of the syringe accompanying the problem).

Example 1: Dr. Bullock has diagnosed a diabetic, Richard Brown. Dr. Bullock wants Mr. Brown to give himself daily injections of Lente U-80 insulin. The order reads: 35 units Lente U-80 q. A.M. SQ. Which of the following syringes should be used? Shade in the correct dosage on that syringe.

Solution:

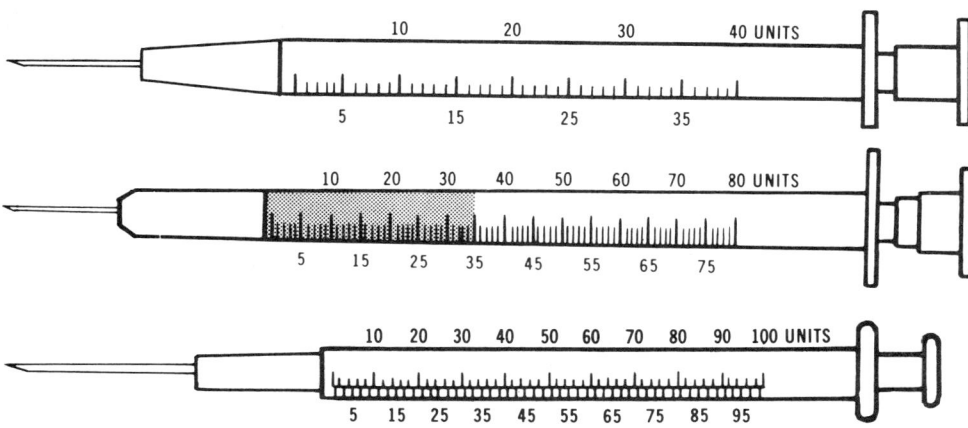

We see that the first syringe (Figure 12-1) is a U-40, the second (Figure 12-2) a U-80, and the third (Figure 12-3) a U-100. Since we are using U-80 insulin, we choose the second syringe (U-80) to give the 35 U of Lente insulin. To show the 35 U of insulin, shade in that portion of the syringe diagram.

Example 2: Last night, the Emergency Room physician admitted a diabetic child in a coma. The doctor ordered 20 units of Regular U-100 insulin stat. Which syringe did the nurse use? Shade in the amount ordered on the correct syringe.

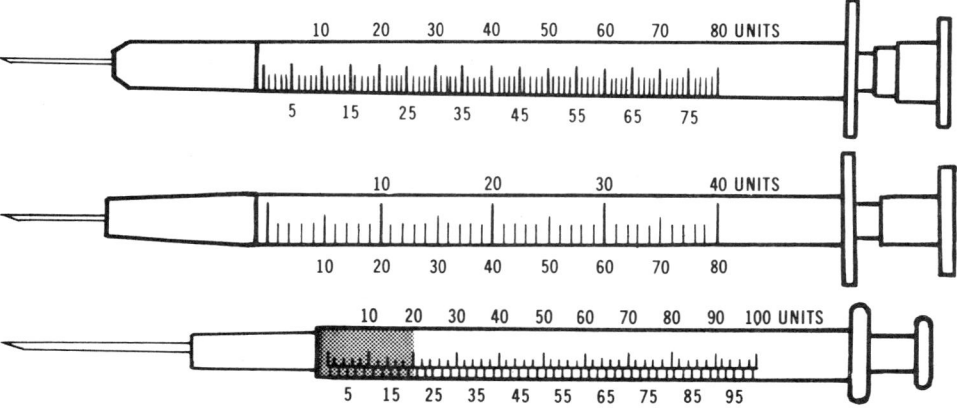

Solution: The first syringe is a U-80, the second is a combination U-40 and U-80, and the third is a U-100. Because we are using the U-100 insulin, we choose the third syringe (U-100) to give the 20 units of Regular U-100 insulin. To show the 20 U of insulin, we shade in that portion of the syringe diagram.

By now you should be able to select the correct syringe and pull back to the desired amount of insulin. However, as we noted earlier, there could be a time when an insulin dose is needed and the proper insulin syringe cannot be found. What can you do in this situation? You may use the kind of insulin syringe that you have on hand, after you calculate the correct amount of insulin to be given with this syringe. Use the following rule to solve the typical insulin-dosage problem involving an insulin syringe not intended to administer the insulin strength you must give the patient.

Rule: Solving Insulin-Dosage Problems
    Method A: Dimensional Analysis
  Step 1A: Read the problem carefully and write down the dose required by the patient.
  Step 2A: Using dimensional analysis, express the concentration of the insulin as one fraction and the type of syringe (when indicated) as another fraction.
  Step 3A: Cancel units and multiply the dose required by the fractions from Step 2A.
  Step 4A: Shade in the correct dosage on the syringe.

Example 3: Use method A to solve the following problem: A patient requires 15 units of insulin from a bottle of U-40 insulin. The nurse has only a U-80 syringe. How many units of U-40 insulin should be given with the U-80 syringe? Shade in the correct dosage.

Solution:
Step 1A: 15 units of insulin are required.

Step 2A: The concentration of the insulin is 40 units/cc. The U-80 syringe can hold a volume of 1 cc.

$$\frac{1 \text{ cc}}{40 \text{ U insulin}}$$

and

$$\frac{80\text{-U syringe}}{1 \text{ cc}}$$

Step 3A: $15 \text{ U} \times \dfrac{1 \cancel{\text{cc}}}{\underset{1}{\cancel{40 \text{ U}}}} \times \dfrac{\overset{2}{\cancel{80 \text{ U}}}}{1 \cancel{\text{cc}}} = \dfrac{30}{1} = 30$ U to be given with the U-80 syringe

Step 4A:

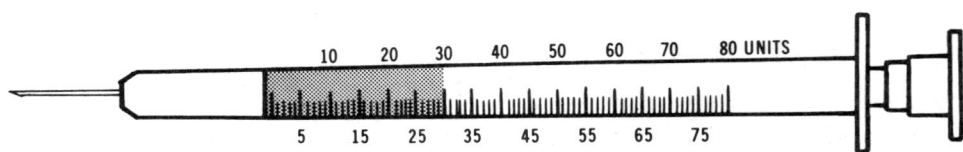

Another kind of insulin-dosage problem involves the situation when *no* insulin syringes are available. In such cases you may substitute a tuberculin syringe or a syringe that is even larger. First, we'll work some examples using tuberculin syringes, then we'll solve some problems involving 3-mℓ (cc) syringes. However, to solve these kinds of insulin problems, we must be familiar with the different concentrations of insulin. Let's review these concentrations, given in Table 12-2.

Table 12-2. Review of insulin concentrations

Strength of insulin has a *concentration of*	
U-40 (all types)	40 units per milliliter of solution
U-80 (all types)	80 units per milliliter of solution
U-100 (all types)	100 units per milliliter of solution
U-500 (Regular—rarely used)	500 units per milliliter of solution

Use the method you just learned to solve insulin-dosage problems to solve the next two problems involving non-insulin syringes. Your answer may be in milliliters or minims; it just depends on the problem. (Remember, Chapter 8 indicated that 16 minims, and occasionally 15 minims, equal 1 mℓ.)

Example 4: Use method A to solve the following problem: The doctor orders 40 units of U-80 insulin to be given to a diabetic. The nurse has only a 1-mℓ (cc) tuberculin syringe. How many milliliters of insulin will contain the 40 units? Shade the correct dosage on the tuberculin syringe.

Solution:
Step 1A: 40 units of insulin are required.
Step 2A: The concentration of the insulin is 80 units/cc.

$$\frac{1 \text{ cc}}{80 \text{ units insulin}} = 1$$

Step 3A: $\overset{1}{\cancel{40 \text{ units}}} \times \frac{1 \text{ cc}}{\underset{2}{\cancel{80 \text{ units}}}} = \frac{1}{2}$ cc, or 0.5 mℓ will contain 40 units.

Step 4A:

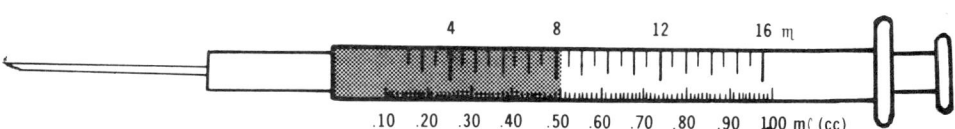

Another way of giving this dose would be to use a 3-mℓ (cc) syringe. The calculation would be the same. The answer of 0.5 mℓ would be measured as follows (shaded area):

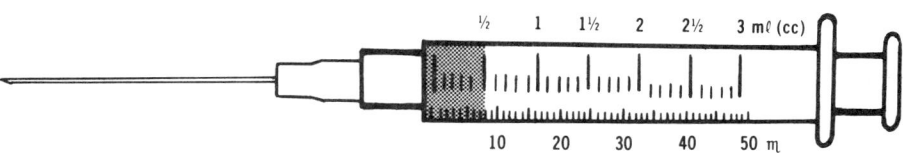

Example 5: Use method A to solve the following problem: The doctor orders 30 units of U-40 insulin to be given to a diabetic. The nurse has only a tuberculin 1-mℓ

(cc) syringe and a 3-mℓ (cc) syringe. Both are graduated in milliliters (cc) and minims (♏). How many minims of insulin will contain the 30 units? Shade in the correct dosage on the tuberculin syringe and the 3-mℓ (cc) syringe.

Solution:

Step 1A: 30 units of insulin are required.

Step 2A: The concentration of the insulin is 40 units/cc. There are 15 or 16 minims/cc.

$$\frac{1 \text{ cc}}{40 \text{ units insulin}} = 1$$

and

$$\frac{15 \text{ or } 16 \text{ minims}}{1 \text{ cc}}$$

Step 3A: $\overset{3}{\cancel{30}} \text{ units} \times \frac{\cancel{1 \text{ cc}}}{\underset{1}{\cancel{\underset{10}{40 \text{ units}}}} \text{ insulin}} \times \frac{\overset{4}{\cancel{16}} \text{ minims}}{\cancel{1 \text{ cc}}} = 12 \text{ minims will contain 30 units.}$

Step 4A:

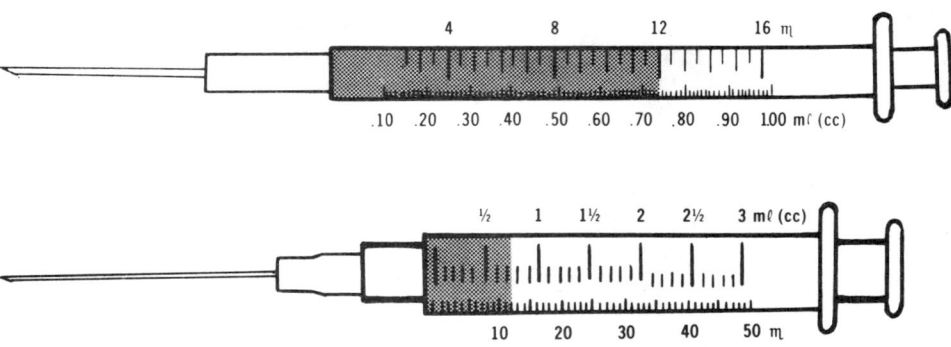

Remember, the problems involving syringe substitutions do not represent the norm. These problems are provided for you to gain experience in insulin administration. Whenever a nurse is on duty, she or he should be prepared to administer insulin with the syringes available. Remember, however, that if substitute syringes are used, the chance of error becomes greater. But if you practice working insulin-dosage problems, you should have no trouble.

Section 12.1 Readiness Review

Work the following problems. If the calculated dose cannot be measured exactly, then shade in an approximate answer on the syringe diagram.

1. A diabetic is to receive 25 units of U-40 insulin. The nurse has only a tuberculin 1-mℓ (cc) syringe and a 3-mℓ (cc) syringe. How many milliliters of insulin will contain 25 units? Shade in the correct dosage on the tuberculin syringe and the 3-mℓ (cc) syringe.

Section 12.1 Insulin 291

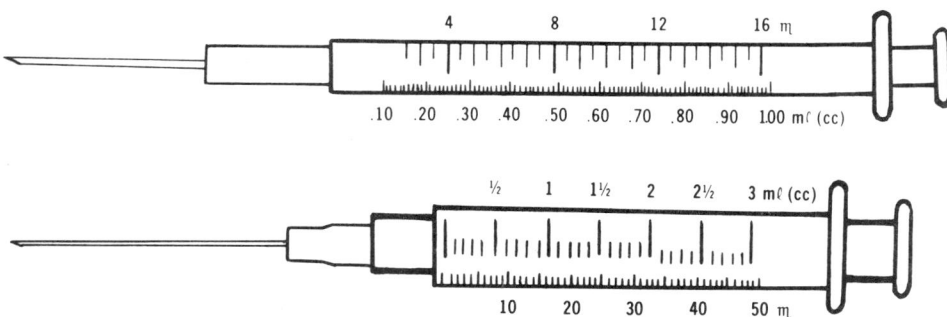

2. A diabetic is to receive 38 units of U-100 insulin. The nurse has only a tuberculin 1-mℓ (cc) syringe and a 3-mℓ (cc) syringe. Both are graduated in milliliters (cc) and minims (♏). How many minims of insulin will contain 38 units? Shade in the correct dosage on the tuberculin syringe and the 3-mℓ syringe.

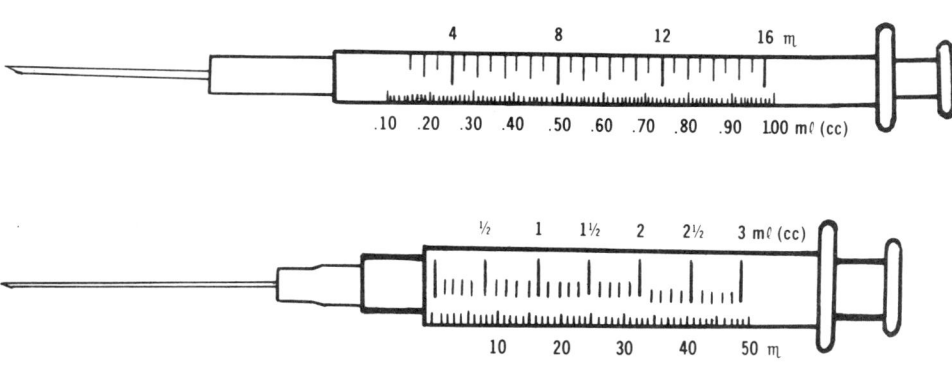

Section readiness review answers:

1. 0.625 mℓ

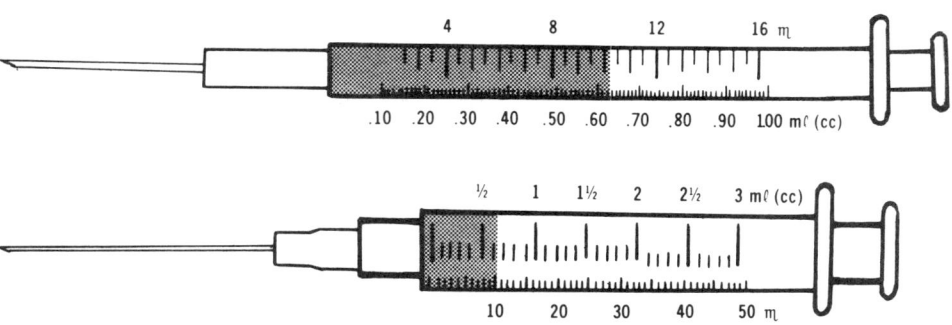

2. 5.7 minims (♏) or 6.08 minims (♏)—(approximately 6 minims)

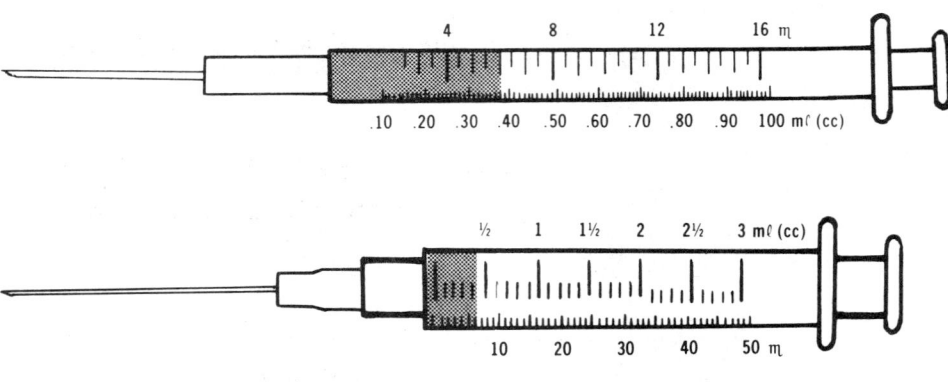

Section 12.1 Review Problems

Solve the following word problems (answers to the odd-numbered problems are at the back of the book).

1. Charles Miller, a diabetic, is to give himself 5 units of Regular U-100 insulin and 20 units of NPH U-100 insulin. Before starting this combination of insulin, the doctor asks him to select the correct syringe. Charles picks a syringe and draws up the 5 units of Regular U-100 insulin and then the 20 units of NPH. He gives himself the injection. Which syringe did he select? Shade in the *total* dosage of insulin that Charles gave himself.

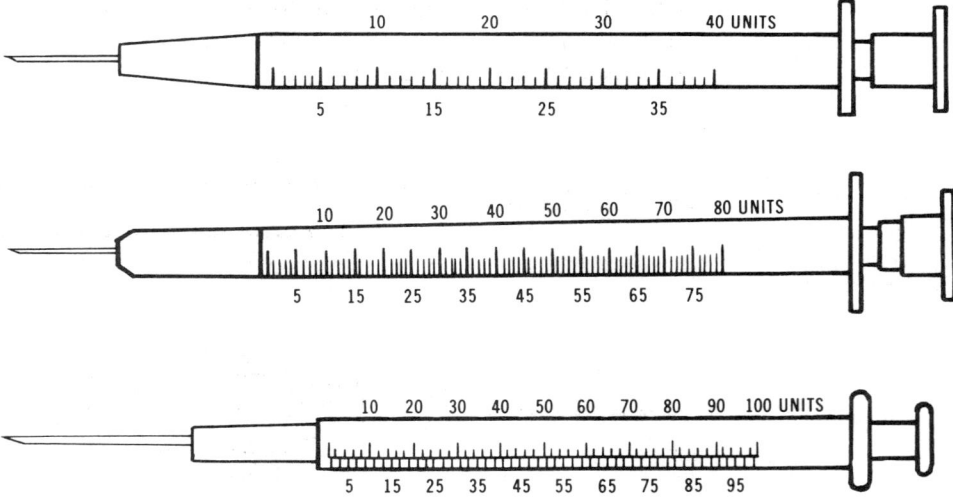

2. A diabetic patient is to be given 9 units of Ultralente U-40 insulin. Which of the following syringes should she use? Shade in the correct dosage on that syringe.

Section 12.1 Insulin 293

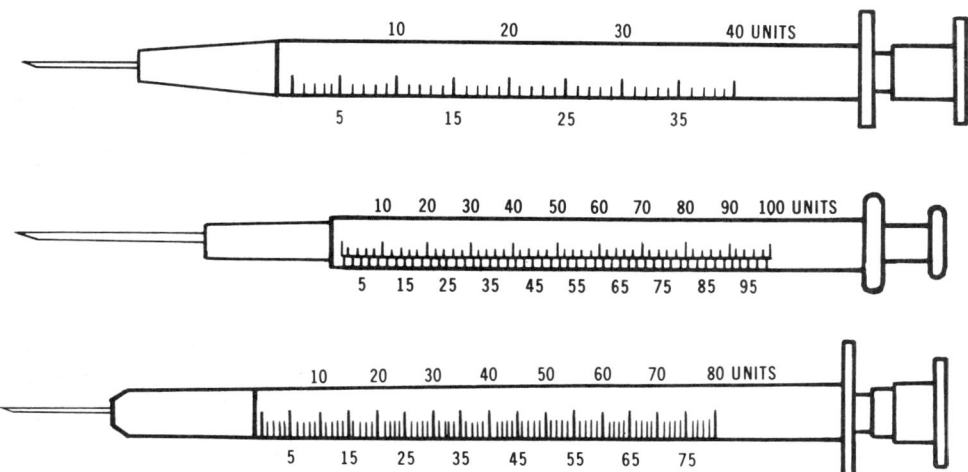

3. The Emergency Room physician treats a patient who is resistant to U-100 Regular insulin and thereby requires higher doses. The doctor orders 240 units of Regular U-500 insulin. The nurse has only a U-100 syringe. How many units of U-500 insulin should be given with the U-100 syringe? Shade in the correct dosage.

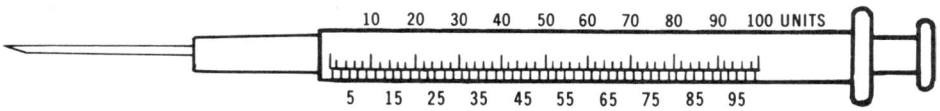

4. A diabetic is to receive 62 units of U-80 insulin. The nurse has only a tuberculin 1-mℓ (cc) syringe and a 3-mℓ (cc) syringe. How many milliliters of insulin will contain 62 units? Shade in the correct dosage on the tuberculin syringe and the 3-mℓ (cc) syringe.

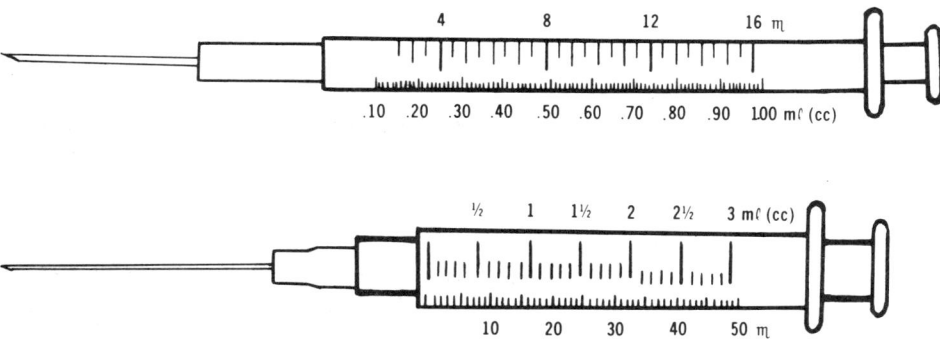

5. A diabetic is to receive 80 units of U-100 insulin. The nurse has a tuberculin 1-mℓ (cc) syringe and a 3-mℓ (cc) syringe on hand. How many milliliters of insulin will contain 80 units? Shade in the correct dosage on the tuberculin syringe and the 3-mℓ (cc) syringe.

294 Chapter 12 Insulin and Heparin

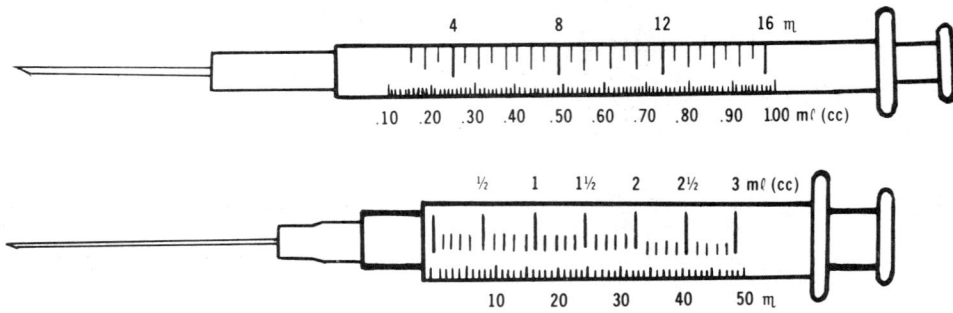

6. A diabetic is to receive 90 units of U-100 insulin. The nurse has only a tuberculin 1-mℓ (cc) syringe and a 3-mℓ (cc) syringe. Both are graduated in milliliters (cc) and minims (♏). How many minims of insulin will contain 90 units? Shade in the correct dosage on the tuberculin syringe and the 3-mℓ (cc) syringe.

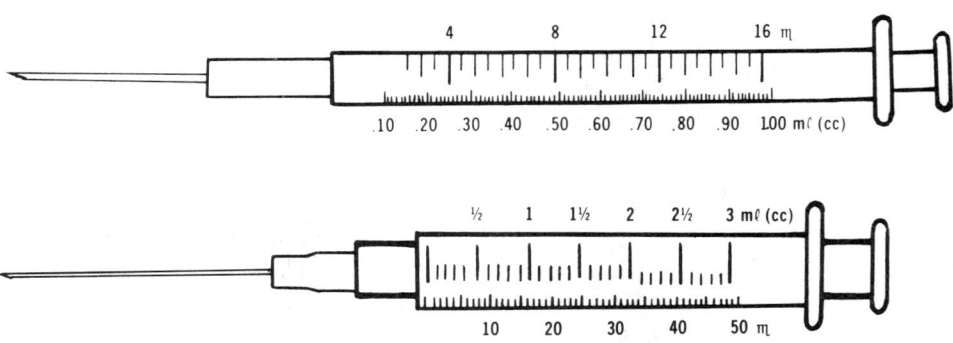

7. A diabetic is to receive 30 units of U-40 insulin. The nurse has a tuberculin 1-mℓ (cc) syringe and a 3-mℓ (cc) syringe. Both are graduated in milliliters (cc) and minims (♏). How many minims of insulin will contain 30 units? Shade in the correct dosage on the tuberculin syringe and the 3-mℓ (cc) syringe.

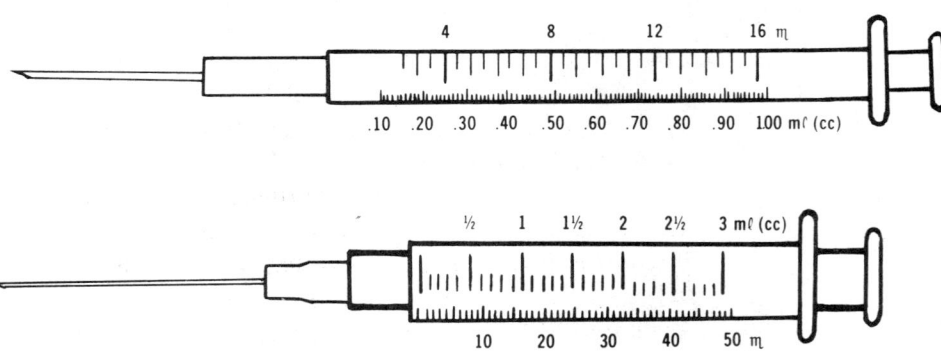

8. A diabetic is to receive 70 units of U-100 insulin. The nurse has only a tuberculin 1-mℓ (cc) syringe and a 3-mℓ (cc) syringe. Both are graduated in milliliters (cc) and minims (♏). How many minims of insulin will contain 70 units? Shade in the correct dosage on the tuberculin syringe and the 3-mℓ (cc) syringe.

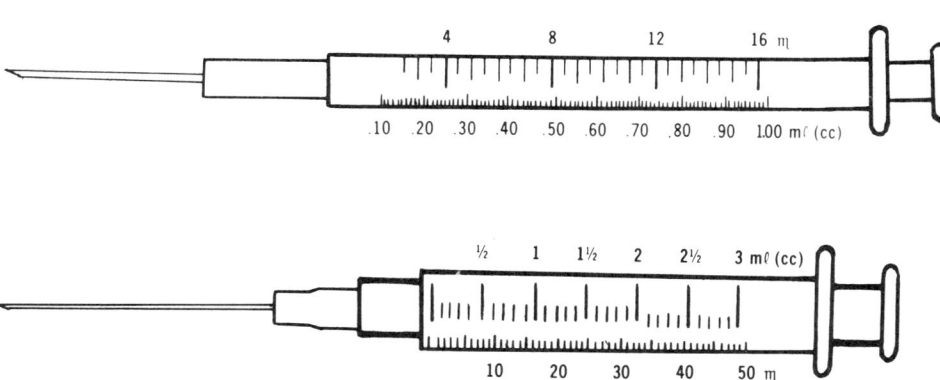

Section 12.2 Heparin

Heparin is a powerful anticoagulant (blood thinner). It is more commonly used than it used to be and is now usually administered by the nurse. It is important, therefore, for nursing students to know how heparin is administered and how to do calculations involving the drug.

Heparin, like insulin, cannot be given orally, because it would be destroyed by stomach acid. It is a natural substance and is obtained commercially from the intestinal mucosa of pigs and the lung tissue of cows. The injectable form of the drug is available as sodium heparin (the salt form). The anticoagulant potency of heparin is expressed in units.

Heparin can be administered subcutaneously, intravenously, or intramuscularly. Intramuscular injection should be avoided, however, because tissue damage may occur. Concentrations commonly range from 1000 units to 40,000 units of heparin sodium per milliliter of solvent (sterile water).

Heparin is used mainly to prevent the formation of thrombi and/or emboli. A *thrombus* is a clot that is formed in an unbroken blood vessel. A thrombus may block the vessel and thereby cut off the oxygen-rich blood supply to the tissues. It may also break off and float through the circulatory system and clog a smaller vessel. In either case, tissues that are not receiving a normal blood supply may be seriously damaged. An *embolus* may be a blood clot, a piece of junk—that is, fat—or an air bubble that is carried by the blood stream. Emboli can also clog blood vessels and cause tissue damage.

The most common medical problems calling for the use of heparin are thrombophlebitis (inflammation of a vein due to the presence of a thrombus) and pulmonary embolism (an embolus that has lodged in the blood vessel[s] supplying the lungs). Heparin is also used to prevent thrombi and/or emboli formation associated with heart and vascular surgery.

Chapter 12 Insulin and Heparin

> **Rule: Solving Heparin-Dosage Problems**
> **Method A:**
> *Step 1A:* Read the problem carefully and write down the dose of heparin ordered by the physician.
> *Step 2A:* Using dimensional analysis, express the concentration of sodium heparin as a fraction.
> *Step 3A:* Cancel units and multiply the dose required by the fraction from Step 2A.
> *Step 4A:* Pull back to the correct number of milliliters on the syringe provided.

Example 1: Use method A to solve the following problem: To prevent the possibility of thromboemboli formation, a patient has been given 5000 units of heparin SQ two hours before surgery and then q. 8 hr SQ. In order to minimize the patient's discomfort, it is recommended that the volume of SQ injection not be more than 0.5 mℓ. (There are exceptions to this, of course.) The heparin available is 40,000 U/mℓ. How many milliliters of the concentration provided will the nurse give? Shade in the correct dosage on the syringe.

Solution:
Step 1A: 5000 units of heparin are required.
Step 2A: The concentration of heparin available is 40,000/mℓ.

$$\frac{1 \text{ m}\ell}{40,000 \text{ units}} = 1$$

Step 3A:

$$\overset{1}{\cancel{5000 \text{ units}}} \times \frac{1 \text{ m}\ell}{\underset{8}{\cancel{40,000 \text{ units}}}} = 0.125 \text{ m}\ell$$

Step 4A: The nurse will give 0.125 mℓ.

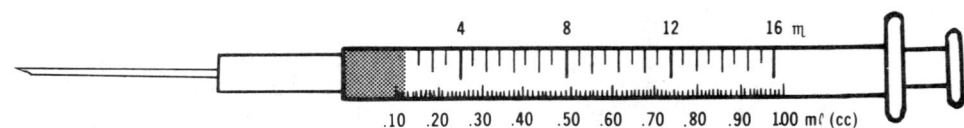

The next problem involves first reading the number of milliliters of heparin in the syringe and then converting them into units of heparin.

Example 2: Use method A to solve the following problem: The heparin concentration available is 20,000 U/mℓ. How many units are contained in the following syringe?

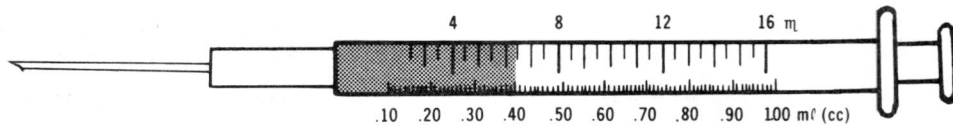

Solution:
Step 1A: The volume contained in the syringe is 0.4 mℓ.

Section 12.2 Heparin   297

Step 2A: The concentration of heparin available is 20,000 U/mℓ.

$$\frac{20{,}000 \text{ U}}{1 \text{ mℓ}} = 1$$

Step 3A: $0.4 \text{ mℓ} \times \dfrac{20{,}000 \text{ U}}{1 \text{ mℓ}} = 8000 \text{ units}$

The syringe contains 8000 units.

Step 4A: The syringe is already shaded in.

Section 12.2 Readiness Review

Solve the following problems:

1. The doctor has discharged Mr. Wong from the hospital. This patient is to continue on his heparin therapy at home for ten days. He is instructed to give himself heparin 5000 U SQ, abdominally q. 6 h. The hospital pharmacy has given Mr. Wong a 10-mℓ multiple-dose vial containing 40,000 U/mℓ. How many milliliters of the concentration will the patient give himself? Shade in the correct dosage on the syringe.

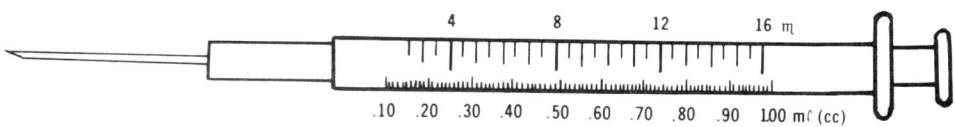

2. The heparin concentration available is 5000 U/mℓ. How many units are contained in the following syringe? _____

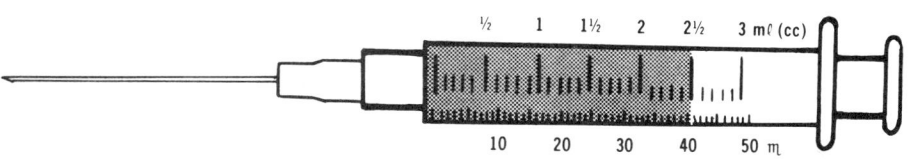

Section readiness review answers (not given in order):

2. 12,500 units

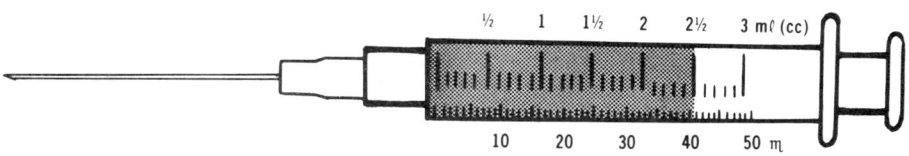

298 Chapter 12 Insulin and Heparin

1. 0.125 mℓ

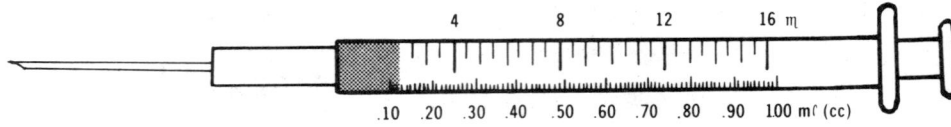

Section 12.2 Review Problems

Solve the following problems (answers to the odd-numbered problems are at the back of the book):

1. A pulmonary embolus has just been diagnosed by Dr. Hoffman. She orders the following treatment for her patient: Heparin 10,000 U I.V. stat. The pharmacy sends a 1-mℓ vial of sodium heparin 20,000 U per milliliter. How many milliliters of the concentration provided will the nurse give? Shade in the correct dosage on the syringe.

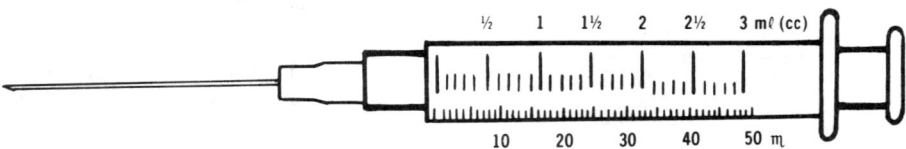

2. Dr. Hoffman's patient was later started on a continuous I.V. infusion of heparin via a pump. (A pump allows a specified amount of drug to be administered evenly in a 24-hour period.) In this case, Dr. Hoffman ordered 48,000 U of heparin to be mixed with 1000 mℓ of D5W. The pharmacy sends a 5-mℓ vial of 20,000 U/mℓ heparin. How many milliliters of the concentration provided will the nurse inject into the liter bottle of D5W? Shade in the correct dosage on the syringe.

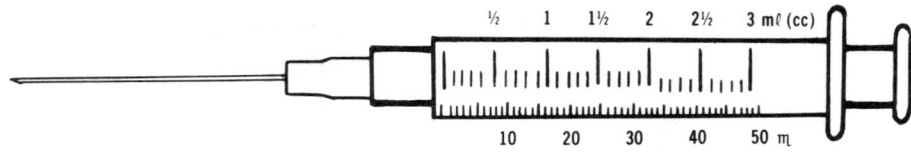

3. The heparin concentration available is 40,000 U/4 mℓ. How many units are contained in the following syringe? _____

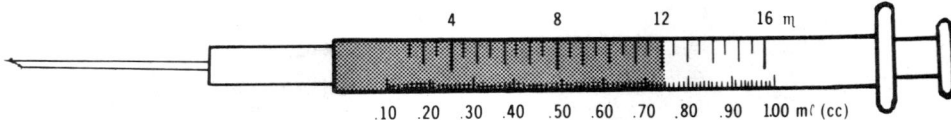

4. The heparin concentration available is 30,000 U/30 mℓ. How many units are contained in the following syringe? _____

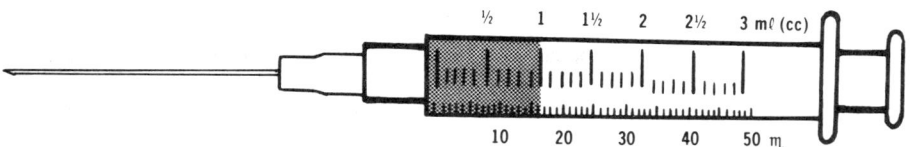

5. The heparin concentration available is 88,000 U/2.2 mℓ. How many units are contained in the following syringe? _____

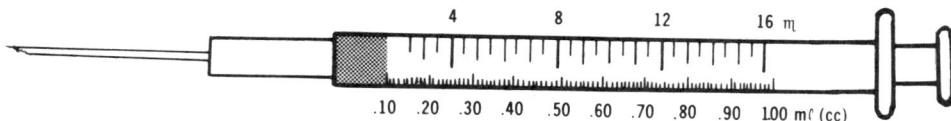

6. The heparin concentration available is 15,000 U/mℓ. How many units are contained in the following syringe? _____

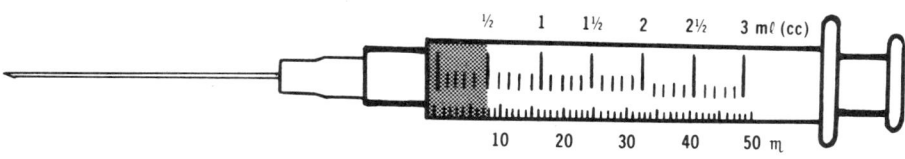

Chapter 12 Readiness Review

1. A diabetic patient is to be given 4 units of Protamine Zinc U-100 insulin. Which of the following syringes should be used? Shade in the correct dosage on that syringe.

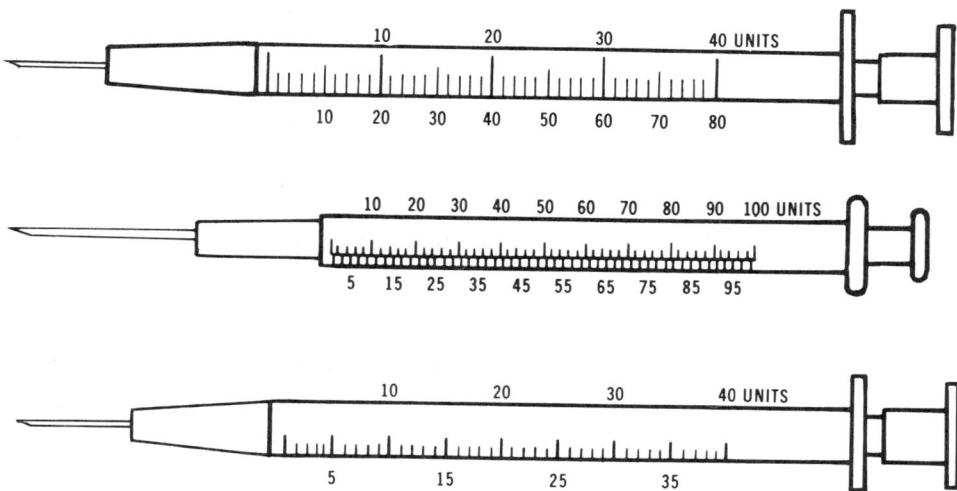

2. A patient requires 38 units of insulin from a bottle of U-40 insulin. The nurse has only a U-100 syringe. How many units of U-40 insulin should be given with the U-100 syringe? Shade in the correct dosage.

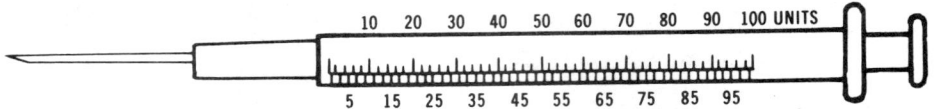

3. A surgical patient is to receive 5000 U of heparin SQ q. 8 hr after surgery. The pharmacy has provided heparin 10,000 U/mℓ. How many milliliters of the concentration provided will the nurse give? Shade in the correct dosage.

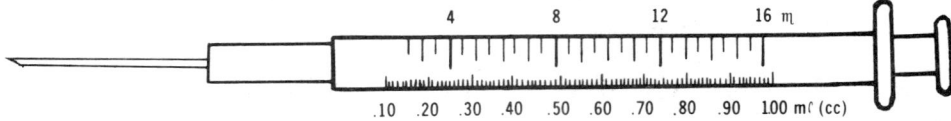

4. The heparin concentration available is 2000 U/2 mℓ. How many units are contained in the following syringe? _____

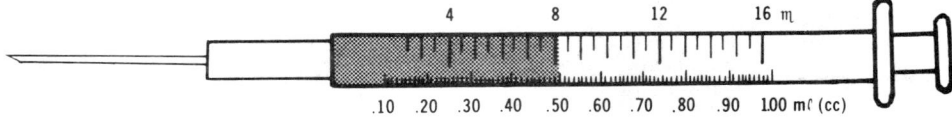

Chapter readiness review answers (not given in order):

4. 500 units

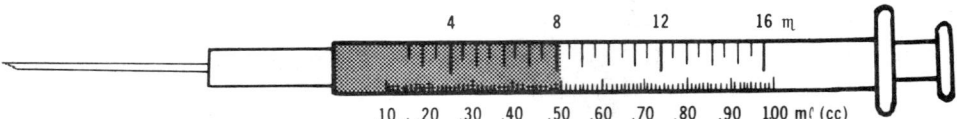

2. 95 units

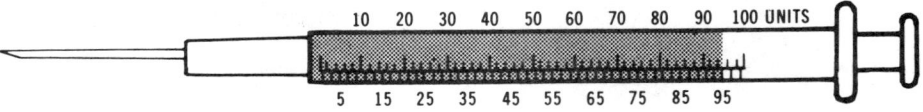

3. 0.5 mℓ

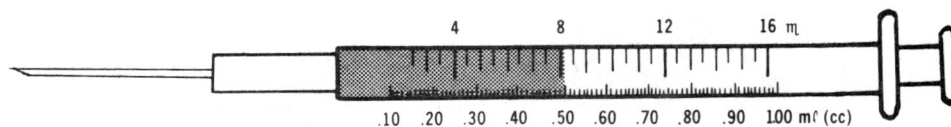

1. The second syringe

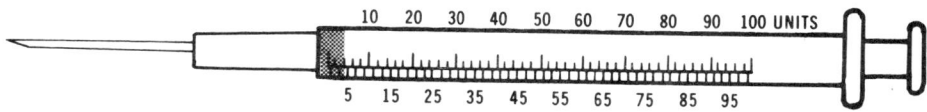

Chapter 12 Summary

Define each item in your own words, then compare your definitions with the text.

Key Words

insulin   (p. 285)
pancreas   (p. 285)
diabetes mellitus   (p. 285)
Regular   (p. 285)
Semilente   (p. 285)
NPH (Isophane)   (p. 285)
Lente   (p. 285)
Protamine Zinc (p. 285)

Ultralente (p. 285)
heparin   (p. 295)
anticoagulant   (p. 295)
clot   (p. 295)
thrombus   (p. 295)
embolus   (p. 295)
thrombophlebitis   (p. 295)
pulmonary embolism (p. 295)

Chapter 12 Review Problems

Solve the following word problems (answers to all the problems are at the back of the book):

1. Helen Drake, a mature-onset diabetic (started at age 42), has been giving herself insulin injections for the past three years. The dose she injects is 42 units of NPH U-100 insulin. Which of the following syringes should she use? Shade in the correct dosage on that syringe.

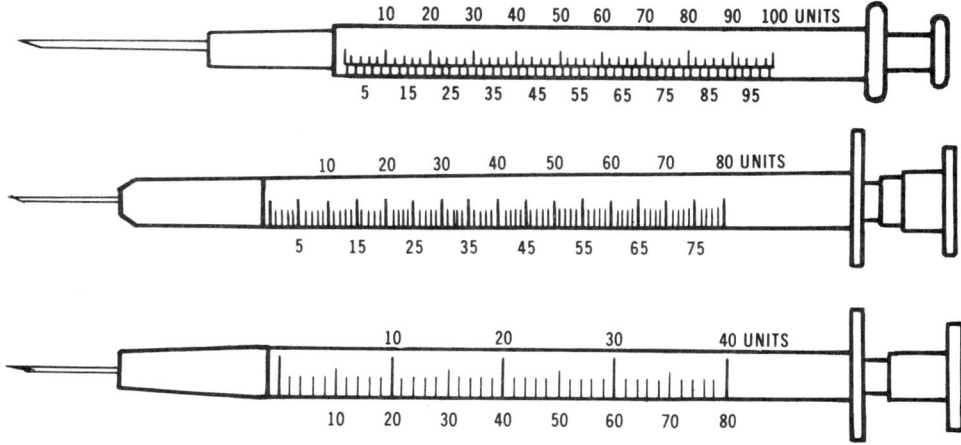

2. A newly diagnosed diabetic is to be given 22 units of U-40 Regular insulin. Which of the following syringes should he use? Shade in the correct dosage on that syringe.

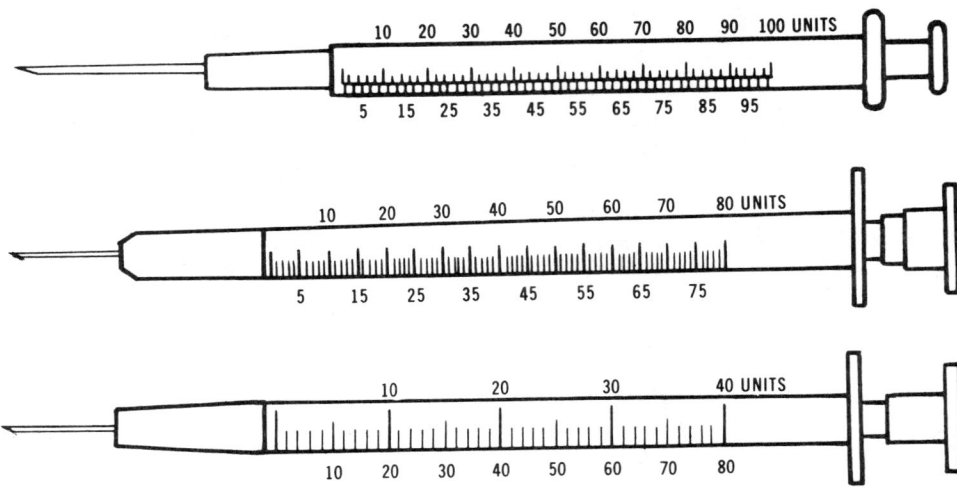

3. Before being discharged from the hospital, Mr. DePree, a diabetic, is asked to demonstrate his injection technique for the nurse. He is to give himself 70 units of Semilente U-80 insulin. The nurse hands him the syringes shown in the following diagram. He selected the correct syringe and gave himself the correct dose. Which syringe did he choose? Shade in the correct dose.

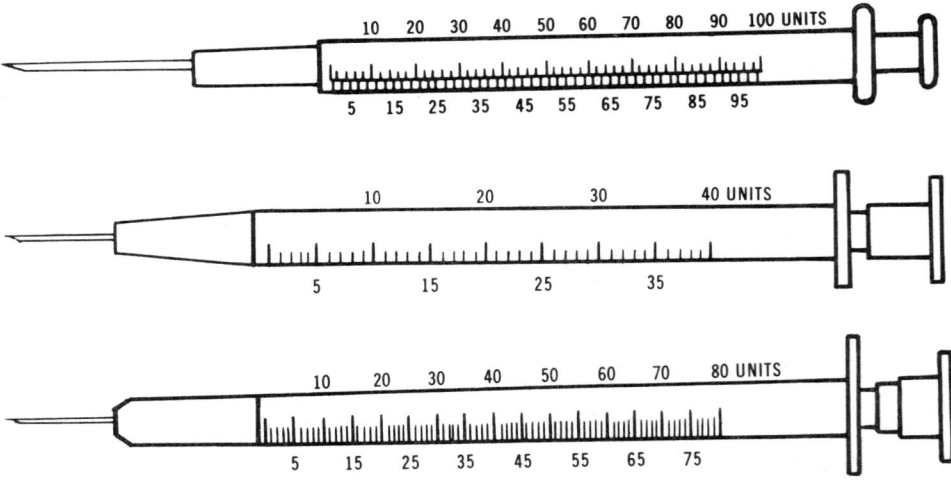

4. A patient requires 10 units of insulin from a bottle of U-80 insulin. The nurse has only a U-40 syringe. How many units of U-80 insulin should be given with the U-40 syringe? Shade in the correct dose.

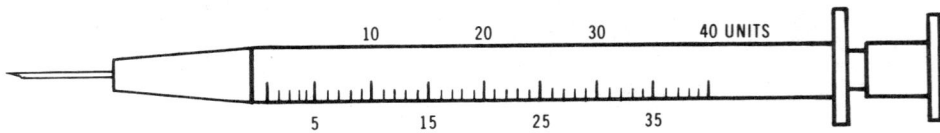

Chapter 12 Review Problems   303

5. A patient requires 4 units of insulin from a bottle of U-100 insulin. The nurse has only a U-40 and a U-80 combination syringe. How many units of U-100 insulin should be given with the U-40 and U-80 combination syringe? (To solve this problem, first use the U-40 scale, then the U-80 scale.) Shade in the correct dose.

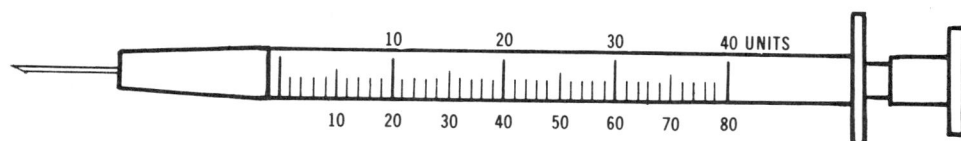

6. A diabetic is to receive 45 units of U-80 insulin. The nurse has only a tuberculin 1-mℓ (cc) syringe and a 3-mℓ (cc) syringe. Both are graduated in milliliters (cc) and minims (♏). How many minims of insulin will contain 45 units? Shade in the correct dose on the tuberculin syringe and the 3-mℓ (cc) syringe.

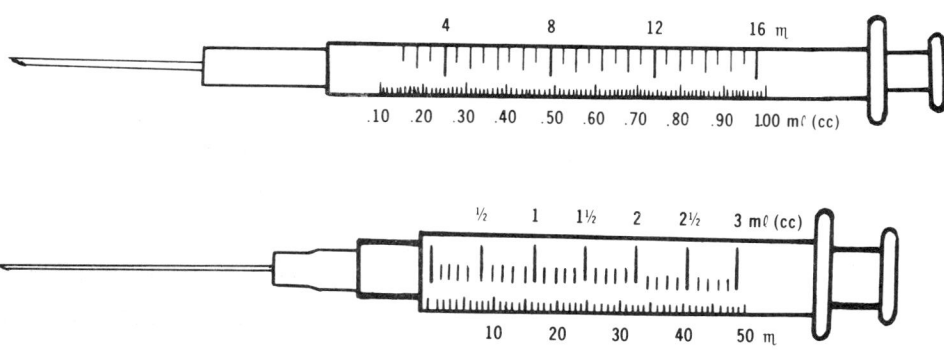

7. A diabetic is to receive 40 units of U-100 insulin. The nurse has only a tuberculin 1-mℓ (cc) syringe and a 3-mℓ (cc) syringe. How many milliliters of insulin will contain 40 units? Shade in the correct dose on the tuberculin syringe and the 3-mℓ (cc) syringe.

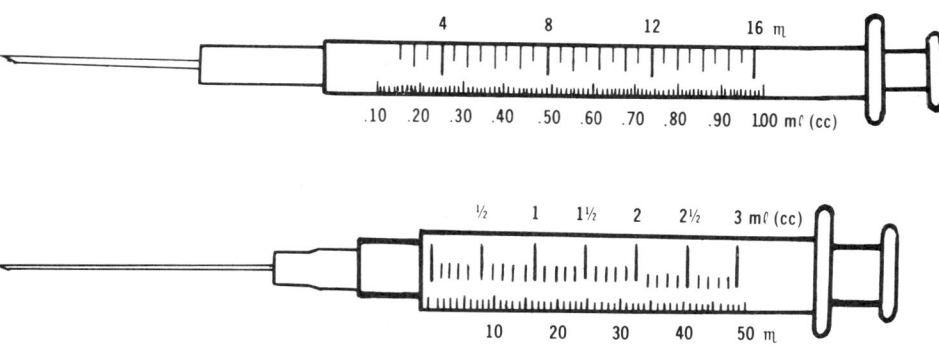

8. A diabetic is to receive 90 units of U-100 insulin. The nurse has only a tuberculin 1-mℓ (cc) syringe and a 3-mℓ (cc) syringe. How many milliliters

of insulin will contain 90 units? Shade in the correct dose on the tuberculin syringe and the 3-mℓ (cc) syringe.

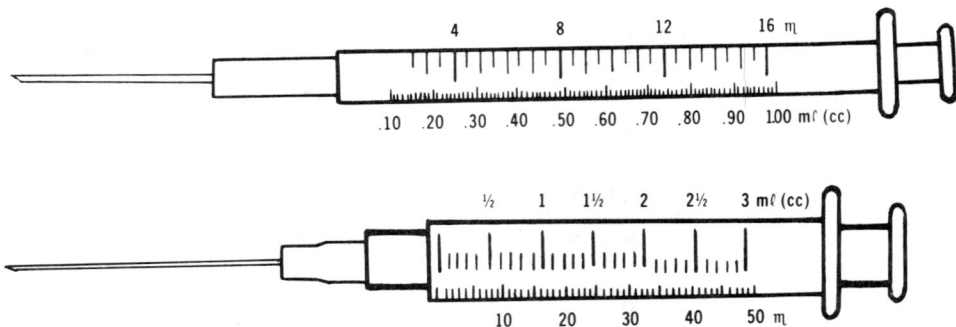

9. Kevin Hale, a post-operative heart surgery patient, is receiving 7500 U SQ heparin q. 6 h. The pharmacy has provided heparin 20,000 U/mℓ. How many milliliters of the concentration provided will the nurse give? Shade in the correct dose on that syringe. _____

10. The heparin concentration available is 1000 U/mℓ. How many units are contained in the following syringe? _____

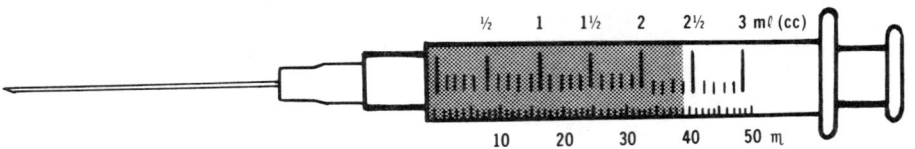

11. The heparin concentration available is 5000 U/mℓ. How many units are contained in the following syringe? _____

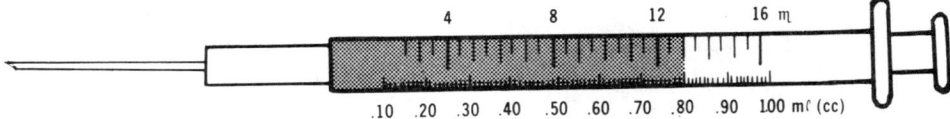

12. The heparin concentration available is 22,000 U/1.1 mℓ. How many units are contained in the following syringe? _____

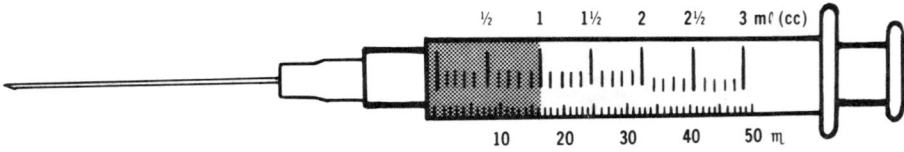

13. The heparin concentration available is 160,000 U/4 mℓ. How many units are contained in the following syringe? _____

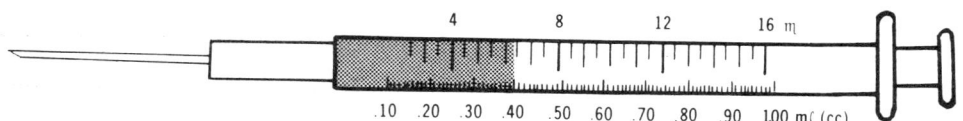

14. The heparin concentration available is 15,000 U/2 mℓ. How many units are contained in the following syringe? _____

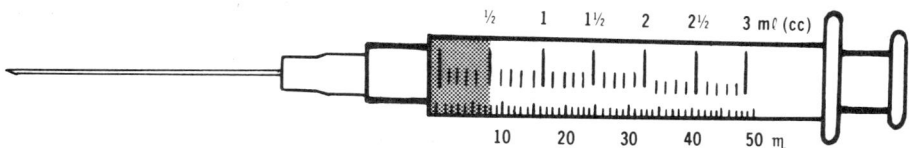

Five
Applications

I
Pediatric and Geriatric Dosages

OBJECTIVES

After studying this application, you should be able to:

1. Solve pediatric-dosage problems involving Fried's Rule, Young's Rule, Clark's Rule, and/or the West Nomogram and Surface Area Formula.
2. Understand how to relate the concept of rate of flow of drops per minute to a rate of microdrops per minute.
3. Understand that, with age, the body undergoes changes and that nurses must be familiar with the unique problems of the elderly.
4. Understand the concept of dose adjustment in treating kidney disease, as an example of a geriatric dosage consideration.
5. Use the Creatinine Clearance Nomogram.

How much medicine should a patient be given? Ideally, it should be the least amount of drug that will produce the desired therapeutic effect. Unfortunately, it is often hard to determine the ideal dose for healthy adult patients, and it can be even harder to do this for children or the elderly. However, nurses have the responsibility to provide quality care, and, therefore, they must have a good idea of what an ideal dose of medicine is for each patient. Nurses may not be required to write original orders for drugs but they *are* required to contact the physician if an order, in their opinion, "doesn't seem just right."

In this application, you'll study the traditional and new methods for calculating pediatric dosages along with some rules for determining dosages for the elderly. In addition to introducing you to some new concepts, this section of the text will provide you with a convenient reference for future use.

A. Pediatric Dosages

It seems that, every time a nursing textbook includes a pediatric-dosage section, the author emphasizes Fried's Rule, Young's Rule, and Clark's Rule. From a nursing point of view, however, these rules alone don't supply enough information to determine a child's dose. Although we'll discuss these rules briefly in order to familiarize you with them, we'll also cover the roles of the West Nomogram, the Surface Area Formula, and microdrop flow in the determination of pediatric dosages.

> **Definition:** *Pediatrics* refers to the branch of medicine that specializes in the care of infants and children.

1. Fried's Rule

Fried's Rule uses the child's age to determine the child's dose. It applies only to children from the time of birth to age 2, and, when you use it, you must give the child's age in months.

> **Formula: Fried's Rule**
> $$\text{Infant's dose} = \frac{\text{Age in months} \times \text{Adult dose}}{150}$$

Example 1: Use the formula for Fried's Rule to solve the following problem:

Child's age: 9 months
Adult dose: Paregoric 5 ml
Infant's dose: ?

Solution: $$\text{Infant's dose} = \frac{9 \text{ (months)} \times 5 \text{ ml}}{150} = \frac{9 \times 5}{150}$$
$$= \frac{45}{150} = 0.3 \text{ ml}$$

310   Application I Pediatric and Geriatric Dosages

Example 2:   Use the formula for Fried's Rule to solve the following problem:

Child's age: 1 year 3 months
Adult dose: Dilantin® 100 mg
Infant's dose: ?

Solution: 1 year 3 months = 15 months

$$\text{Infant's dose} = \frac{15 \text{ (months)} \times 100 \text{ mg}}{150} = \frac{15 \times 100}{150}$$

$$= \frac{1500}{150} = 10 \text{ mg}$$

2. Young's Rule

Young's Rule is also based on the child's age. It applies to children between approximately the ages of 1 and 12; give the child's age in years.

Formula: Young's Rule

$$\text{Child's dose} = \frac{\text{Age of child (years)} \times \text{Adult dose}}{\text{Age of child (years)} + 12}$$

Example 3:   Use the formula for Young's Rule to solve the following problem:

Child's age: 7 years
Adult dose: Colace® 100 mg
Child's dose: ?

Solution:
$$\text{Child's dose} = \frac{7 \text{ (years)} \times 100 \text{ mg}}{7 \text{ (years)} + 12} = \frac{7 \times 100}{7 + 12}$$

$$= \frac{700}{19} = 36.8 \text{ mg}$$

Example 4:   Use the formula for Young's Rule to solve the following problem:

Child's age: 22 months
Adult dose: Penicillin VK 250 mg
Child's dose: ?

Solution: 22 months = 1 5/6 years or 1.83 years

$$\frac{1.83 \text{ (years)} \times 250 \text{ mg}}{1.83 \text{ (years)} + 12} = \frac{1.83 \times 250}{1.83 + 12}$$

$$= \frac{457.5}{13.83} = 33.1 \text{ mg}$$

3. Clark's Rule

Clark's Rule is based on the child's weight in pounds. It is used for children who are 2 years of age or older.

Formula: Clark's Rule

$$\text{Child's dose} = \frac{\text{Weight in pounds} \times \text{Adult dose}}{150}$$

A. Pediatric Dosages 311

Example 5: Use the formula for Clark's Rule to solve the following problem:

Child's weight: 50 lb
Adult dose: Mycostatin® Oral Susp. 6 mℓ
Child's dose: ?

Solution:
$$\frac{50 \text{ (lb)} \times 6 \text{ mℓ}}{150} = \frac{50 \times 6}{150} = \frac{300}{150} = 2 \text{ mℓ}$$

Example 6: Use the formula for Clark's Rule to solve the following problem:

Child's weight: 33 lb
Adult dose: Nembutal® 100 mg
Child's dose: ?

Solution:
$$\frac{33 \text{ (lb)} \times 100 \text{ mg}}{150} = \frac{33 \times 100}{150} = \frac{3300}{150} = 22 \text{ mg}$$

4. Surface Area Formula—West Nomogram

As you read earlier, the preceding three rules are not adequate ways of determining pediatric dosages. The main problem stems from the fact that not all children of the same age are the same size. This problem can be avoided by expressing the child's size in terms of body-surface area. To do this, use the West Nomogram (see Figure I-1).

> **Definition:** The *West Nomogram* is a chart that relates the height and weight of a child in terms of body-surface area.

> **Rule: Determining Body-Surface Area—Children**
> *Step 1:* Place a straight edge (for example, a ruler, a tongue depressor, or the like) at the level of the child's height in the left-hand column. Line the straight edge up with the child's weight in the right-hand column.
> *Step 2:* Read the surface-area (SA) measurement at the point where the straight edge intersects the SA column. This will give you the estimated surface area in m^2 (square meters).

Example 7: Use the West Nomogram to determine the child's body-surface area in the following problem:

Child's height: 34 in.
Child's weight: 28 lb
Body-surface area: ?

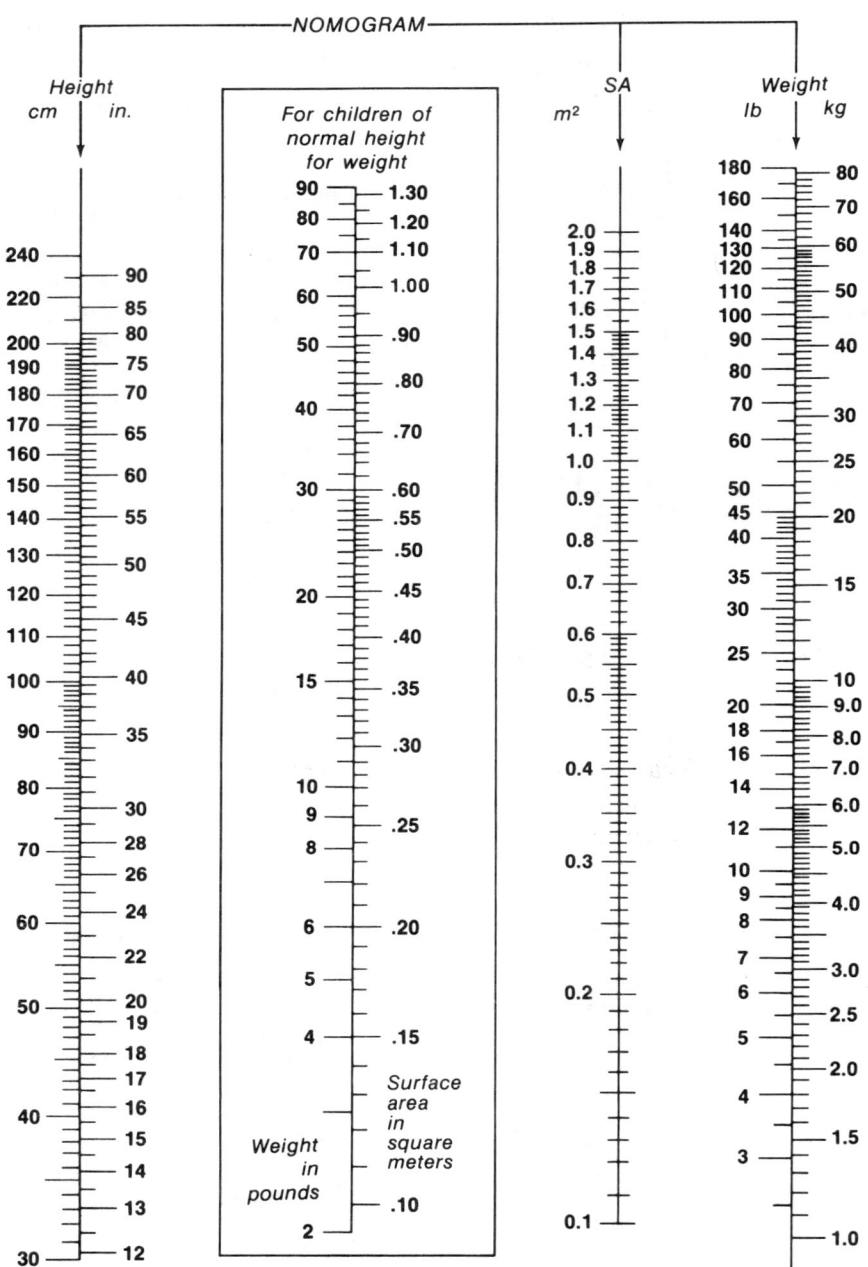

Figure I-1. West Nomogram for estimating body-surface area. A straight edge placed from height in the left column to weight in the right column will intersect the body-surface column (SA) at the number indicating the child's body-surface area. The enclosed column will estimate the body-surface of children of average build using weight alone. (Figure of West Nomogram modified from *Textbook of Pediatrics* (8th ed.) by W. E. Nelson (Ed.). Copyright 1964 by W. B. Saunders Company, Philadelphia. This and all other West Nomograms reprinted by permission of the publisher.)

Solution:
Step 1: Place a straight edge at the level of the child's height in the left-hand column. Line the straight edge up with the child's weight in the right-hand column. (This text uses a line to represent the straight edge.)

A. Pediatric Dosages   313

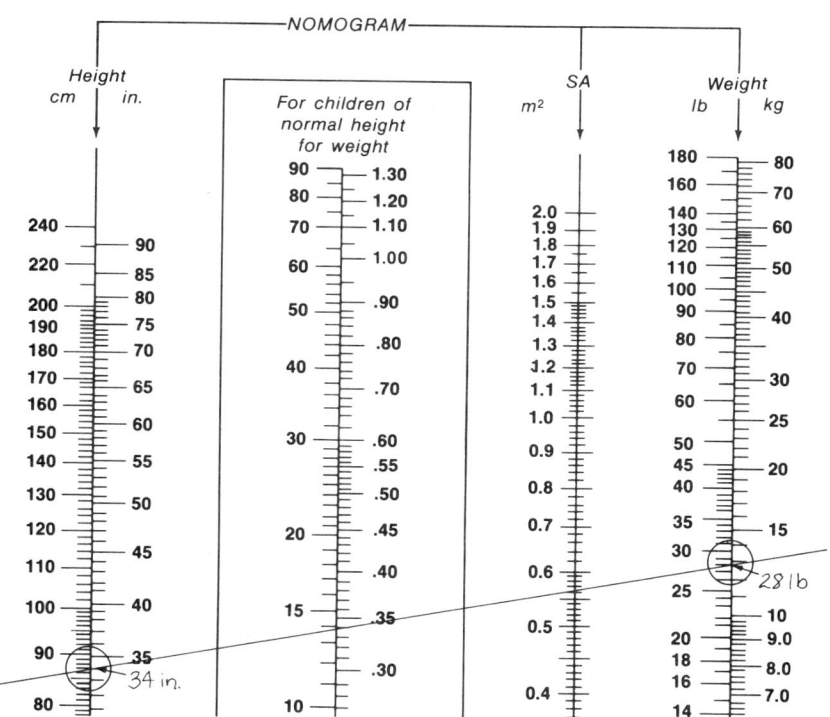

Step 2:   Read the surface-area (SA) measurement at the point where the straight edge intersects the SA column. This will give you the estimated surface area in m² (square meters).

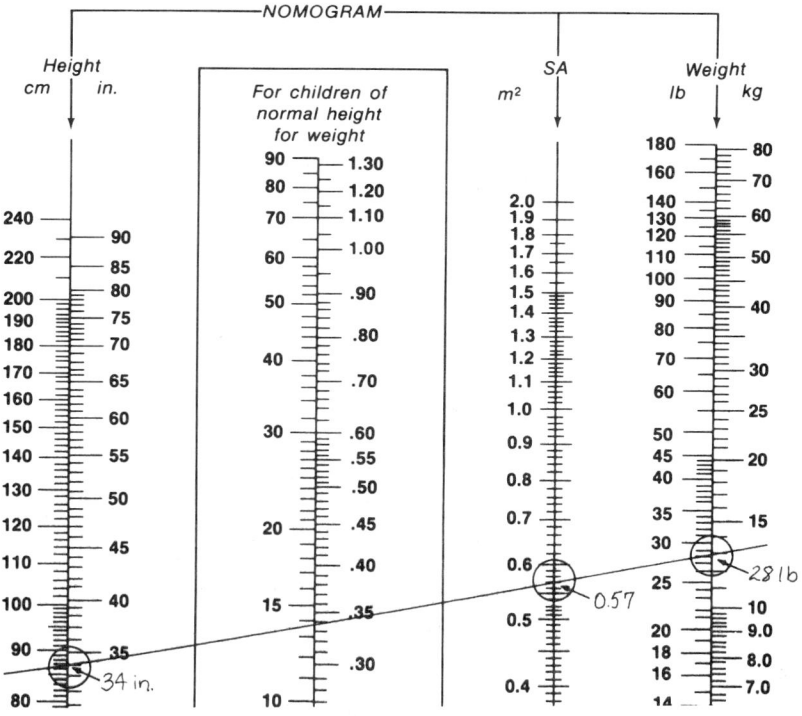

The West Nomogram shows that this child's body-surface area is 0.57 m².

Application I Pediatric and Geriatric Dosages

Example 8: Use the West Nomogram to determine the child's body-surface area in the following problem:

Child's height: 80 cm
Child's weight: 10 kg
Body-surface area: ?

Solution:

Step 1: Place the straight edge at the child's height (note that this time the height is in cm and the weight is in kg). Line up the straight edge on the left-hand side with the child's weight on the right-hand side.

Step 2: Read the surface-area (SA) measurement.

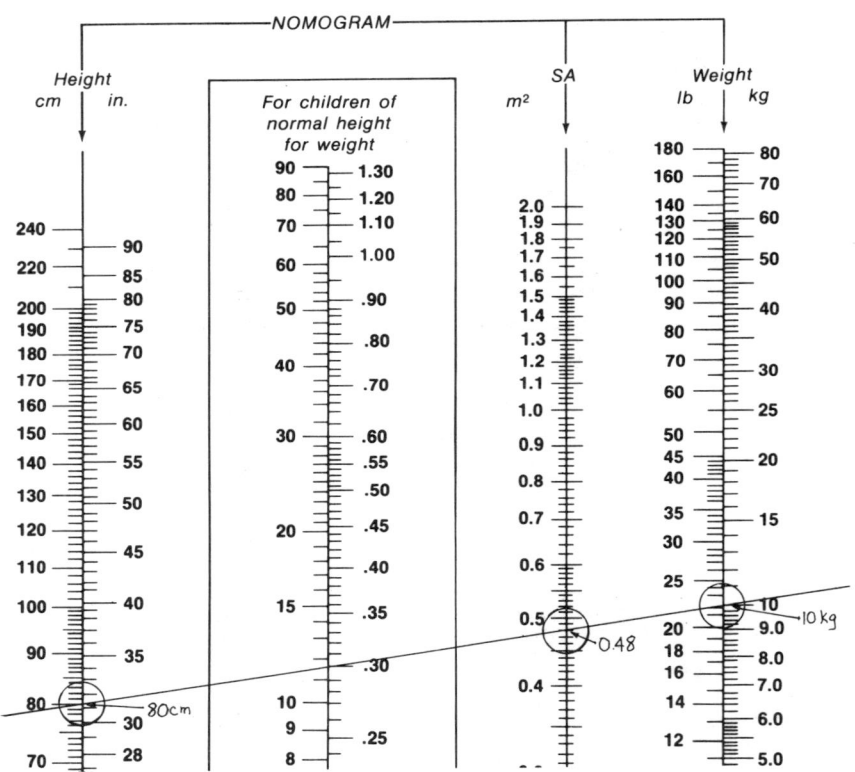

The child's body-surface area is 0.48 m².

The Surface Area Formula is used along with the West Nomogram to determine pediatric dosages. This is by far the most accurate method. The West Nomogram is used to find the child's body-surface area, which is then used in the Surface Area Formula. The Surface Area Formula divides the child's surface area by the average adult surface area (1.7 m²) and then multiplies the result by the adult drug dose.

A. Pediatric Dosages   315

> **Formula: Surface Area Formula**
> $$\text{Child's dose} = \frac{\text{Surface area of child (m}^2)}{1.7 \text{ m}^2} \times \text{Adult dose}$$

Example 9: Use the West Nomogram and the Surface Area Formula to solve the following problem:

Child's height: 70 in.
Child's weight: 120 lb
Adult dose: Theophylline 200 mg
Child's dose: ?

Solution:
Step 1: Height: 70 in.
Weight: 120 lb

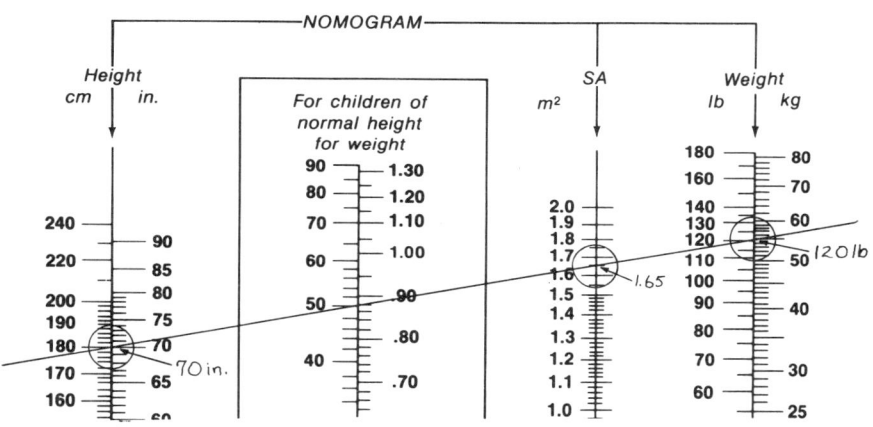

Step 2: On the West Nomogram, we see that SA ≈ 1.65 m². Using the Surface Area Formula, the child's dose is calculated to be 194.1 mg.

$$\text{Child's dose} = \frac{\text{Surface area of child (m}^2)}{1.7 \text{ m}^2} \times \text{Adult dose}$$
$$= \frac{1.65 \text{ m}^2}{1.7 \text{ m}^2} \times 200 \text{ mg} = 194.1 \text{ mg}$$

Example 10: Use the West Nomogram and the Surface Area Formula to solve the following problem:

Child's height: 106 cm
Child's weight: 9.8 kg
Adult dose: Ampicillin 250 mg
Child's dose: ?

Solution:
Step 1: Height: 106 cm
Weight: 9.8 kg

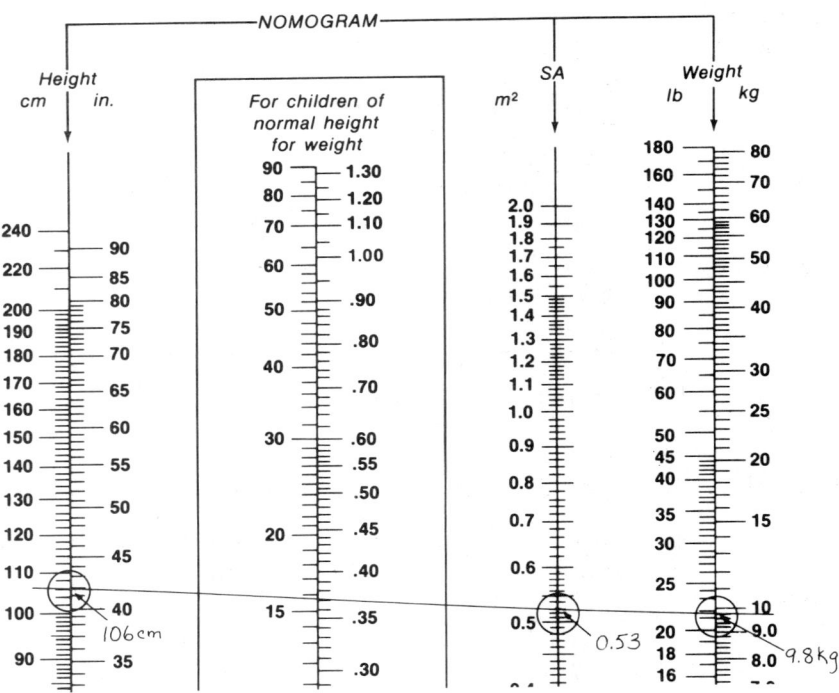

Step 2: On the West Nomogram, we see that SA ≈ 0.53 m². Use the Surface Area Formula to calculate the child's dose of 77.9 mg.

$$\text{Child's dose} = \frac{\text{Surface area of child (m}^2)}{1.7 \text{ m}^2} \times \text{Adult dose}$$

$$= \frac{0.53 \text{ m}^2}{1.7 \text{ m}^2} \times 250 \text{ mg} = 77.9 \text{ mg}$$

5. Microdrop—Pediatric I.V. Fluids

Dehydration in infants and children can have serious consequences. When a child is ill, it is essential to restore and maintain his or her fluid and electrolyte balances without overloading the child with fluid. In the case of adults, the rate of flow of I.V. fluid infusions is given in terms of drops per minute. For infants and many children, however, the rate of flow is measured in microdrops per minute. Microdrops are much smaller than drops.

> **Definition:** 60 *microdrops* equal 1 milliliter of fluid.

In Section 11.1, General Concepts and Drops Per Minute, you learned how to solve rate-of-flow problems. Use that knowledge to solve microdrops-per-minute problems, as shown in the following examples.

A. Pediatric Dosages

Example 11: Find the rate of flow for the following pediatric-fluid order:

Ordered: 100 mℓ LR (lactated Ringer's) over 60 min
Administration set: 60 microdrops/mℓ

Solution:

Step 1A:
   a. 100 mℓ
   b. 60 min
   c. 60 microdrops/mℓ

Step 2A: $\dfrac{60 \text{ microdrops}}{1 \text{ mℓ}}$

Step 3A: $\dfrac{100 \text{ mℓ}}{60 \text{ min}}$

Step 4A: $\dfrac{\cancel{60}^{1} \text{ microdrops}}{1 \cancel{\text{ mℓ}}} \times \dfrac{100 \cancel{\text{ mℓ}}}{\underset{1}{\cancel{60}} \text{ min}} = 100 \text{ microdrops/min}$

Step 5A: 100 microdrops/min

Example 12: Find the rate of flow for the following pediatric-fluid order:

Ordered: 50 mℓ D5LR (dextrose 5% in lactated Ringer's) I.V. over eight hours
Administration set: 60 microdrops/mℓ

Solution:

Step 1A:
   a. 50 mℓ
   b. $8 \cancel{\text{ hr}} \times \dfrac{60 \text{ min}}{1 \cancel{\text{ hr}}} = 480 \text{ min}$
   c. 60 microdrops/mℓ

Step 2A: $\dfrac{60 \text{ microdrops}}{1 \text{ mℓ}}$

Step 3A: $\dfrac{50 \text{ mℓ}}{480 \text{ min}}$

Step 4A: $\dfrac{\cancel{60}^{1} \text{ microdrops}}{1 \cancel{\text{ mℓ}}} \times \dfrac{50 \cancel{\text{ mℓ}}}{\underset{8}{\cancel{480}} \text{ min}} = 6.25 \text{ microdrops/min}$

Step 5A: Round off answer to nearest whole number: 6.25 = 6 microdrops/min

---

Section I-A Readiness Review

Solve each of the following problems using the rule indicated:

1. *Clark's Rule:* (see Examples 5 & 6)
   Child's weight: 80 lb
   Adult dose: Atropine sulfate 0.4 mg
   Child's dose: ?

2. *Young's Rule:* (see Examples 3 & 4)
   Child's age: 12 years
   Adult dose: Castor oil 15 mℓ
   Child's dose: ?

318   Application I   Pediatric and Geriatric Dosages

3. *Fried's Rule:* (see Examples 1 & 2)
   Child's age: 18 months
   Adult dose: Aspirin 5 gr
   Infant's dose: ?

Use the West Nomogram to find the child's body-surface area in the following problem (see Examples 7 & 8):

4. *Child's height:* 53 in.
   Child's weight: 70 lb
   Body-surface area: ?

In the next example, use the West Nomogram and the Surface Area Formula to determine the child's dose (see Examples 9 & 10):

5. *Child's height:* 84 cm
   Child's weight: 6.6 kg
   Adult dose: Erythromycin 250 mg
   Child's dose: ?

Find the rate of flow for the following pediatric-fluid order (see Examples 11 & 12):

6. *Ordered:* 50 mℓ D5W I.V. over six hours
   Administration set: 60 microdrops per milliliter

Section readiness review answers (not given in order):

| 5. 57.4 mg | 6. 8 microdrops/min | 3. 0.6 gr |
| 1. 0.2 mg | 2. 7.5 mℓ | 4. 1.08 m^2 |

Section I-A   Review Problems

Use Fried's Rule to solve the following problems (answers to the odd-numbered problems are at the back of the book):

1. *Child's age:* 8 months
   Adult dose: Amoxicillin 250 mg
   Infant's dose: ?

2. *Child's age:* 1 year
   Adult dose: Benadryl® 25 mg
   Infant's dose: ?

3. *Child's age:* 5 months
   Adult dose: Ferrous sulfate 300 mg
   Infant's dose: ?

4. *Child's age:* 7 months
   Adult dose: Milk of magnesia 30 mℓ
   Infant's dose: ?

Use Young's Rule to solve the following problems:

5. *Child's age:* 4 years
   Adult dose: Ampicillin 250 mg
   Child's dose: ?

6. *Child's age:* 5 years
   Adult dose: Robitussin® 10 mℓ
   Child's dose: ?

7. *Child's age:* 18 months
   Adult dose: Sudafed® 60 mg
   Child's dose: ?

8. *Child's age:* 3 years
   Adult dose: Gantrisin® 1 g
   Child's dose: ?

Use Clark's Rule to solve the following problems:

9. *Child's weight:* 20 lb
   Adult dose: Mysoline® 250 mg
   Child's dose: ?

10. *Child's weight:* 25 lb
    Adult dose: Novafed® syrup 10 mℓ
    Child's dose: ?

11. *Child's weight:* 31 lb
    Adult dose: Somophyllin® oral liquid 15 mℓ
    Child's dose: ?

12. *Child's weight:* 40 lb
    Adult dose: Polaramine® syrup 5 mℓ
    Child's dose: ?

Use the West Nomogram to determine the body-surface area in each of the following examples:

13. *Child's height:* 28 in.
    Child's weight: 20 lb
    Body-surface area: ?

14. *Child's height:* 112 cm
    Child's weight: 28 kg
    Body-surface area: ?

15. *Child's height:* 95 cm
    Child's weight: 14 kg
    Body-surface area: ?

16. *Child's height:* 43 in.
    Child's weight: 40 lb
    Body-surface area: ?

17. *Child's height:* 32 in.
    Child's weight: 26 lb
    Body-surface area: ?

18. *Child's height:* 104 cm
    Child's weight: 19 kg
    Body-surface area: ?

Use the West Nomogram and the Surface Area Formula to solve the following problems:

19. *Child's height:* 21 in.
    Child's weight: 8 lb
    Adult dose: Atarax® 25 mg
    Child's dose: ?

20. *Child's height:* 50 in.
    Child's weight: 50 lb
    Adult dose: Neomycin 500 mg
    Child's dose: ?

21. *Child's height:* 17.5 in.
    Child's weight: 2.2 kg
    Adult dose: Tempra® 325 mg
    Child's dose: ?

22. *Child's height:* 39 in.
    Child's weight: 10 kg
    Adult dose: Amoxil® 500 mg
    Child's dose: ?

23. *Child's height:* 100 cm
    Child's weight: 10 kg
    Adult dose: Chloral hydrate 500 mg
    Child's dose: ?

24. *Child's height:* 129 cm
    Child's weight: 57 kg
    Adult dose: Dicloxacillin 250 mg
    Child's dose: ?

25. *Child's height:* 56 cm
    Child's weight: 3 kg
    Adult dose: Colace® 100 mg
    Child's dose: ?

Find the rate of flow for the following pediatric-fluid orders:

26. *Ordered:* 100 mℓ D5W I.V. over 1.8 hr
    Administration set: 60 microdrops/mℓ

27. *Ordered:* 100 mℓ D5W I.V. over 6 hr
    Administration set: 60 microdrops/mℓ

28. *Ordered:* 100 mℓ D5NS (dextrose 5% in normal saline) I.V. over seven hr
    Administration set: 60 microdrops/mℓ

29. *Ordered:* 50 mℓ D5LR (dextrose 5% in lactated Ringer's) I.V. over 0.8 hr
    Administration set: 60 microdrops/mℓ

30. *Ordered:* 60 mℓ NS I.V. over 12 hr
    Administration set: 60 microdrops/mℓ

B. Geriatric Dosages

It is estimated that, by the year 2000, sixty million Americans will be over the age of 65. Unfortunately, the unique physical problems of this segment of the population have often been ignored when it comes to dosage calculation. However, times are changing, and nurses must learn about these problems and keep up-to-date on geriatric medicine in order to provide quality care for the elderly.

> **Definition:** *Geriatrics* refers to the branch of medicine that specializes in the care of the elderly. North American society has established the age of 65 as the cut-off point between middle and old age.

As the body grows older, changes occur in specific body organs. For example:

Heart — The heart may not pump as efficiently as it did when the body was younger. This can lead to a decrease in blood flow, which in turn reduces the blood supply to other organs.

Kidney — The ability of the kidneys to eliminate drugs from the body may be impaired.

Liver — There may be a decrease in the ability of the liver to metabolize drugs—that is, to change drugs into forms that the body can get rid of.

Gastrointestinal (G.I.) Tract — The ability of the stomach and intestines to absorb drugs may decline because of changes in G.I. motility.

Central Nervous System — The brain's blood supply may decrease, thus possibly leading to an increased risk of stroke.

As a nurse, you will learn a great deal about the body's organ systems. Just keep in mind that, as age increases, certain organ changes may occur and that these changes will influence dosage determinations.

There has been a recent increase in the amount of information available on how to calculate specific drug dosages when the patient has renal (kidney) impairment. Of course, this information applies not just to elderly patients but also to all other adult patients with decreased kidney function. The method for determining geriatric dosages based on renal function that we are about to study concerns creatinine clearance.

> **Definition:** *Creatinine* is a product of normal muscle breakdown.

Creatinine is normally carried in the bloodstream. The concentration of creatinine in the blood is commonly referred to as the *serum creatinine level*. The kidneys filter a measured, constant amount of this substance out of the blood and into the urine over a certain amount of time. This amount is called *creatinine clearance*. Thus, if the normal value for creatinine clearance is approximately 100 mℓ/min, this means that the kidneys are able to "clear" 100 mℓ of blood plasma of a constant amount of creatinine per minute and excrete it into the urine. If, then, the creatinine clearance is only 50 mℓ per minute, the kidneys may be diseased, or, perhaps, as in the case of an elderly patient, they simply may not be working as well as they used to. (Measuring the serum creatinine level is also helpful in evaluating kidney function.) The rate of creatinine clearance affects certain drug dosages; if it is not within normal limits, then the dose may have to be adjusted.

1. Creatinine Clearance Nomogram

There are several ways to estimate creatinine clearance. For our purposes we will use a nomogram that relates serum creatinine, age, and weight to give us the amount of creatinine clearance (see Figure I-2).

Rule: Determining Creatinine Clearance (Ccr) by Nomogram

Step 1: Use a straight edge to line up the patient's weight to his or her age* (be sure the straight edge doesn't move off line R). Mark where the straight edge crosses line R.

Step 2: Now use the right side of the straight edge to line up the given serum creatinine value in the right-hand column with the mark on the R line. The left end of the straight edge will automatically line up with the creatinine clearance value. Creatinine clearance (Ccr) will be in milliliters of blood cleared per minute.

*The age in years is listed according to sex: male = ♂; female = ♀.

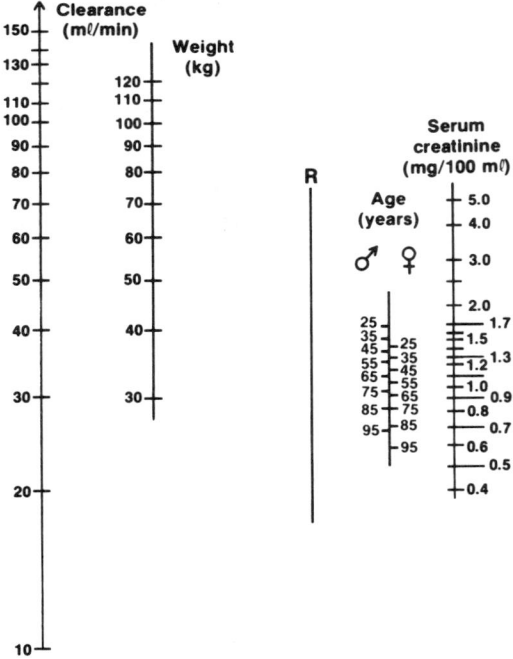

Figure I-2. Creatinine Clearance Nomogram. (Figure of Creatinine Clearance Nomogram courtesy of *Lancet,* May 29, 1971, p. 1134, and K. Siersbaek-Nelson, Medical Department B, Frederiksberg Hospital, Copenhagen, Denmark. This and all other Creatinine Clearance Nomograms reprinted by permission.)

B. Geriatric Dosages   323

Example 1:   Determine the creatinine clearance using the Creatinine Clearance Nomogram and the following information:

Patient's age/sex: 80 yr/male
Patient's weight: 70 kg
Serum creatinine: 2.0 mg%
Creatinine clearance: ?

Solution:
Step 1:   Using a straight edge, join the patient's weight to age (be sure the straight edge doesn't move off line R). Mark where the straight edge crosses line R.

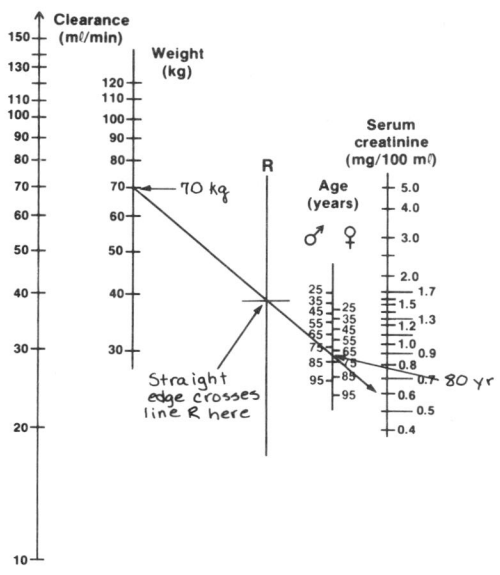

Step 2:   Move the right end of the straight edge to the given serum creatinine value and line it up with the mark on the R line. The left end of the straight edge will line up with the creatinine clearance value. Creatinine clearance is stated as milliliters of blood cleared per minute.

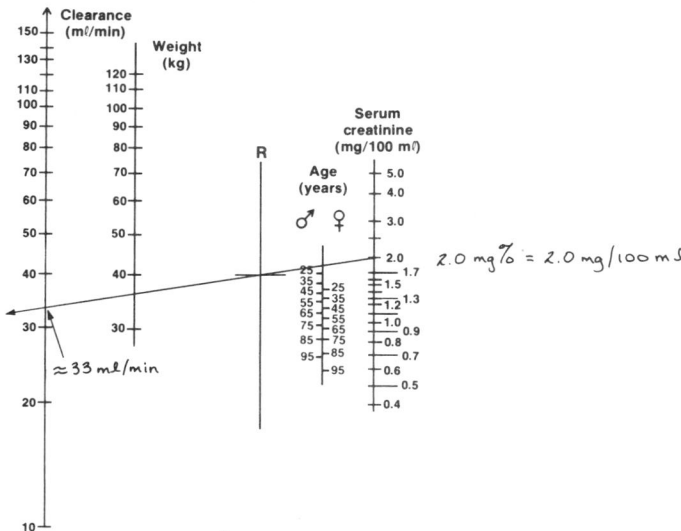

Serum creatinine: 2.0 mg%. According to the Nomogram, the creatinine clearance is approximately 33 ml/min.

324 Application I Pediatric and Geriatric Dosages

Example 2:   Determine the creatinine clearance using the Creatinine Clearance Nomogram and the following information:

Patient's age/sex: 75 yr/female
Patient's weight: 55 kg
Serum creatinine: 1.2 mg%
Creatinine clearance: ?

Solution:
Step 1:   Use a straight edge to line the patient's weight up with his or her age (be sure the straight edge doesn't move off line R). Mark where the straight edge crosses line R.

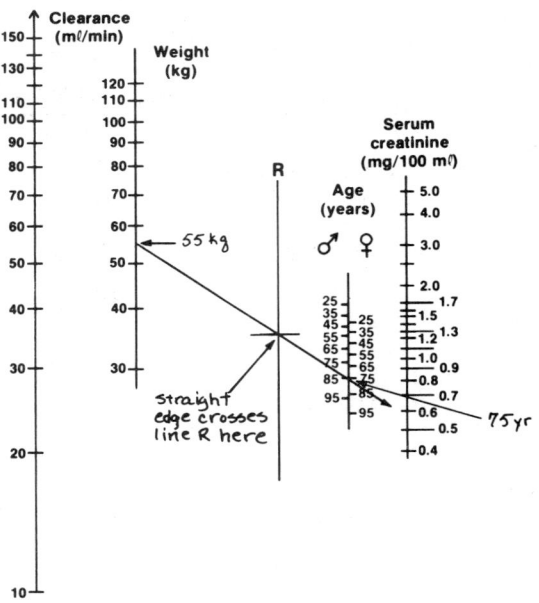

Step 2:   Move the right end of the straight edge to the given serum creatinine value and line it up with the mark on the R line. The left end of the straight edge will line up with the creatinine clearance value.

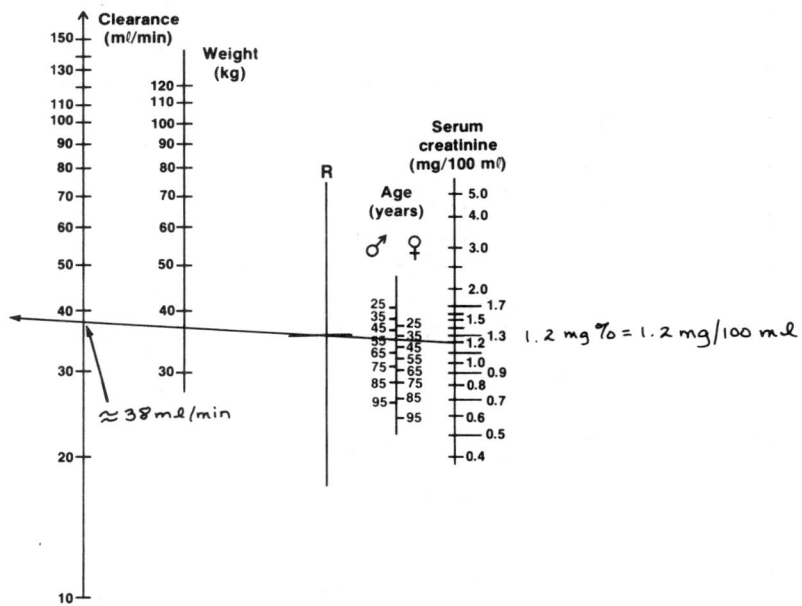

Serum creatinine: 1.2 mg%. According to the Nomogram, the creatinine clearance is approximately 38 mℓ/min.

Once you have determined the creatinine clearance, you may use it to determine the correct drug dosage. Many drug manufacturers now include dosage schedules for patients with renal impairment. Let's study several different schedule tables for specific antibiotics and then calculate doses.

Example 3: Using the following table for the drug Mandol® and the other information provided, determine the correct dosage(s) based on Ccr. (Hint: first determine the Ccr.)

Maintenance Dosage Guide for Patients with Renal Impairment[1]

Renal-function impairment	Ccr (mℓ/min)	Life-threatening infections (maximum dosage)	Less severe infections
Normal	>80	2 g q. 4 h.	1-2 g q. 6 h.
Mild	80-50	1.5 g q. 4 h. or 2 g q. 6 h.	0.75-1.5 g q. 6 h.
Moderate	50-25	1.5 g q. 6 h. or 2 g q. 8 h.	0.75-1.5 g q. 8 h.
Severe	25-10	1 g q. 6 h. or 1.25 g q. 8 h.	0.5-1 g q. 8 h.
Marked	10-2	0.67 g q. 8 h. or 1 g q. 12 h.	0.5-0.75 g q. 12 h.
None	<2	0.5 g q. 8 h. or 0.75 g q. 12 h.	0.25-0.5 g q. 12 h.

Patient's age/sex: 80 yr/male
Patient's weight: 70 kg
Serum creatinine: 2.0 mg%
Creatinine clearance (Ccr): ?
Type of infection: Less severe
Dosage: ?

Solution:
Step 1: Weight: 70 kg
Sex: male
Age: 80 yr

[1] This and all other Reduced Renal Function charts reprinted with permission from *Facts and Comparisons*, © 1980.

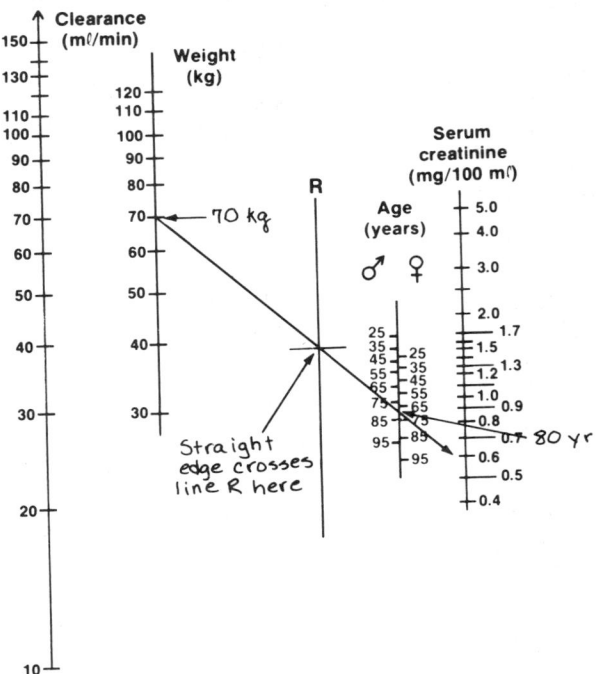

Step 2: Serum creatinine: 2.0 mg%. According to the Nomogram, the creatinine clearance is approximately 33 mℓ/min.

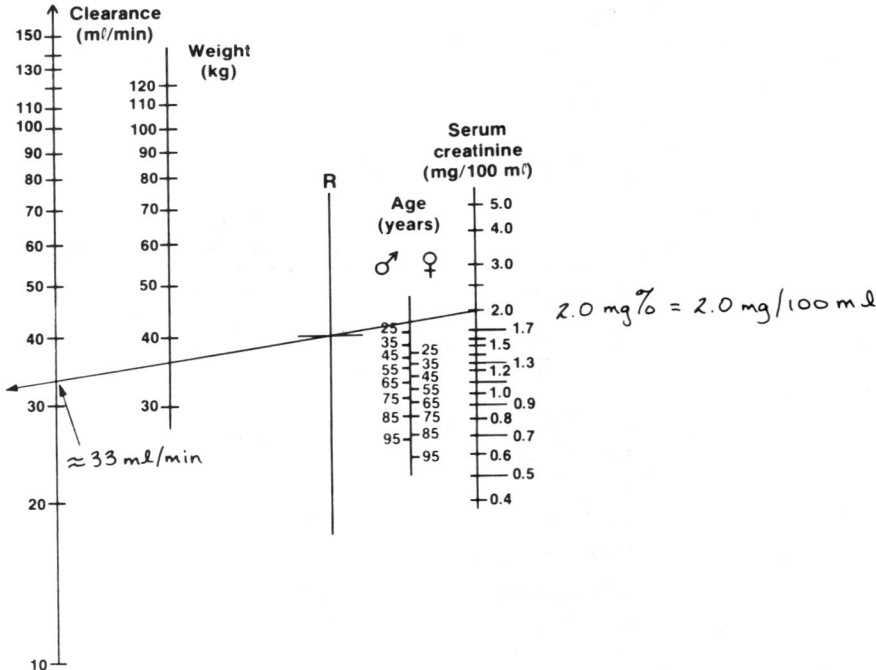

Step 3: The table indicates that 33 mℓ/min is considered moderate renal impairment. The infection is less severe; therefore, the dose of Mandol® is 0.75 to 1.5 g q. 8 h. (Compare this to the normal renal-function value: in the latter case, a greater amount of drug is given more frequently.)

B. Geriatric Dosages

Maintenance Dosage Guide for Patients with Renal Impairment

Renal-function impairment	Ccr (mℓ/min)	Life-threatening infections (maximum dosage)	Less severe infections
Normal	>80	2 g q. 4 h.	1–2 g q. 6 h.
Mild	80–50	1.5 g q. 4 h. or 2 g q. 6 h.	0.75–1.5 g q. 6 h.
Moderate	50–25 (33 mℓ/min)	1.5 g q. 6 h. or 2 g q. 8 h.	0.75–1.5 g q. 8 h.
Severe	25–10	1 g q. 6 h. or 1.25 g q. 8 h.	0.5–1 g q. 8 h.
Marked	10–2	0.67 g q. 8 h. or 1 g q. 12 h.	0.5–0.75 g q. 12 h.
None	<2	0.5 g q. 8 h. or 0.75 g q. 12 h.	0.25–0.5 g q. 12 h.

Example 4: Using the following table for the drug Ancef® and the other information provided, determine the correct dosage(s) based on Ccr. (Hint: first determine Ccr.)

Maintenance Dosage in Adults with Reduced Renal Function

Renal-function impairment	BUN (mg%)	Ccr (mℓ/min)	Dosage Mild to moderate infection (mg)	Dosage Moderate to severe infection (mg)	Dosage interval (hr)
Mild	20–34	70–40	250 to 500	500 to 1250	12
Moderate	35–49	40–20	125 to 250	250 to 600	12
Severe	50–75	20–5	75 to 150	150 to 400	24
Essentially no function	>75	<5	37.5 to 75	75 to 200	24

Patient's age/sex: 75 yr/female
Patient's weight: 55 kg
Serum creatinine: 1.2 mg%
Creatinine clearance (Ccr): ?
Type of infection: Severe
Dosage: ?

Solution:
Step 1: Weight: 55 kg
Sex: female
Age: 75 yr

328   Application I Pediatric and Geriatric Dosages

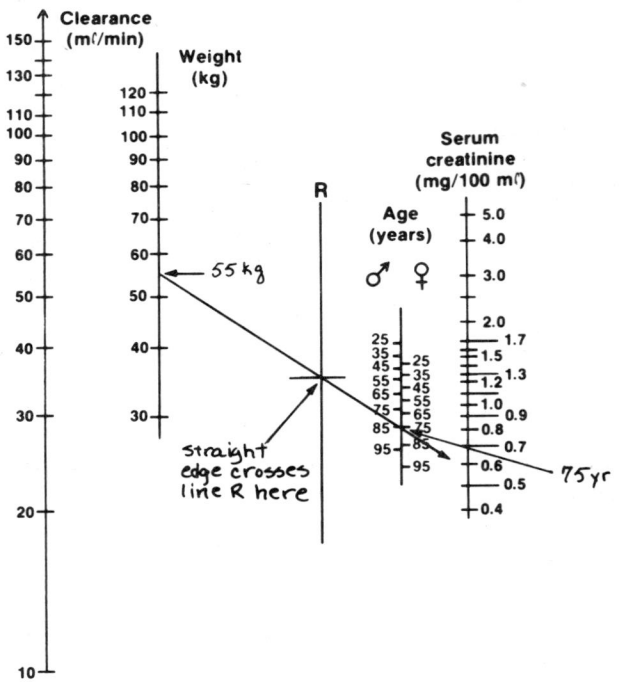

Step 2:   Serum creatinine: 1.2 mg%. According to the Nomogram, the creatinine clearance is approximately 38 mℓ/min.

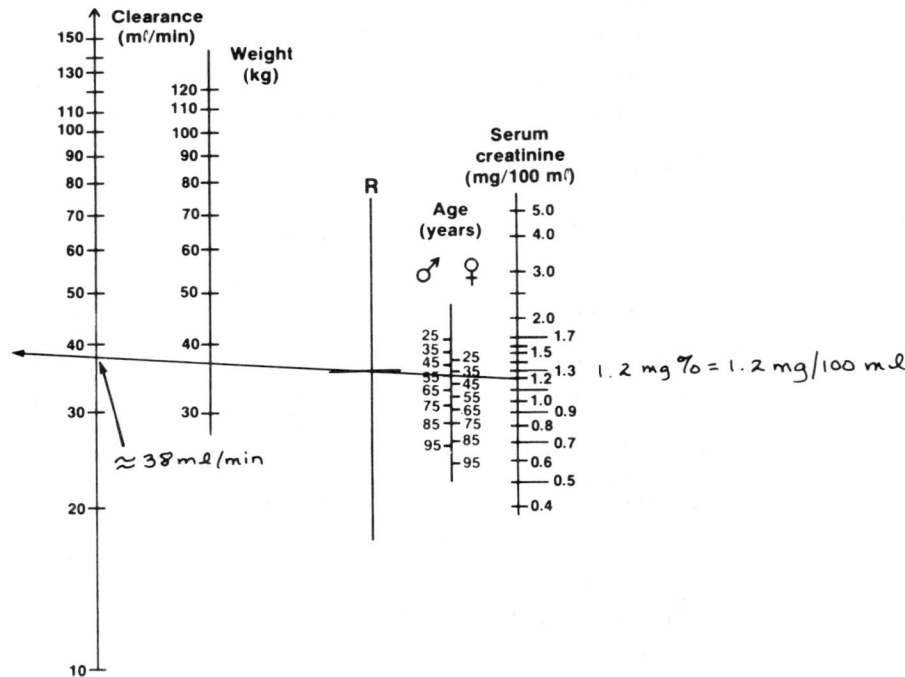

Step 3:   The table indicates that 38 mℓ/min is considered moderate renal impairment. The infection is severe; therefore, the dose of Ancef® is 250 to 600 mg q. 12 h.

B. Geriatric Dosages   329

Maintenance Dosage in Adults with Reduced Renal Function

Renal-function impairment	BUN (mg%)	Ccr (mℓ/min)	Dosage		Dosage interval (hrs)
			Mild to moderate infection (mg)	Moderate to severe infection (mg)	
Mild	20–34	70–40	250 to 500	500 to 1250	12
Moderate	35–49	(38) 40–20	125 to 250	250 to 600	12
Severe	50–75	20–5	75 to 150	150 to 400	24
Essentially no function	>75	<5	37.5 to 75	75 to 200	24

Section I-B Readiness Review

Use the Creatinine Clearance Nomogram to determine the creatinine clearance in each of the following examples:

1. *Patient's age/sex:* 75 yr/male   (see Example 1)
   Patient's weight: 65 kg
   Serum creatinine: 3.0 mg%
   Creatinine clearance: ?

2. *Patient's age/sex:* 85 yr/female   (see Example 2)
   Patient's weight: 70 kg
   Serum creatinine: 1.2 mg%
   Creatinine clearance: ?

In Problems 3 and 4, use the following table for the drug Mefoxin® and the other information provided to determine the correct dosage(s) based on Ccr. Be sure to first determine Ccr.

Maintenance Dosage in Adults with Reduced Renal Function

Renal function	Creatinine clearance (mℓ/min)	Dose (grams)	Frequency
Mild impairment	50–30	1–2	every 8–12 hours
Moderate impairment	29–10	1–2	every 12–24 hours
Severe impairment	9–5	0.5–1	every 12–24 hours
Essentially no function	<5	0.5–1	every 24–48 hours

3. *Patient's age/sex:* 70 yr/male   (see Example 3)
   Patient's weight: 75 kg
   Serum creatinine: 1.5 mg%
   Creatinine clearance (Ccr): ?
   Type of infection: Not mentioned
   Dosage: ?

Application I Pediatric and Geriatric Dosages

4. *Patient's age/sex:* 65 yr/female   (see Example 4)
   Patient's weight: 55 kg
   Serum creatinine: 3.0 mg%
   Creatinine clearance (Ccr): ?
   Type of infection: Not mentioned
   Dosage: ?

Section readiness review answers (not given in order):

1. 22 mℓ/min
4. Ccr = 18 mℓ/min: Dose—Mefoxin® 1-2 g q. 12-24 hours
3. Ccr = 50 mℓ/min: Dose—Mefoxin® 1-2 g q. 8-12 hours
2. 40 mℓ/min

Section I-B Review Problems

Use the Creatinine Nomogram to determine the creatinine clearance in each of the following examples (answers to the odd-numbered problems are at the back of the book):

1. *Patient's age/sex:* 65 yr/female
   Patient's weight: 60 kg
   Serum creatinine: 3.0 mg%
   Creatinine clearance: ?

2. *Patient's age/sex:* 70 yr/male
   Patient's weight: 75 kg
   Serum creatinine: 0.6 mg%
   Creatinine clearance: ?

3. *Patient's age/sex:* 65 yr/male
   Patient's weight: 70 kg
   Serum creatinine: 1.3 mg%
   Creatinine clearance: ?

4. *Patient's age/sex:* 70 yr/male
   Patient's weight: 80 kg
   Serum creatinine: 1.2 mg%
   Creatinine clearance: ?

5. *Patient's age/sex:* 70 yr/female
   Patient's weight: 60 kg
   Serum creatinine: 1.5 mg%
   Creatinine clearance: ?

6. *Patient's age/sex:* 65 yr/female
   Patient's weight: 55 kg
   Serum creatinine: 1.7 mg%
   Creatinine clearance: ?

B. Geriatric Dosages   331

7. *Patient's age/sex:* 65 yr/male
   Patient's weight: 80 kg
   Serum creatinine: 1.5 mg%
   Creatinine clearance: ?

8. *Patient's age/sex:* 70 yr/female
   Patient's weight: 50 kg
   Serum creatinine: 0.9 mg%
   Creatinine clearance: ?

In each of the following problems, use the tables provided to determine the correct dosage(s) based on Ccr.

Mandol®—Maintenance Dosage Guide for Patients with Renal Impairment

Renal-function impairment	Ccr (mℓ/min)	Life-threatening infections (maximum dosage)	Less severe infections
Normal	>80	2 g q. 4 h.	1–2 g q. 6 h.
Mild	80–50	1.5 g q. 4 h. or 2 g q. 6 h.	0.75–1.5 g q. 6 h.
Moderate	50–25	1.5 g q. 6 h. or 2 g q. 8 h.	0.75–1.5 g q. 8 h.
Severe	25–10	1 g q. 6 h. or 1.25 g q. 8 h.	0.5–1 g q. 8 h.
Marked	10–2	0.67 g q. 8 h. or 1 g q. 12 h.	0.5–0.75 g q. 12 h.
None	<2	0.5 g q. 8 h. or 0.75 g q. 12 h.	0.25–0.5 g q. 12 h.

9. *Patient's age/sex:* 70 yr/male
   Patient's weight: 75 kg
   Serum creatinine: 0.6 mg%
   Creatinine clearance (Ccr): ?
   Type of infection: Life threatening
   Dosage: ?

10. *Patient's age/sex:* 80 yr/female
    Patient's weight: 40 kg
    Serum creatinine: 0.5 mg%
    Creatinine clearance (Ccr): ?
    Type of infection: Less severe
    Dosage: ?

Application I Pediatric and Geriatric Dosages

Ancef® — Maintenance Dosage in Adults with Reduced Renal Function

Renal-function impairment	BUN (mg%)	Ccr (mℓ/min)	Dosage		Dosage interval (hr)
			Mild to moderate infection (mg)	Moderate to severe infection (mg)	
Mild	20–34	70–40	250 to 500	500 to 1250	12
Moderate	35–49	40–20	125 to 250	250 to 600	12
Severe	50–75	20–5	75 to 150	150 to 400	24
Essentially no function	>75	<5	37.5 to 75	75 to 200	24

11. *Patient's age/sex:* 70 yr/male
    Patient's weight: 80 kg
    Serum creatinine: 1.2 mg%
    Creatinine clearance (Ccr): ?
    Type of infection: Moderate to mild
    Dosage: ?

12. *Patient's age/sex:* 65 yr/male
    Patient's weight: 80 kg
    Serum creatinine: 1.5 mg%
    Creatinine clearance (Ccr): ?
    Type of infection: Moderate to severe
    Dosage: ?

13. *Patient's age/sex:* 70 yr/female
    Patient's weight: 50 kg
    Serum creatinine: 0.9 mg%
    Creatinine clearance (Ccr): ?
    Type of infection: Mild to moderate
    Dosage: ?

Mefoxin® — Maintenance Dosage in Adults with Reduced Renal Function

Renal function	Creatinine clearance (mℓ/min)	Dose (grams)	Frequency
Mild impairment	50–30	1–2	every 8–12 hours
Moderate impairment	29–10	1–2	every 12–24 hours
Severe impairment	9–5	0.5–1	every 12–24 hours
Essentially no function	<5	0.5–1	every 24–48 hours

14. *Patient's age/sex:* 70 yr/female
    Patient's weight: 60 kg
    Serum creatinine: 1.5 mg%
    Creatinine clearance (Ccr): ?
    Type of infection: Not mentioned
    Dosage: ?

15. *Patient's age/sex:* 65 yr/female
    Patient's weight: 55 kg
    Serum creatinine: 1.7 mg%
    Creatinine clearance (Ccr): ?
    Type of infection: Not mentioned
    Dosage: ?

II
Electrolyte Solutions

OBJECTIVES After studying this application, you should be able to:
1. Name some electrolytes normally found in the body.
2. Solve problems involving milliequivalents per liter.

A. Electrolytes

As a nurse, you may have to add potassium chloride to an I.V. or start I.V. fluids that contain any number of different electrolytes; therefore, you must know how to do calculations involving electrolyte solutions. In Application II you'll find many examples and problems to help you learn about electrolytes and give you practice and confidence in doing electrolyte-solution calculations.

> **Definition:** An atom or group of atoms that carries an electrical charge is called an *ion*.

> **Definition:** A substance that dissociates into its component ions when in solution is called an *electrolyte*.

What is an electrolyte? In chemistry we are taught that some chemical compounds stay together—that is, remain intact—whereas others come apart, or dissociate. The particles that become separate carry a charge, either positive or negative. These charged particles are called *ions* (see Figure II-1).

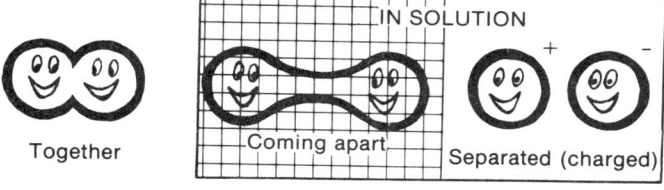

Figure II-1. Charged particles in solution—ions

You might correctly expect that, if there are electrolytes in the body, there must also be nonelectrolytes. Nonelectrolytes don't dissociate in solution, as shown in Figure II-2. An example of a nonelectrolyte is glucose (a sugar source of energy).

Figure II-2. Nonelectrolytes

Common examples of electrolytes found in the body are sodium, chloride, and bicarbonate ions. If sodium chloride (NaCl) is placed in a water solution, it dissociates into Na^+ (sodium) and Cl^- (chloride) (see Figures II-3 and II-4).

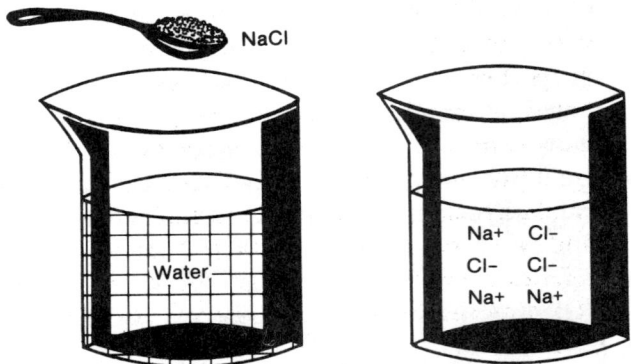

Figure II-3. Sodium chloride placed in solution to form sodium (Na^+) and chloride (Cl^-) ions

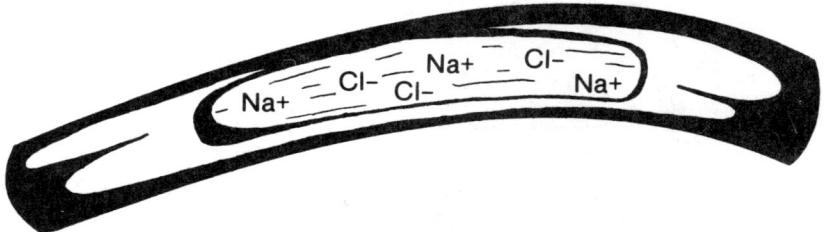

Figure II-4. Sodium ions (Na^+) and chloride ions (Cl^-) in the bloodstream

Blood plasma contains the ions Na^+ (sodium), K^+ (potassium), Ca^{++} (calcium), Mg^{++} (magnesium), Cl^- (chloride), HCO_3^- (bicarbonate), HPO_4^{--} (monohydrogen phosphate), SO_4^{--} (sulfate), organic acid$^-$ (organic acid anions, which come from the breakdown of organic acids within the cells), and protein$^-$ (protein anions, which come from the breakdown of proteins in the body). When they are in correct balance, the body's electrolytes provide a healthy setting for maintaining normal body functions.

Many medical problems become seriously complicated because of an imbalance of electrolytes—for example, heart attacks, severe dehydration, high blood pressure, respiratory failure, burns, kidney failure, and so on. In any case, when electrolytes in the body go out of balance, therapy must be started to correct the situation. Electrolyte solutions are available for injection and are used frequently to replace fluids and electrolytes that have been lost.

B. Valence and Milliequivalents

Concentrations of electrolytes are usually expressed in milliequivalents. Milliequivalents are based on how many negatively and positively charged ions are present rather than how much they weigh. We'll cover the significance of this concept in more detail at the end of this section. In the meantime, we'll review the concept of valence and how it applies to milliequivalent calculations.

> **Definition:** *Valence* is the capacity of atoms to bind together.

B. Valence and Milliequivalents

You'll recall from your chemistry studies that the valence of an atom is determined by the number of orbital electrons that influence the chemical properties of the atom; that is to say, each atom has a certain bonding power based on its valence number. From every atom in a structural formula, there must extend as many bonds as correspond to the valence number of the atom.

For our purposes, we will use the valence concept as it applies to structural formulas of atoms and molecules. This is a simplified approach, however, and barely scratches the surface of valence bond theory. Therefore, you should refer to your nursing chemistry text for more information on this theory. Table II-1 provides the valence of selected electrolytes for your convenience.

Table II-1. Valence of Selected Electrolytes

Name	Symbol	Valence
Acetate	$C_2H_3O_2^-$	1
Ammonium	NH_4^+	1
Bicarbonate	HCO_3^-	1
Calcium	Ca^{++}	2
Chloride	Cl^-	1
Fluoride	F^-	1
Hydrogen	H^+	1
Lactate	$C_3H_5O_3^-$	1
Magnesium	Mg^{++}	2
Phosphate	PO_4^{---}	3
Potassium	K^+	1
Sodium	Na^+	1
Sulfate	SO_4^{--}	2

First let's look at a molecule of water represented by structural formula.

H_2O (water)

```
           O
bonds → / \ ← bonds
       H   H
```

Atoms
H — Hydrogen
O — Oxygen

Look at the atom of oxygen. It has two bonds extending from it.

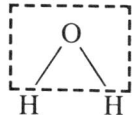

Therefore, the valence of oxygen is 2.

Now look at one of the atoms of hydrogen. It has one bond extending from it.

The valence of hydrogen is 1.

Next let's examine another molecule commonly found in the body, carbon dioxide (the gas you breathe out).

CO_2 (carbon dioxide)

O=C=O

Atoms
C — Carbon
O — Oxygen

The atom of carbon has four bonds extending from it.

O⎡=C=⎤O

Therefore, the valence of carbon is 4.
Now look at the oxygen atoms. Each oxygen atom has two bonds extending from it.

⎡O=⎤C=O   O=C⎡=O⎤

The valence of oxygen is always 2.
The acetate ion also includes an oxygen atom, but it has a more complicated structure.

$C_2H_3O_2^-$ (Acetate ion)

```
    H   O
    |   ‖
  H-C-C
    |   \
    H   O⁻
        ↑
   Business end of
    acetate ion
```

Atoms
C — Carbon
H — Hydrogen
O — Oxygen

You'll notice that the acetate-ion structure is a little different from the structure of the water and carbon-dioxide molecules you just looked at: in the acetate ion, one of the two oxygen bonds is all by itself. This is the "business end" of the acetate ion: a positive ion may bind to it at that site.
Now that we've reviewed valence and structural formula a bit, see if you can determine the valences of C, H, and O from the acetate-ion structure.

```
    H    O              H     O
    |    ‖              |     ‖
  H-C-  C             H-C  -C
    |    \              |     \
    H    O⁻             H     O⁻
```

Carbon has a valence of 4.

```
    H    O              H     O
    |    ‖              |     ‖
  H-C-C                H-C-C
    |    \              |     \
    H    O⁻             H     O⁻
```

Oxygen has a valence of 2.

```
    H   O              H  O              H  O
   ┌─┐  ║              │  ║              │  ║
   │H│  //         ┌──┐ │ //             │ //
   │ │                            
H─C┼─┼C          H─│C─│C         H─C─C
   │ │  \           │  │  \         │  │  \
   H  O─           H  O─       ┌──┐ O─
                               │H │
                               └──┘
```

Hydrogen has a valence of 1.

The valence numbers you'll need to work problems in this book are in Table II-1, Valences of Selected Electrolytes.

It's obvious that, to use the structural approach in solving electrolyte-solution problems, you must know how to write the structural formula of the electrolyte you are interested in. If you don't know how to write the formula, you can look it up in your chemistry book. However, nurses don't spend their time at the hospital drawing chemical structures. The structural approach just gives nurses something to fall back on, if they need to check whether they've correctly remembered the valence of a certain electrolyte. This structural approach to determining valence is not infallible, but it does give students a good working model for solving electrolyte problems.

The *equivalent weight* of any element is calculated by dividing the atomic weight by the valence. A *milliequivalent weight* is equal to the equivalent weight divided by 1000.

Milliequivalents measure electrolytes in terms of their combining power instead of their weight. It is generally accepted that one mEq of any electrolyte has the same chemical combining power as 1 mg of hydrogen. That is, 1 mg of hydrogen exerts 1 mEq of chemical activity; 23 mg of sodium, 39 mg of potassium, and 35 mg of chloride *also* exert 1 mEq of chemical activity or combining power.

Definition: One milliequivalent of a positive ion is equivalent chemically to one milliequivalent of a negative ion.

The following formula is a simplified way to convert milligrams per liter to milliequivalents per liter. (We will not go into the derivation of this formula in this text. If you are interested in pursuing it, consult your nursing chemistry book.) The abbreviation for milliequivalents is mEq.

$$\frac{mEq}{\ell} = \frac{\left(\frac{mg}{\ell}\right) \times \text{Valence}}{\text{Atomic or molecular weight}}$$

Table II-2 will provide you with the atomic weights you'll need to work milliequivalents-per-liter problems in this text.

Table II-2. Approximate Atomic Weights of Selected Elements

Name	Symbol	Atomic Weight
Aluminum	Al	27
Calcium	Ca	40
Carbon	C	12
Chlorine	Cl	35
Fluorine	F	19
Hydrogen	H	1
Lithium	Li	7
Magnesium	Mg	24
Nitrogen	N	14
Oxygen	O	16
Phosphorus	P	31
Potassium	K	39
Sodium	Na	23
Sulfur	S	32

Rule: Solving Milliequivalents-Per-Liter Problems

Step 1: Determine the atomic weight of each ion from Table II-2, Approximate Atomic Weights of Selected Elements. (The molecular weight is found by adding these together.)

Step 2: Find the valence of each ion using Table II-1, Valences of Selected Electrolytes.

Step 3: Use the concentration of substance given in the problem, the valence, and the atomic or molecular weight in the following formula:

$$\frac{mEq}{\ell} = \frac{(\frac{mg}{\ell}) \times \text{Valence}}{\text{Atomic or molecular weight}}$$

Complete multiplication and division as indicated. Your answer is in mEq/ℓ.

Example 1: Solve the following problem: A solution of potassium chloride (KCl) has a concentration of 1490 mg per liter. How many milliequivalents (mEq) of potassium and chloride ions are contained in each liter of solution?

Solution:

Step 1: The atomic weights of K and Cl are:

K = 39 and Cl = 35

Thus the molecular weight of KCl is:

39 + 35 = 74

Step 2: The valence of K^+ and Cl^- is 1.

Step 3:

$$\frac{mEq}{\ell} = \frac{(\frac{mg}{\ell}) \times Valence}{Atomic\ or\ molecular\ weight}$$

$$\frac{mEq}{\ell} = \frac{1490 \times 1}{74} = \frac{1490}{74} = 20.1$$

There are 20 mEq each of potassium and chloride ions per liter of solution.

Example 2: Solve the following problem: A solution contains sodium bicarbonate $NaHCO_3$ with a concentration of 84 mg per liter. How many milliequivalents (mEq) of bicarbonate (HCO_3^-) and sodium ions are contained in each liter of solution? How many milliequivalents (mEq) of Na^+ are contained per liter of solution?

Solution:

Step 1: The atomic weights of Na and HCO_3 are:

$$Na = 23\ and\ HCO_3 = 61$$

Thus the molecular weight of $NaHCO_3$ is:

$$23 + 61 = 84$$

Step 2: The valence of Na^+ and HCO_3^- is 1.

Step 3: $\frac{mEq}{\ell} = \frac{84 \times 1}{84} = \frac{84}{84} = 1$

There is 1 mEq each of sodium and bicarbonate ion per liter of solution.

Section II-B Readiness Review

Solve the following problems (see Examples 1 & 2):

1. A solution of sodium lactate ($NaC_3H_5O_3$) has a concentration of 560 mg/ℓ. How many milliequivalents (mEq) of lactate ($C_3H_5O_3$—valence 1) and sodium ions are contained in each liter of solution?

2. A solution of sodium chloride (NaCl) has a concentration of 50 g/ℓ (50,000 mg/ℓ). How many milliequivalents of sodium and chloride ions are contained in each liter of solution?

Section readiness review answers:

1. There are 5 mEq each of sodium and lactate ions per liter of solution.
2. There are 862 mEq each of sodium and chloride ions per liter of solution.

Application II Review Problems

Solve the following problems (answers to the odd-numbered problems are at the back of the book):

1. A solution of potassium acetate ($KC_2H_3O_2$) has a concentration of 2.45 g/ℓ (2450 mg/ℓ). How many milliequivalents (mEq) of acetate ($C_2H_3O_2$—valence 1) and potassium are contained in each liter of solution?

2. A solution of potassium chloride (KCl) has a concentration of 3000 mg/ℓ. How many milliequivalents (mEq) of potassium and chloride ions are contained in each liter of fluid?

3. A solution of sodium chloride (NaCl) has a concentration of 4.5 g/ℓ (4500 mg/ℓ). How many milliequivalents (mEq) of sodium and chloride ions are contained in each liter of solution?

4. A solution of ammonium chloride (NH_4Cl) is used as a source of chloride ion for a patient who is lacking this electrolyte. If the physician orders ammonium chloride solution, 9 g/ℓ (9000 mg/ℓ), how many milliequivalents of chloride and ammonium (NH_4^+—valence 1) ions will each liter of solution contain?

5. Liz Bryte has had diarrhea for some time. Her physician has decided to admit her to the hospital for treatment. Dr. Sharp has ordered an I.V. that contains sodium lactate, knowing that the sodium lactate will be changed in the body to bicarbonate. Dr. Sharp has ordered sodium lactate ($NaC_3H_5O_3$) with a concentration of 19 g/ℓ (19,000 mg/ℓ). How many milliequivalents (mEq) of lactate ($C_3H_5O_3^-$—valence 1) and sodium will the patient receive per liter of solution?

6. Dr. Weston has ordered an I.V. that contains potassium chloride (KCl) 2250 mg/ℓ. How many milliequivalents (mEq) of potassium and chloride ions are contained in each liter of fluid?

7. A solution of potassium chloride (KCl) has a concentration of 750 mg/ℓ. How many milliequivalents (mEq) of potassium and chloride ions are contained in each liter of solution?

8. A solution of sodium chloride (NaCl) has a concentration of 9.0 g/ℓ (9000 mg/ℓ). How many milliequivalents of sodium and chloride ions are contained in each liter of solution?

III
Total Parenteral Nutrition

OBJECTIVES After studying this application, you should be able to:

1. Understand the role of total parenteral nutrition in specific disease states.
2. Solve problems involving T.P.N solutions and extra electrolytes.

Overview

Electrolyte solutions are given to correct electrolyte imbalances in the body. These solutions also maintain electrolyte balance in patients undergoing total parenteral nutrition (T.P.N.) or hyperalimentation (H.A.).

Total parenteral nutrition is the administration of nutrients to the body through a large vein. The nutrients are in a highly concentrated solution containing glucose, proteins, electrolytes, vitamins, and so on. Because the solution is too concentrated (hypertonic) to be given through an arm vein (peripherally), it's fed through a carefully placed catheter in the superior vena cava (see Figure III-1; the superior vena cava is the large vessel through which oxygen-poor blood returns to the heart.) The solution travels from the bottle to the heart, where it is diluted immediately in a large volume of blood. The nutrient solution is eventually carried, by way of the bloodstream, to all parts of the body (see Figure III-2).

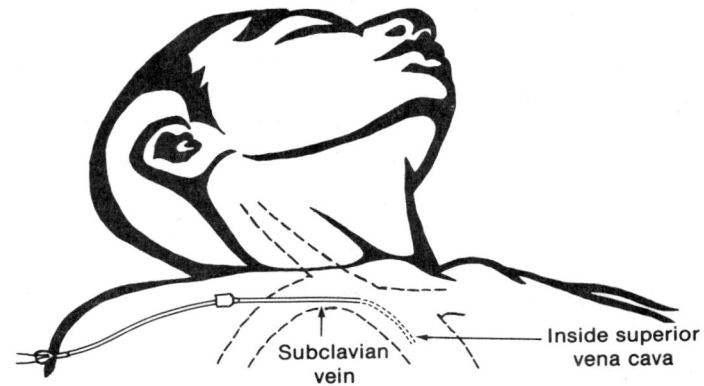

Figure III-1. The superior vena cava

Patients who need total parenteral nutrition usually cannot take food by mouth or through a tube in the stomach. However, they obviously still require a source of carbohydrates and protein to provide the building blocks for repair of damaged tissues and to prevent starvation. The type of patient that is placed on total parenteral nutrition may suffer, for example, from diseases of the bowel and stomach, uncontrollable diarrhea, burns, or cancer.

The nursing role in total parenteral nutrition is expanding. Among other things, the patient-care nurse monitors the T.P.N. patient for:

1. fever,
2. excess sugar in the urine,
3. redness around catheter site,
4. body weight,
5. daily input and output of fluids,
6. constant infusion rate of the T.P.N. solution,
7. vital signs,
8. electrolyte imbalances.

It is essential, therefore, that nurses given the responsibility to care for the T.P.N. patient understand all aspects of the therapy and be able to solve

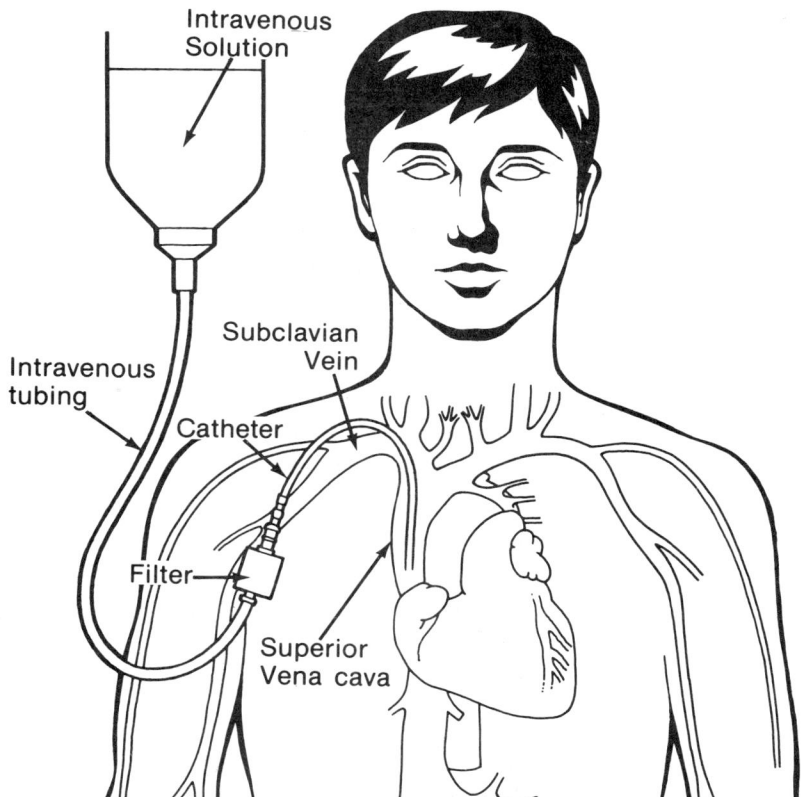

Figure III-2. The nutrient solution is carried by the bloodstream

problems involving total parenteral nutrition. This section will give you an opportunity to apply your knowledge of milliequivalents to solving T.P.N. problems.

> **Rule: Solving T.P.N. Problems**
> *Step 1:* Read the entire problem and try to get a general idea of what it's about. Keep in mind that the problem may include some "extra" information, as is often the case in real-life situations.
> *Step 2:* Work the problem.
> *Step 3:* Check your answer by any method you have learned in the past. If you can't solve the problem, read it once more and try again. If, after several tries, you are still having trouble, please get help from the instructor. It is extremely important that you get help if you do not understand a problem; you owe it to yourself and you owe it to your future patients.

After the physician has ordered total parenteral nutrition for a patient, the pharmacist makes the necessary calculations and prepares the T.P.N. solution. The pharmacist must calculate how much commercial T.P.N. solution to use based on the doctor's orders for a certain amount of grams of protein and dextrose required by the patient. Any item added to the T.P.N. solution must be ordered by the physician. When the T.P.N. solution arrives at the nursing station, it will be labeled in the following manner:

```
┌─────────────────────────────────────────────┐
│           INTRAVENOUS SOLUTION ADDITIVES    │
├──────────────────────┬──────────────────────┤
│ Room_____        │ Name_____    │
│ I.V. No._____      │ Doctor_____    │
│ Flo-rate_____      │ Date _____    │
│ Started _____      │                      │
│ By_____       │ _____ Bottle    │
├──────────────────────┴──────────────────────┤
│           DRUGS ADDED this bottle           │
├─────────────────────────────────┬───────────┤
│ Protein                         │ R  A  N   │
│ Dextrose                        │ E  D  O   │
│   ___mEq Na+    ___mEq K+       │ T  M  T   │
│   ___mEq Mg++   ___mEq CL-      │ U  I      │
│   ___mEq HPO4-- ___mEq Ca++     │ R  N  B   │
│   ___mEq SO4--                  │ N  I  E   │
│   ___mEq Acetate___ ___         │    S  G   │
│   ___ml MVI___Folbesyn          │ T  T  U   │
│            (vitamins)           │ O  R  N   │
│ Total Volume_____ml          │    A      │
│ Refrigerate qs ad sterile water │ P  T  B   │
│                                 │ H  I  Y:  │
│                                 │ A  O      │
│                                 │ R  N      │
│                                 │ M         │
│                                 │ A  H      │
│                                 │ C  A      │
│                                 │ Y  S      │
│                                 │ IF        │
└─────────────────────────────────┴───────────┘
```

The type of label used may differ from hospital to hospital. However, the basic information will be the same; that is, the label will show the electrolytes and other ingredients added to each bottle. It will also show the amount of protein and dextrose contained as well as the total volume of the bottle.

How can the information provided on the label assist the nurse in caring for the T.P.N. patient? First of all, the label tells the nurse exactly what is contained in the bottle—that is, protein, dextrose, electrolytes, and so on. Second, it tells the nurse the volume of fluid contained in the bottle. By knowing the volume of fluid at the start of infusion, the nurse can monitor fluid input over a 24-hour period and be sure that a constant rate of fluid is infused into the patient per hour. In addition, the expiration date on the label of the T.P.N. solution can be checked so that a fresh solution can be used if needed. The label must also be numbered to insure that the solutions are hung in the correct order.

As mentioned previously, the pharmacist determines the volume of commercial protein solution to obtain the amount of protein grams ordered by the physician. The pharmacist also calculates the volume of dextrose 50% in water required to provide the number of calories ordered by the doctor. Commercial T.P.N. solutions don't always contain enough electrolytes to fulfill a doctor's order; therefore, extra electrolytes occasionally must be added. In this case, the extra electrolytes will normally be added to the T.P.N. bottles in the sterile area of the pharmacy.

Now that we've discussed some aspects of total parenteral nutrition, use your knowledge of electrolyte calculations and intravenous fluids to solve the following T.P.N. problems.

Example 1: Dr. Hata orders a 40-g-protein, 400-g dextrose-per-day T.P.N. solution for his patient, Francisco Arce. The commercial product Aminosyn® 7% is used as the source of protein and dextrose 50% is used as the energy source. Three bottles of T.P.N. solution will be hung daily. Each bottle will have a total volume of 1000 mℓ. The infusion rate will be 125 mℓ per hour.

To get 40 g of *protein per day,* 571 mℓ of Aminosyn® 7% is needed. That is,

$$40 \text{ g} \times \frac{100 \text{ m}\ell}{7 \text{ g}} = 571 \text{ m}\ell$$

To get 13.3 g of *protein per bottle,* 190 mℓ of Aminosyn® 7% is needed. That is,

$$13.3 \text{ g} \times \frac{100 \text{ m}\ell}{7 \text{ g}} = 190 \text{ m}\ell$$

The electrolytes obtained from 190.48 mℓ of Aminosyn® 7% solution are as follows:

Electrolyte	Milliequivalents per bottle
Na^+	0 mEq
K^+	1 mEq
Mg^{++}	0 mEq
Cl^-	0 mEq
Acetate$^-$	17 mEq

The following labels show the electrolyte concentrations requested by Dr. Hata:

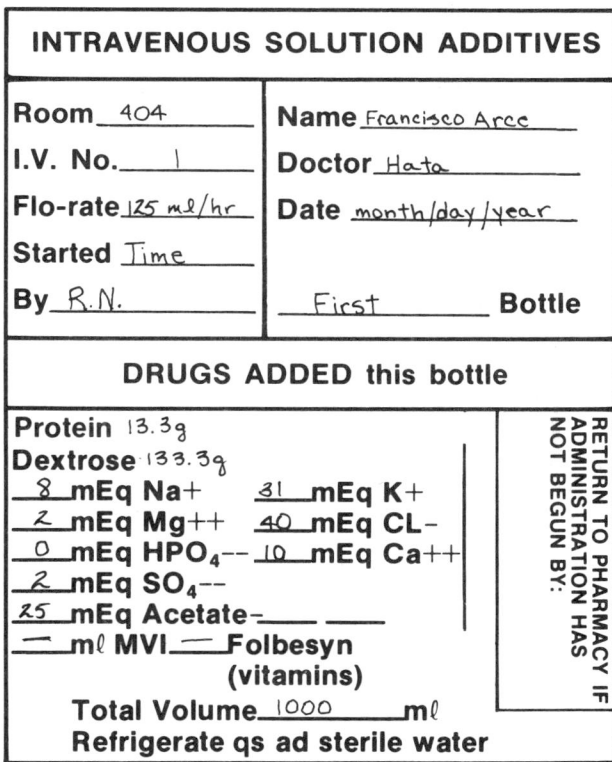

```
INTRAVENOUS SOLUTION ADDITIVES

Room  404          Name Francisco Arce
I.V. No.  1        Doctor Hata
Flo-rate 125 mℓ/hr  Date month/day/year
Started Time
By R.N.              First      Bottle

DRUGS ADDED this bottle

Protein 13.3g
Dextrose 133.3g
  8 mEq Na+     31 mEq K+
  2 mEq Mg++   40 mEq CL-
  0 mEq HPO4-- 10 mEq Ca++
  2 mEq SO4--
 25 mEq Acetate-
  — mℓ MVI — Folbesyn
          (vitamins)
  Total Volume 1000 mℓ
Refrigerate qs ad sterile water

RETURN TO PHARMACY IF
ADMINISTRATION HAS
NOT BEGUN BY:
```

Application III Total Parenteral Nutrition

INTRAVENOUS SOLUTION ADDITIVES

Room __404__	Name __Francisco Arce__
I.V. No. __2__	Doctor __Hata__
Flo-rate __125 ml/hr__	Date __month/day/year__
Started __Time__	
By __R.N.__	__Second__ Bottle

DRUGS ADDED this bottle

Protein 13.3 g
Dextrose 133.3 g
__8__ mEq Na+   __31__ mEq K+
__2__ mEq Mg++ __40__ mEq CL-
__0__ mEq HPO$_4$-- __10__ mEq Ca++
__2__ mEq SO$_4$--
__25__ mEq Acetate ___ ___
__10__ ml MVI ___ Folbesyn
            (vitamins)
Total Volume __1000__ ml
Refrigerate qs ad sterile water

RETURN TO PHARMACY IF ADMINISTRATION HAS NOT BEGUN BY:

INTRAVENOUS SOLUTION ADDITIVES

Room __404__	Name __Francisco Arce__
I.V. No. __3__	Doctor __Hata__
Flo-rate __125 ml/hr__	Date __month/day/year__
Started __Time__	
By __R.N.__	__Third__ Bottle

DRUGS ADDED this bottle

Protein 13.3 g
Dextrose 133.3 g
__8__ mEq Na+   __31__ mEq K+
__2__ mEq Mg++ __40__ mEq CL-
__0__ mEq HPO$_4$-- __10__ mEq Ca++
__2__ mEq SO$_4$--
__25__ mEq Acetate ___ ___
___ ml MVI ___ Folbesyn
            (vitamins)
Total Volume __1000__ ml
Refrigerate qs ad sterile water

RETURN TO PHARMACY IF ADMINISTRATION HAS NOT BEGUN BY:

The following chart shows the levels of the electrolytes in the commercial solution and the final infusion, after extra electrolytes have been added.

Electrolyte	From Aminosyn®	Added	Total amount per bottle
Na^+	0 mEq	8 mEq	8 mEq
Cl^-	0 mEq	[30 mEq + 10 mEq]	40 mEq
K^+	1 mEq	30 mEq	31 mEq
Ca^{++}	0 mEq	10 mEq	10 mEq
Mg^{++}	0 mEq	2 mEq	2 mEq
SO_4^{--}	0 mEq	2 mEq	2 mEq
$Acetate^-$	17 mEq	8 mEq	25 mEq

Given the preceding information, how many milliliters of:

a. KCl-2 mEq/mℓ were added to the T.P.N. solution to supply the K^+ as ordered? _____ mℓ

b. $CaCl_2$-1.36 mEq/mℓ were added to the T.P.N. solution to supply the Ca^{++} as ordered? _____ mℓ

c. $MgSO_4$-4 mEq/mℓ were added to the T.P.N. solution to supply the Mg^{++} as ordered? _____ mℓ

d. Na Acetate-2 mEq/mℓ were added to the T.P.N. solution to supply the Na^+ as ordered? _____ mℓ

e. How long will one bottle of T.P.N. run? _____

Solution:

a. To find the number of milliliters of KCl added to the T.P.N. solution, use dimensional analysis. Locate the number of milliequivalents of K^+ from the *Added* column on the electrolyte chart and multiply it by the KCl available.

Electrolyte	From Aminosyn®	Added	Total amount per bottle
Na^+	0 mEq	8 mEq	8 mEq
Cl^-	0 mEq	[30 mEq + 10 mEq]	40 mEq
K^+	1 mEq	☐ 30 mEq ☐	31 mEq
Ca^{++}	0 mEq	10 mEq	10 mEq
Mg^{++}	0 mEq	2 mEq	2 mEq
SO_4^{--}	0 mEq	2 mEq	2 mEq
$Acetate^-$	17 mEq	8 mEq	25 mEq

$$\underset{\downarrow}{K^+\ \text{required}} \quad \underset{\downarrow}{\text{KCl available}}$$

$$\overset{15}{\cancel{30}}\ \text{mEq} \times \frac{1\ \text{mℓ}}{\underset{1}{\cancel{2\ \text{mEq}}}} =$$

$$\frac{15\ \text{mℓ}}{1} = 15\ \text{mℓ of KCl 2 mEq/mℓ}$$

b. Use the same method as you did in (a) to find out how many milliliters of $CaCl_2$ were added to the T.P.N. solution.

Electrolyte	From Aminosyn®	Added	Total amount per bottle
Na^+	0 mEq	8 mEq	8 mEq
Cl^-	0 mEq	[30 mEq + 10 mEq]	40 mEq
K^+	1 mEq	30 mEq	31 mEq
Ca^{++}	0 mEq	[10 mEq]	10 mEq
Mg^{++}	0 mEq	2 mEq	2 mEq

Ca^{++} required $\quad$ $CaCl_2$ available

$$10 \text{ mEq} \times \frac{1 \text{ ml}}{1.36 \text{ mEq}} =$$

$$\frac{10 \text{ ml}}{1.36} = 7.4 \text{ ml of } CaCl_2 \text{ 1.36 mEq/ml}$$

c. The $MgSO_4$ quantity is found the same way.

Electrolyte	From Aminosyn®	Added	Total amount per bottle
Na^+	0 mEq	8 mEq	8 mEq
Cl^-	0 mEq	[30 mEq + 10 mEq]	40 mEq
K^+	1 mEq	30 mEq	31 mEq
Ca^{++}	0 mEq	10 mEq	10 mEq
Mg^{++}	0 mEq	[2 mEq]	2 mEq
SO_4^{--}	0 mEq	2 mEq	2 mEq

Mg^{++} required $\quad$ $MgSO_4$ available

$$\overset{1}{\cancel{2} \text{ mEq}} \times \frac{1 \text{ ml}}{\underset{2}{\cancel{4} \text{ mEq}}} =$$

$$\frac{1 \text{ ml}}{2} = 0.5 \text{ ml of } MgSO_4 \text{ 4 mEq/ml}$$

d. Finally, use dimensional analysis once again to find the amount of Na Acetate required.

Electrolyte	From Aminosyn®	Added	Total amount per bottle
Na^+	0 mEq	[8 mEq]	8 mEq
Cl^-	0 mEq	[30 mEq + 10 mEq]	40 mEq

Na^+ required $\quad$ Na Acetate available

$$\overset{4}{\cancel{8} \text{ mEq}} \times \frac{1 \text{ ml}}{\underset{1}{\cancel{2} \text{ mEq}}} =$$

$$\frac{4 \text{ ml}}{1} = 4 \text{ ml of Na Acetate 2 mEq/ml}$$

e. The volume of the T.P.N. bottle is 1000 mℓ (from label). The rate of administration is 125 mℓ/hr.

$$\overset{8}{\cancel{1000\text{ mℓ}}} \times \frac{1\text{ hr}}{\underset{1}{\cancel{125\text{ mℓ}}}} = 8 \text{ hours}$$

The one bottle of T.P.N. infusion will run 8 hours.

In summary, the answers to the questions in Example 1 are:

a. How many KCl-2 mEq/mℓ were added to the T.P.N. solution to supply the K$^+$ as ordered? ___15___ mℓ

b. How many CaCl$_2$-1.36 mEq/mℓ were added to the T.P.N. solution to supply the Ca^{++} as ordered? ___7.4___ mℓ

c. How many MgSO$_4$-4 mEq/mℓ were added to the T.P.N. solution to supply the Mg^{++} as ordered? ___0.5___ mℓ

d. How many Na Acetate-2 mEq/mℓ were added to the T.P.N. solution to supply the Na$^+$ as ordered? ___4___ mℓ

e. How long will one bottle of T.P.N. infusion run? ___8 hours___

Let's solve another T.P.N. problem, using the commercial product Travisol® 5.5%.

Example 2: Dr. Marks has decided to place her patient, Shannon O'Farrell, on a 40-g-protein, 400-g-dextrose-per-day T.P.N. solution. The pharmacist decides to use Travisol® 5.5% as the protein source and dextrose 50% as the energy source. Three bottles of T.P.N. solution will be hung daily. Each bottle will have a total volume of 1000 mℓ. The infusion rate will be 125 mℓ per hour.

To get 40 g of *protein per day,* 727 mℓ of Travisol® 5.5% is needed.

To get 13.3 g of *protein per bottle,* 242 mℓ of Travisol® 5.5% is needed.

The electrolytes obtained from 242 mℓ of Travisol® 5.5% are as follows:

Electrolyte	*Milliequivalents per bottle*
Na$^+$	0 mEq
K$^+$	0 mEq
Mg^{++}	0 mEq
Acetate$^-$	8 mEq
Cl$^-$	5 mEq
HPO$_4^{--}$	0 mEq

The labels in this example would be the same as those in the first example, except that the number of milliequivalents of electrolytes would be different. Dr. Marks has requested the electrolytes shown on the following labels.

Electrolyte	*From* Travisol®	*Added*	*Total amount per bottle*
Na$^+$	0 mEq	15 mEq	15 mEq
K$^+$	0 mEq	40 mEq	40 mEq
Mg^{++}	0 mEq	0 mEq	0 mEq
Acetate$^-$	8 mEq	15 mEq	23 mEq
Cl$^-$	5 mEq	[40 mEq + 8 mEq]	40 mEq
HPO$_4^{--}$	0 mEq	0 mEq	53 mEq
Ca^{++}	0 mEq	8 mEq	8 mEq

Application III Total Parenteral Nutrition

INTRAVENOUS SOLUTION ADDITIVES

Room **307**   Name **Shannon O'Farrell**
I.V. No. **1**   Doctor **Marks**
Flo-rate **125 ml/hr**   Date **month/day/year**
Started **Time**
By **R.N.**   **First** Bottle

DRUGS ADDED this bottle

Protein **13.3 g**
Dextrose **133.3 g**
15 mEq Na+   **40** mEq K+
0 mEq Mg++   **53** mEq CL-
0 mEq HPO$_4$--   **8** mEq Ca++
0 mEq SO$_4$--
23 mEq Acetate= ___ ___
___ ml MVI ___ Folbesyn
        (vitamins)
Total Volume **1000** ml
Refrigerate qs ad sterile water

RETURN TO PHARMACY IF ADMINISTRATION HAS NOT BEGUN BY:

INTRAVENOUS SOLUTION ADDITIVES

Room **307**   Name **Shannon O'Farrell**
I.V. No. **2**   Doctor **Marks**
Flo-rate **125 ml/hr**   Date **month/day/year**
Started **Time**
By **R.N.**   **Second** Bottle

DRUGS ADDED this bottle

Protein **13.3 g**
Dextrose **133.3 g**
15 mEq Na+   **40** mEq K+
0 mEq Mg++   **53** mEq CL-
0 mEq HPO$_4$--   **8** mEq Ca++
0 mEq SO$_4$--
23 mEq Acetate= ___ ___
___ ml MVI **4 ml** Folbesyn
        (vitamins)
Total Volume **1000** ml
Refrigerate qs ad sterile water

RETURN TO PHARMACY IF ADMINISTRATION HAS NOT BEGUN BY:

INTRAVENOUS SOLUTION ADDITIVES

Room **307** Name **Shannon O'Farrell**
I.V. No. **3** Doctor **Marks**
Flo-rate **125 ml/hr** Date **month/day/year**
Started **Time**
By **R.N.** **Third** Bottle

DRUGS ADDED this bottle

Protein **13.3 g**
Dextrose **133.3 g**
15 mEq Na+   **40** mEq K+
0 mEq Mg++   **53** mEq CL−
0 mEq HPO$_4$−−   **8** mEq Ca++
0 mEq SO$_4$−−
23 mEq Acetate
___ ml MVI ___ Folbesyn
(vitamins)
Total Volume **1000** ml
Refrigerate qs ad sterile water

RETURN TO PHARMACY IF ADMINISTRATION HAS NOT BEGUN BY:

Given the preceding information, how many milliliters of:

a. KCl-2 mEq/ml were added to the T.P.N. solution to supply the K$^+$ as ordered? _____ ml

b. CaCl$_2$-1.36 mEq/ml were added to the T.P.N. solution to supply the Ca^{++} as ordered? _____ ml

Solution:

a. Use dimensional analysis to find the number of milliliters of KCl added to the T.P.N. solution.

Electrolyte	From Travisol®	Added	Total amount per bottle
Na$^+$	0 mEq	15 mEq	15 mEq
K$^+$	0 mEq	**40 mEq**	40 mEq
Mg^{++}	0 mEq	0 mEq	0 mEq

K$^+$ required   KCl available
↓   ↓

$$\overset{20}{\cancel{40} \text{ mEq}} \times \frac{\text{ml}}{\cancel{2 \text{ mEq}}_1} =$$

$$\frac{20 \text{ ml}}{1} = 20 \text{ ml of KCl 2 mEq/ml}$$

b. To find the number of milliliters of CaCl$_2$ added to the T.P.N. solution, first determine how much extra calcium is needed.

Electrolyte	From Travisol®	Added	Total amount per bottle
Na⁺	0 mEq	15 mEq	15 mEq
K⁺	0 mEq	40 mEq	40 mEq
Mg⁺⁺	0 mEq	0 mEq	0 mEq
Acetate⁻	8 mEq	15 mEq	23 mEq
Cl⁻	5 mEq	[40 mEq + 8 mEq]	53 mEq
HPO₄⁻⁻	0 mEq	0 mEq	0 mEq
Ca⁺⁺	0 mEq	8 mEq	8 mEq

$$\underset{\text{Ca}^{++}\text{ required}}{8 \text{ mEq}} \times \underset{\text{CaCl}_2\text{ available}}{\frac{1 \text{ m}\ell}{1.36 \text{ mEq}}} =$$

$$\frac{8 \text{ m}\ell}{1.36} = 5.9 \text{ m}\ell \text{ of CaCl}_2 \text{ 1.36 mEq/m}\ell$$

In summary, the answers to the questions in Example 2 are:

a. How many KCl-2 mEq/mℓ were added to the T.P.N. solution to supply the K⁺ as ordered? ____20____ mℓ

b. How many CaCl₂-1.36 mEq/mℓ were added to the T.P.N. solution to supply the Ca⁺⁺ as ordered? ____5.9____ mℓ

Application III Readiness Review

1. Dr. Sullivan orders a 40-g-protein, 400-g-dextrose-per-day T.P.N. solution for his patient, Laurie Bettag-Bain. The commercial protein source is FreAmine III® 8.5%, and dextrose 50% is the energy source. Three bottles of T.P.N. solution will be hung daily. Each bottle will have a total volume of 1000 mℓ. The infusion rate will be 125 mℓ per hour.

   To get 40 g of *protein per day,* we need 471 mℓ of FreAmine III® 8.5%.

   To get 13.3 g of *protein per bottle,* we need 156 mℓ of FreAmine III® 8.5%.

   The electrolytes obtained from 156 mℓ of FreAmine III® 8.5% solution are as follows:

Electrolyte	Milliequivalents per bottle
Na⁺	2 mEq
Acetate⁻	12 mEq
HPO₄⁻⁻	2 mEq

   The following label shows the electrolyte concentrations requested by Dr. Sullivan. (Only one label is shown to save space. The two remaining labels would be identical to the first, except the label for I.V. No. 2 would have a vitamin additive.)

INTRAVENOUS SOLUTION ADDITIVES

Room __106__ Name __Bettag-Bain__
I.V. No. __1__ Doctor __Sullivan__
Flo-rate __125 ml/hr__ Date __month/day/year__
Started __Time__
By __R.N.__ __First__ Bottle

DRUGS ADDED this bottle

Protein __13.3 g__
Dextrose __133.3 g__
__40__ mEq Na+   __30__ mEq K+
__4__ mEq Mg++ __35__ mEq CL-
__40__ mEq HPO₄-- __5__ mEq Ca++
__4__ mEq SO₄--
__12__ mEq Acetate- ___ ___
___ ml MVI ___ Folbesyn
(vitamins)
Total Volume __1000__ ml
Refrigerate qs ad sterile water

RETURN TO PHARMACY IF ADMINISTRATION HAS NOT BEGUN BY:

Electrolyte	From FreAmine III®	Added	Total amount per bottle
Na^+	2 mEq	38 mEq	40 mEq
K^+	0 mEq	30 mEq	30 mEq
Mg^{++}	0 mEq	4 mEq	4 mEq
Ca^{++}	0 mEq	5 mEq	5 mEq
Cl^-	0 mEq	[30 mEq + 5 mEq]	35 mEq
HPO_4^{--}	2 mEq	38 mEq	40 mEq
SO_4^{--}	0 mEq	4 mEq	4 mEq
$Acetate^-$	12 mEq	0 mEq	12 mEq

Given the preceding information, how many milliliters of:

a. KCl-2 mEq/ml were added to the T.P.N. solution to supply the K^+ as ordered? _____ ml

b. CaCl₂-1.36 mEq/ml were added to the T.P.N. solution to supply the Ca^{++} as ordered? _____ ml

c. MgSO₄-4 mEq/ml were added to the T.P.N. solution to supply the Mg^{++} as ordered? _____ ml

d. Na₂HPO₄-4 mEq/ml were added to the T.P.N. solution to supply the Na^+ as ordered? _____ ml

---

Readiness review answers:

a. 15 ml **b.** 3.7 ml **c.** 1 ml **d.** 9.5 ml

Application III Review Problems

All of the following problems have a few things in common. In each case, you can assume that three bottles of T.P.N. solution will be hung daily. Each bottle will have a total volume of 1000 mℓ. The infusion rate will be 125 mℓ per hour. The nutrition level will be stated at the beginning of each problem. Answer the questions at the end of each problem (answers to the odd-numbered questions are at the back of the book).

1. 40 g of protein—400 g of dextrose 50% per day
   Doctor: Edgar
   Patient: Amy Braddock
   Room: 101

   To get 40 g of *protein per day*, 571 mℓ of Aminosyn® 7% is needed.

   To get 13.3 g of *protein per bottle*, 190 mℓ of Aminosyn® 7% is needed.

   The electrolytes obtained from 190 mℓ of Aminosyn® 7% solution are as follows:

Electrolyte	Milliequivalents per bottle
Na^+	0 mEq
K^+	1 mEq
Mg^{++}	0 mEq
Cl^-	0 mEq
$Acetate^-$	17 mEq

Electrolyte	From Aminosyn® 7%	Added	Total amount per bottle
Na^+	0 mEq	30 mEq	30 mEq
K^+	1 mEq	16.5 mEq	17.5 mEq
Mg^{++}	0 mEq	2 mEq	2 mEq
Ca^{++}	0 mEq	6.3 mEq	6.3 mEq
Cl^-	0 mEq	6.3 mEq	6.3 mEq
HPO_4^{--}	0 mEq	16.5 mEq	16.5 mEq
$Acetate^-$	17 mEq	30 mEq	47 mEq
SO_4^{--}	0 mEq	2 mEq	2 mEq

   Given the preceding information, how many milliliters of:

   a. Na Acetate 2 mEq/mℓ were added to the T.P.N. solution to supply the Na^+ as ordered? _____ mℓ
   b. K_2HPO_4 4.4 mEq/mℓ were added to the T.P.N. solution to supply the K^+ as ordered? _____ mℓ
   c. $MgSO_4$ 4 mEq/mℓ were added to the T.P.N. solution to supply the Mg^{++} as ordered? _____ mℓ
   d. $CaCl_2$ 1.36 mEq/mℓ were added to the T.P.N. solution to supply the Ca^{++} as ordered? _____ mℓ

2. 60 g of protein—600 g of dextrose 50% per day
Doctor: Anthony
Patient: Jim Kinney
Room: 213

To get 60 g of *protein per day*, 706 mℓ of FreAmine III® 8.5% is needed.

To get 20 g of *protein per bottle*, 235 mℓ of FreAmine III® 8.5% is needed.

The electrolytes obtained from 235 mℓ of FreAmine III® 8.5% solution are as follows:

Electrolyte	Milliequivalents per bottle
Na^+	2 mEq
HPO_4^{--}	2 mEq
Acetate⁻	17 mEq

Electrolyte	From FreAmine III®	Added	Total amount per bottle
Na^+	2 mEq	28 mEq	30 mEq
K^+	0 mEq	40 mEq	40 mEq
Mg^{++}	0 mEq	2 mEq	2 mEq
Ca^{++}	0 mEq	3.4 mEq	3.4 mEq
Cl^-	0 mEq	[40 mEq + 3.4 mEq]	43.4 mEq
HPO_4^{--}	2 mEq	28 mEq	30 mEq
SO_4^{--}	0 mEq	2 mEq	2 mEq
Acetate⁻	17 mEq	0 mEq	17 mEq

Given the preceding information, how many milliliters of:

a. Na_2HPO_4 4 mEq/mℓ were added to the T.P.N. solution to supply the Na^+ as ordered? _____ mℓ

b. KCl 2 mEq/mℓ were added to the T.P.N. solution to supply the K^+ as ordered? _____ mℓ

c. $MgSO_4$ 4 mEq/mℓ were added to the T.P.N. solution to supply the Mg^{++} as ordered? _____ mℓ

d. $CaCl_2$ 1.36 mEq/mℓ were added to the T.P.N. solution to supply the Ca^{++} as ordered? _____ mℓ

3. 60 g of protein—600 g of dextrose 50% per day
Doctor: Spoto
Patient: Susan Davison
Room: 211

To get 60 g of *protein per day*, 1091 mℓ of Travisol® 5.5% is needed.

To get 20 g of *protein per bottle*, 364 mℓ of Travisol® 5.5% is needed.

Application III Total Parenteral Nutrition

The electrolytes obtained from 364 mℓ of Travisol® 5.5% solution are as follows:

Electrolyte	Milliequivalents per bottle
Na$^+$	0 mEq
K$^+$	0 mEq
Mg^{++}	0 mEq
Acetate$^-$	13 mEq
Cl$^-$	8 mEq
HPO$_4^{--}$	0 mEq

Electrolyte	From Travisol® 5.5%	Added	Total amount per bottle
Na$^+$	0 mEq	30 mEq	30 mEq
K$^+$	0 mEq	30 mEq	30 mEq
Mg^{++}	0 mEq	0 mEq	0 mEq
Ca^{++}	0 mEq	3 mEq	3 mEq
Cl$^-$	8 mEq	[8 mEq + 3 mEq]	11 mEq
HPO$_4^{--}$	0 mEq	0 mEq	0 mEq
Acetate$^-$	13 mEq	0 mEq	13 mEq

Given the preceding information, how many milliliters of:

a. KCl 2 mEq/mℓ were added to the T.P.N. solution to supply the K$^+$ as ordered? _____ mℓ

b. CaCl$_2$ 1.36 mEq/mℓ were added to the T.P.N. solution to supply the Ca^{++} as ordered? _____ mℓ

Appendices

Appendix 1 Abbreviations Used in Drug Administration

Abbreviation	Meaning
a.c.	before meals
aq.	water
b.i.d.	twice a day
$\bar{c}$	with
h.s.	at bedtime
O.D.	right eye
O.S.	left eye
O.U.	both eyes
p.c.	after meals
p.o.	by mouth
p.r.n.	as needed
q.d.	every day
q.i.d.	four times daily
stat.	immediately
t.i.d.	three times daily

Appendix 2 Unit Abbreviations

Abbreviation	Meaning
cc	cubic centimeter
mℓ	milliliter
ℓ	liter
℥	ounce
pt	pint
qt	quart
gr	grain
ʒ	dram
♏	minim
gtt	drop
tsp	teaspoonful
tbsp	tablespoonful
mcg	microgram
mg	milligram
g	gram
kg	kilogram

Appendix 3 Approximate Equivalents

	Metric		Apothecaries'	English
Weight	0.065 g or 65 mg	=	1 gr	
	1 g	=	15 gr	
	28.35 g	=		1 oz
	1 kg	=		2.2 lb
Length	1 m	=		39.37 in.
	2.54 cm	=		1 in.
Volume	1 mℓ	=		16 ♏
	30 mℓ	=		1 f℥
	480 mℓ	=		1 pt

Appendix 4 Table of Approximate Atomic Weights of Selected Elements

Name	Symbol	Atomic Weight
Aluminum	Al	27
Calcium	Ca	40
Carbon	C	12
Chlorine	Cl	35
Fluorine	F	19
Hydrogen	H	1
Lithium	Li	7
Magnesium	Mg	24
Nitrogen	N	14
Oxygen	O	16
Phosphorus	P	31
Potassium	K	39
Sodium	Na	23
Sulfur	S	32

Answers

Chapter 1

Section 1.1 Review Problems (p. 5)

    1. 29   **3.** 42   **5.** 45
    7. 51   **9.** 538   **11.** 1334
    13. 1891   **15.** 1160   **17.** 4400
    19. 9970

Section 1.2 Review Problems (p. 8)

    1. 11   **3.** 9   **5.** 9
    7. 266   **9.** 148   **11.** 90
    13. 218   **15.** 8889   **17.** 26,589
    19. 9999

Section 1.3 Review Problems (p. 11)

    1. 54   **3.** 336   **5.** 840
    7. 11,234   **9.** 81,213   **11.** 33,666
    13. 18,676   **15.** 185,496   **17.** 22,804,600
    19. 205,600

Section 1.4 Review Problems (p. 15)

    1. 41   **3.** 101   **5.** 89
    7. $13\frac{3}{4}$ **9.** $411\frac{7}{16}$ **11.** $66\frac{5}{12}$
    13. $121\frac{33}{40}$ **15.** $360\frac{11}{25}$ **17.** $1375\frac{9}{10}$
    19. $1881\frac{184}{481}$

362   Answers

Section 1.5 Review Problems (p. 17)

  1. 82 blood samples will be drawn.
  3. He needs to work 328 more hours.
  5. There will be 123 containers left to count for inventory.
  7. He will administer 54 capsules.
  9. There are 280 individual drawers in the hospital.

Chapter 1 Review Problems (p. 19)

1. 17	2. 85	3. 53
4. 88	5. 47	6. 68
7. 628	8. 227	9. 1270
10. 1012	11. 693	12. 1962
13. 3500	14. 5877	15. 3092
16. 37,662	17. 39	18. 36
19. 39	20. 75	21. 227
22. 160	23. 70	24. 366
25. 136	26. 166	27. 7778
28. 6662	29. 28,149	30. 6559
31. 23,988	32. 9181	33. 104
34. 51	35. 816	36. 798
37. 350	38. 714	39. 7874
40. 13,965	41. 17,523	42. 7095
43. 7776	44. 322,905	45. 19,548
46. 315,198	47. 120,276	48. 400,000
49. 61	50. 106	51. 101
52. 170	53. 87	54. 9
55. $8\frac{4}{5}$	56. $32\frac{1}{2}$	57. $57\frac{8}{13}$
58. $342\frac{3}{4}$	59. $191\frac{3}{5}$	60. $487\frac{24}{31}$
61. $1365\frac{9}{10}$	62. $18\frac{267}{911}$	63. $2159\frac{86}{421}$
64. $391\frac{87}{151}$	65. 504	66. 8
67. 1204	68. 75	69. 2079
70. 2321	71. 29	72. 139
73. 88	74. 64,405	75. 802
76. 68	77. 114	78. 24,510
79. 980	80. 78	81. $11\frac{1}{7}$

82. 28   **83.** 1018   **84.** 9

85. $26\frac{875}{958}$ **86.** 864   **87.** 14,664

88. 215   **89.** $13\frac{8}{9}$ **90.** 3508

91. 24,998   **92.** 144   **93.** $230\frac{59}{258}$

94. 28,980

95. For two days he requires 48 drops.

96. The treadmill will be in operation for 545 minutes.

97. 846 units are left.

98. There were 25 capsules in each vial.

99. Each one will handle 5 surgeries.

100. There will be 16 milliliters left in the bottle.

Chapter 2

Section 2.1 Review Problems (p. 28)

1. proper   **3.** mixed number   **5.** complex

7. improper   **9.** proper   **11.** $\frac{3}{2}$

13. $\frac{143}{12}$ **15.** $\frac{665}{13}$ **17.** $1\frac{1}{2}$

19. $3\frac{1}{25}$ **21.** $\frac{2}{3}$ **23.** $\frac{1}{5}$

25. $\frac{1}{9}$

Section 2.2 Review Problems (p. 36)

1. $\frac{7}{9}$ **3.** $\frac{7}{16}$ **5.** $2\frac{1}{4}$

7. $1\frac{3}{4}$ **9.** $\frac{24}{34}$ **11.** $\frac{56}{245}$

13. $\frac{900}{1000}$ **15.** 210   **17.** 264

19. $1\frac{881}{2100}$ **21.** $1\frac{3}{280}$ **23.** $2\frac{187}{780}$

25. $2\frac{31}{110}$ **27.** $2\frac{2}{5}$ **29.** $2\frac{1}{24}$

31. $2\frac{2}{21}$ **33.** $1\frac{7}{96}$ **35.** $1\frac{11}{28}$

37. $1\frac{1}{4}$ **39.** $\frac{19}{56}$ **41.** $\frac{107}{126}$

43. $\frac{31}{42}$ **45.** $1\frac{253}{840}$

Answers

47. Dr. Dugan spent $2\frac{2}{15}$ hours with his patients (or 2 hours, 8 minutes).

49. The partially filled jars will add up to $\frac{23}{24}$ of one jarful.

Section 2.3 Review Problems (p. 39)

1. $\frac{1}{2}$
3. $\frac{1}{2}$
5. $\frac{11}{21}$
7. $\frac{1}{8}$
9. $\frac{9}{16}$
11. $\frac{5}{8}$
13. $2\frac{1}{2}$
15. $6\frac{1}{6}$
17. $\frac{10}{21}$
19. $\frac{48}{65}$

Section 2.4 Review Problems (p. 43)

1. yes
3. no
5. yes
7. yes
9. yes
11. no
13. $\frac{1}{4}, \frac{1}{3}, \frac{1}{2}$
15. $\frac{1}{2}, \frac{5}{6}, \frac{11}{12}$
17. $\frac{1}{4}, \frac{1}{3}, \frac{7}{12}$
19. $\frac{5}{12}, \frac{1}{2}, \frac{2}{3}$
21. $\frac{11}{12}, \frac{5}{6}, \frac{3}{4}$
23. $\frac{3}{4}, \frac{1}{2}, \frac{3}{10}$
25. $\frac{4}{5}, \frac{3}{4}, \frac{3}{10}$

Section 2.5 Review Problems (p. 49)

1. $7\frac{1}{2}$
3. $36\frac{2}{3}$
5. $10\frac{1}{3}$
7. $14\frac{41}{42}$
9. $16\frac{5}{12}$
11. $18\frac{17}{33}$
13. $112\frac{3}{8}$
15. $38\frac{3}{5}$
17. $22\frac{1}{3}$
19. $43\frac{1}{4}$
21. $5\frac{1}{4}$

23. $23\frac{1}{2}$ teaspoonsful are left in the bottle.

25. He rode $11\frac{1}{2}$ miles.

Section 2.6 Review Problems (p. 51)

1. $\frac{18}{35}$
3. $\frac{11}{16}$
5. $\frac{5}{11}$
7. $\frac{91}{160}$
9. 11
11. $26\frac{2}{3}$
13. 19
15. $62\frac{2}{5}$
17. $9\frac{3}{8}$

19. $17\frac{3}{5}$ 21. $20\frac{3}{20}$ 23. $135\frac{27}{32}$

25. $4\frac{8}{9}$

Section 2.7 Review Problems (p. 54)

1. $\frac{3}{7}$ 3. $\frac{14}{23}$ 5. $\frac{5}{8}$

7. $11\frac{2}{3}$ 9. $2\frac{1}{32}$ 11. $1\frac{3}{8}$

13. $11\frac{1}{3}$ 15. $1\frac{16}{17}$ 17. $\frac{17}{27}$

19. $1\frac{1}{3}$ 21. $\frac{5}{9}$ 23. $3\frac{13}{32}$

25. $2\frac{89}{260}$

Chapter 2 Review Problems (p. 58)

1. complex 2. proper 3. proper
4. improper 5. mixed number 6. complex
7. improper 8. mixed number 9. $\frac{29}{6}$
10. $\frac{59}{8}$ 11. $\frac{148}{11}$ 12. $\frac{344}{7}$
13. $2\frac{1}{3}$ 14. $3\frac{1}{7}$ 15. $2\frac{5}{18}$
16. $2\frac{8}{45}$ 17. $\frac{1}{4}$ 18. $\frac{1}{3}$
19. $\frac{1}{5}$ 20. $\frac{1}{9}$ 21. $\frac{1}{3}$
22. $\frac{1}{6}$ 23. no 24. no
25. no 26. yes 27. yes
28. yes 29. yes 30. no
31. yes 32. no 33. $\frac{3}{10}, \frac{1}{2}, \frac{3}{4}$
34. $\frac{1}{8}, \frac{5}{32}, \frac{3}{16}$ 35. $\frac{3}{16}, \frac{3}{4}, \frac{7}{8}$ 36. $\frac{5}{16}, \frac{1}{2}, \frac{7}{12}$
37. $\frac{1}{8}, \frac{18}{64}, \frac{5}{16}$ 38. $\frac{1}{9}, \frac{10}{27}, \frac{2}{3}$ 39. $\frac{1}{3}, \frac{1}{2}, \frac{3}{4}$
40. $\frac{3}{5}, \frac{75}{100}, \frac{21}{25}$ 41. $\frac{1}{16}, \frac{3}{8}, \frac{1}{2}$ 42. $\frac{3}{11}, \frac{1}{3}, \frac{8}{9}$
43. $\frac{7}{10}, \frac{2}{5}, \frac{1}{5}$ 44. $\frac{2}{5}, \frac{3}{10}, \frac{4}{15}$ 45. $\frac{2}{3}, \frac{1}{2}, \frac{1}{24}$

46. $\frac{2}{3}, \frac{5}{21}, \frac{1}{7}$ 47. $\frac{7}{8}, \frac{5}{6}, \frac{3}{4}$ 48. $\frac{1}{2}, \frac{3}{7}, \frac{5}{14}$

49. $\frac{17}{30}, \frac{5}{10}, \frac{6}{15}$ 50. $\frac{2}{3}, \frac{11}{100}, \frac{1}{10}$ 51. $\frac{3}{4}, \frac{7}{12}, \frac{1}{2}$

52. $\frac{21}{25}, \frac{82}{100}, \frac{3}{5}$ 53. $1\frac{1}{8}$ 54. $1\frac{4}{5}$

55. $1\frac{4}{13}$ 56. $1\frac{1}{3}$ 57. $1\frac{5}{7}$

58. $1\frac{1}{2}$ 59. $\frac{21}{35}$ 60. $\frac{143}{156}$

61. $\frac{285}{625}$ 62. $\frac{28}{60}$ 63. 12

64. 168   65. 4080   66. 660

67. $1\frac{923}{1386}$ 68. $2\frac{523}{700}$ 69. $2\frac{5}{112}$

70. $\frac{35}{176}$ 71. $7\frac{4}{5}$ 72. $17\frac{4}{15}$

73. $29\frac{8}{15}$ 74. $23\frac{1}{6}$ 75. $12\frac{19}{56}$

76. $435\frac{25}{84}$ 77. $\frac{2}{9}$ 78. $\frac{5}{12}$

79. $7\frac{11}{26}$ 80. $39\frac{201}{232}$ 81. $293\frac{29}{48}$

82. $8\frac{2}{3}$ 83. $\frac{15}{56}$ 84. $\frac{7}{17}$

85. $83\frac{4}{7}$ 86. $3\frac{41}{63}$ 87. $16\frac{3}{16}$

88. $15\frac{1}{5}$ 89. $\frac{14}{25}$ 90. $2\frac{1}{8}$

91. $31\frac{1}{2}$ 92. $1\frac{19}{22}$ 93. $\frac{14}{17}$

94. $2\frac{3}{8}$

95. He used $10\frac{11}{24}$ inches of Testape®.

96. Dr. Phillips gave $1\frac{13}{24}$ jarfuls to her patients.

97. The patient will have taken $52\frac{1}{2}$ tablets.

98. $20\frac{19}{24}$ inches were not used.

99. There are $100\frac{1}{2}$ inches left over.

100. Baby Franklin received a total of $1\frac{2}{3}$ teaspoonsful.

Chapter 3

Section 3.1 Review Problems (p. 66)

1. two and five tenths
3. eighteen and seven tenths
5. eight and fourteen hundredths
7. one and one hundred twenty-three thousandths
9. twelve and one thousandth
11. eighty three millionths

13. 3.6	15. 19.3	17. 4.21
19. 2.345	21. 13.013	23. 0.000011
25. 0.005		

Section 3.2 Review Problems (p. 71)

1. $\frac{4}{5}$	3. $\frac{8}{25}$	5. $\frac{1}{4}$
7. $\frac{627}{1000}$	9. $\frac{3}{8}$	11. $\frac{9}{16}$
13. 0.2	15. 0.4	17. 0.75
19. 0.8125	21. 0.1666	23. 0.3437
25. 0.4166		

Section 3.3 Review Problems (p. 77)

1. 104.668	3. 36.7406	5. 1604.184
7. 222.52758	9. 83.922147	11. 2.89
13. 282.1541	15. 0.747743	17. 120.999
19. 0.005, 0.001, 0.006	21. 0.175 mg	23. 25.75 mg
25. 12.6 mg		

Section 3.4 Review Problems (p. 82)

1. 4.1	3. 21.6	5. 15.86
7. 39.21	9. 10	11. 211.007
13. 316.142	15. 475.012	17. 0.8, 0.845
19. 5.7, 5.716	21. 0.5, 0.501	23. 3.3, 3.325
25. 0.0, 0.005		

Section 3.5 Review Problems (p. 90)

1. 31.95	3. 773.64	5. 1393.02
7. 0.0116	9. 0.750	11. 555,220
13. 115.9	15. 2.667	17. 0.194
19. 547.5 g	21. 35 days	23. 0.5 mℓ
25. 166.25 g		

Answers

Chapter 3 Review Problems (p. 94)

1. two and four tenths
2. two and three tenths
3. seventeen and eight tenths
4. thirty-seven and eighty-eight hundredths
5. fifteen and twenty-six hundredths
6. four and nine hundred eighty-seven thousandths
7. four and four ten-thousandths
8. one thousandth
9. three hundred seventy-five hundred-thousandths
10. 4.8
11. 20.1
12. 3.24
13. 44.44
14. 8.567
15. 3.107
16. 14.014
17. 0.0075
18. 0.000012
19. $\frac{2}{5}$
20. $\frac{1}{10}$
21. $\frac{13}{20}$
22. $\frac{22}{25}$
23. $\frac{11}{50}$
24. $\frac{463}{1000}$
25. $\frac{5}{8}$
26. $\frac{7}{8}$
27. $\frac{11}{16}$
28. 0.6
29. 0.5
30. 0.7
31. 0.9375
32. 0.3125
33. 0.1875
34. 0.6667
35. 0.0909
36. 0.0313
37. 0.2188
38. 12.7108
39. 153.546
40. 25.451
41. 67.666
42. 79.727
43. 308.909
44. 957.332
45. 95.125
46. 933.768
47. 219.702
48. 23.123
49. 90.4102
50. 2.35
51. 3.61
52. 1.88
53. 12.31
54. 46.17
55. 44.038
56. 28.272
57. 113.518
58. 107.148
59. 3
60. 20
61. 51.3
62. 67.2
63. 18.58
64. 41.68
65. 13.30
66. 155.913
67. 121.606
68. 5.199
69. 2.3, 2.316
70. 0.2, 0.154
71. 6.7, 6.715
72. 6.4, 6.446
73. 0.1, 0.120
74. 4.3, 4.254
75. 32.96
76. 0.9
77. 8.832
78. 298.422
79. 2042.87
80. 0.0106
81. 0.021
82. 0.15384
83. 1.5
84. 0.1395

Chapter 4

85. 384.1   86. 425   87. 0.042
88. 2.692   89. 10   90. 0.089
91. 3.333   92. 0.0231   93. 0.0006
94. 4.8035   95. 24.545 μ 96. 0.03 mg
97. 16.5 cm   98. 5.59 IU (vit. D), 9.4 mg (vit. C)
99. four times per day   100. 3.5 mg

Chapter 4

Section 4.1 Review Problems (p. 100)

1. 1 : 6   3. 13 : 7   5. 1 : 4
7. 1 : 10,000   9. 4 : 3   11. $\frac{33}{25}$
13. $\frac{1}{8}$ 15. $\frac{203}{18}$ 17. $\frac{1}{2}$
19. $\frac{101}{6}$ 21. $\frac{13}{20}$ 23. $\frac{121}{13}$ (is already reduced)
25. $\frac{15}{16}$ (is already reduced)

Section 4.2 Review Problems (p. 107)

1. $N = 8$ 3. $G = 8$ 5. $I = 33\,3/4$
7. $K = 12$ 9. $M = 10$ 11. $P = 2$
13. $R = 10$ 15. $T = 2.5$ 17. $V = 20$
19. $X = 0.033$
21. There were 10 R.N.s on duty.
23. The diabetic used 250 units of insulin X.
25. Burn Patient B will use $1\,1/3$ jarfuls of burn cream per day.

Section 4.3 Review Problems (p. 112)

1. 720 hr   3. 60,480 sec   5. 4032 min
7. $12\,4/5$ cups   9. 32 fl oz   11. 57.6 oz
13. 224 fl oz   15. 10,800 sec   17. 55 min
19. 3.5 d

Chapter 4 Review Problems (p. 114)

1. 1 : 5   2. 2 : 8   3. 13 : 10
4. 12 : 144   5. 3 : 13   6. 10 : 1
7. 100 : 1   8. 1 : 200   9. 1 : 130
10. 1 : 15,000   11. $\frac{12}{25}$ 12. $\frac{1}{2}$

13. $\frac{18}{37}$ 14. $\frac{63}{188}$ 15. $\frac{9}{1}$

16. $\frac{34}{7}$ 17. $\frac{1}{25}$ 18. $\frac{23}{14}$

19. $B = 21$ 20. $C = \frac{15}{32}$ 21. $D = 4\frac{1}{2}$

22. $G = 24$ 23. $H = 1$ 24. $L = 11\frac{2}{3}$

25. $M = 1$ 26. $N = 13\frac{1}{3}$ 27. $P = 31\frac{1}{2}$

28. $Q = 12$ 29. $R = 1000$ 30. $S = 2\frac{1}{3}$

31. $T = 27$ 32. $U = 0.05$ 33. $V = 0.017$

34. He took 72 green tablets.

35. She should order 96 acetominophen suppositories.

36a. There are 143 nursing students in the program.

36b. There are 34 inhalation-therapy students in the program.

36c. There are 21 medical students involved in the program.

37. 604,800 sec   38. 7560 min   39. 80,640 min

40. 43,200 sec   41. 6400 oz   42. 4 pt

43. 48 cups   44. 24 cups   45. 8 fl oz

46. 6.4 fl oz   47. 2 pt   48. 24 fl oz

Chapter 5

Section 5.1 Review Problems (p. 120)

1. 18%   3. 12.5%   5. 62.7%

7. 817%   9. 10%   11. 0.31

13. 0.275   15. 0.04   17. 1.02

19. 0.552

Section 5.2 Review Problems (p. 123)

1. $92\frac{4}{13}\%$ 3. $706\frac{2}{3}\%$ 5. $966\frac{2}{3}\%$

7. 10%   9. $83\frac{1}{3}\%$ 11. $73\frac{1}{3}\%$

13. $\frac{37}{100}$ 15. $\frac{43}{100}$ 17. $\frac{3}{50}$

19. $\frac{1}{24}$ 21. $\frac{43}{240}$ 23. $\frac{1}{600}$

25. $3\frac{7}{10}$

Chapter 5   371

Section 5.3 Review Problems (p. 126)

1. 44.2 **3.** 1.83 **5.** 5/16
7. 100 **9.** 6.944% or ≈ 7% **11.** 516.129% or ≈ 516%
13. 81.818% or ≈ 82% **15.** 27.778% or ≈ 28%

Section 5.4 Review Problems (p. 130)

1. 3.75 g. Weigh 3.75 g of drug and place it in a graduate that can hold 125 mℓ. Fill the graduate with distilled water to the 125-mℓ mark. Stir well.

3. 0.15 g. Weigh 0.15 g of gentian violet crystals and place it in a graduate that can hold 15 mℓ. Fill the graduate with rubbing alcohol to the 15-mℓ mark. Stir well.

5. 7.5 g. Weigh 7.5 g of neomycin sulfate powder and place it in a 1500-mℓ graduate. Fill the graduate with normal saline to the 1500-mℓ mark. Stir well.

Section 5.5 Review Problems (p. 135)

1. 0.015 g. Weigh 0.015 g of potassium permanganate crystals and place it in a 100-mℓ graduate. Fill the graduate with distilled water to the 75-mℓ mark. Stir well.

3. 2.5 mℓ. Measure 2.5 mℓ of stock solution and place it in a graduate that can hold 125 mℓ. Fill the graduate with alcohol to the 125-mℓ mark. Stir well.

5. 0.2 mℓ. Measure 0.2 mℓ and place it in a 250-mℓ graduate. Fill the graduate with tap water to the 250-mℓ mark. Stir well.

Chapter 5 Review Problems (p. 137)

1. 55% **2.** 23% **3.** 13.8%
4. 12.3% **5.** 675% **6.** 154%
7. 428% **8.** 372% **9.** 60%
10. 50% **11.** 0.36 **12.** 0.20
13. 0.815 **14.** 0.05 **15.** 0.073
16. 0.144 **17.** 0.008 **18.** 2.01
19. 1.23 **20.** 5.65 **21.** $81\frac{9}{11}\%$
22. $12\frac{1}{2}\%$ **23.** $83\frac{1}{3}\%$ **24.** 350%
25. 725% **26.** $833\frac{1}{3}\%$ **27.** $37\frac{1}{2}\%$
28. 90% **29.** 60% **30.** 65%
31. $\frac{3}{4}$ **32.** $\frac{51}{100}$ **33.** $\frac{1}{25}$
34. $\frac{1}{50}$ **35.** $\frac{27}{500}$ **36.** $\frac{59}{800}$
37. $\frac{1}{175}$ **38.** $\frac{1}{150}$ **39.** $\frac{51}{400}$
40. $\frac{2}{225}$ **41.** 21 **42.** 7.29 or ≈ 7.3

372 Answers

43. 60 **44.** $\frac{2}{3}$ **45.** 720

46. $16\frac{2}{3}$ **47.** 520 **48.** 345.45% or ≈ 345%

49. 8.33% or ≈ 8% **50.** 70% **51.** 4250 patients

52. 5.56% or ≈ 6%

53. 4.5 g. Weigh 4.5 g of sodium chloride and place it in a 500-mℓ graduate. Fill the graduate with distilled water to the 500-mℓ mark. Stir well.

54. 5.6 g. Weigh 5.6 g of magnesium sulfate crystals and place it in a graduate that will hold 112 mℓ. Fill the graduate with distilled water to the 112-mℓ mark. Stir well.

55. 1.2 g. Weigh 1.2 g of hydrocortisone powder and place it in a graduate that will hold 120 mℓ. Fill the graduate with lotion to the 120-mℓ mark. Stir well.

56. 0.37 g. Weigh 0.37 g of potassium permanganate crystals and place it in a 100-mℓ graduate. Fill the graduate with distilled water to the 55-mℓ mark. Stir well.

57. 4 mℓ. Measure 4 mℓ of stock solution and place it in a 1000-mℓ graduate. Fill the graduate with distilled water to the 1000-mℓ mark. Stir well.

58. 2.4 mℓ. Measure 2.4 mℓ of stock solution and place it in a 50-mℓ graduate. Fill the graduate with alcohol to the 45-mℓ mark. Stir well.

59. 1.03 mℓ. Measure 1.03 mℓ of stock solution and place it in a graduate that will hold 175 mℓ. Fill the graduate with alcohol to the 175-mℓ mark.

60. 3.33 mℓ. Measure 3.33 mℓ of stock solution and place it in a 100-mℓ graduate. Fill the graduate with distilled water to the 76-mℓ mark. Stir well.

Chapter 6

Section 6.1 Review Problems (p. 145)

1. 0.000082 m **3.** 0.000076 m **5.** 0.004 mm
7. 0.00003 cm **9.** 0.00005 cm **11.** 0.000000141 km
13. 0.657 m **15.** 0.477 cm **17.** 0.00003281 km
19. 8.109 m **21.** 5.09 m **23.** 0.0625 m
25. 0.0050394 km **27.** 0.04931 km **29.** 24.8 cm
31. 70.7 cm **33.** 10,303 mm **35.** 3281 mm
37. 325,000 m **39.** 473 cm

Section 6.2 Review Problems (p. 151)

1. 0.075 ℓ **3.** 50 mℓ **5.** 0.48 ℓ
7. 0.002 ℓ **9.** 0.35 ℓ **11.** 3.15 ℓ
13. 11 hℓ **15.** 1.3 kℓ **17.** 423.4 mℓ
19. 0.0004 ℓ **21.** 1 ℓ **23.** 317 cc
25. 320 cc **27.** 270 mℓ **29.** 420 cc
31. 63,000 cc **33.** 10,000 mℓ **35.** 101 mℓ
37. 25,000 cc **39.** 7960 cc

Section 6.3 Review Problems (p. 156)

1. 80,000 mcg	3. 370 mcg	5. 0.001439 g
7. 0.000005 g	9. 50 g	11. 1400 g
13. 0.00019 kg	15. 0.000014 kg	17. 0.00003 kg
19. 0.000018 g	21. 0.063 g	23. 1,052,000 g
25. 18,500 mg	27. 5050 dg	29. 2460 dag
31. 0.000182 kg	33. 0.343 g	35. 0.03 kg
37. 484 g	39. 0.025 kg	

Section 6.4 Review Problems (p. 159)

1. 48.9° C	3. 28° C	5. 71.1° C
7. 100° C	9. 68° F	11. 110.1° F
13. 181.9° F	15. 204.4° F	

Chapter 6 Review Problems (p. 161)

1. 0.000031 m	2. 0.000048 m	3. 0.006 mm
4. 0.005 mm	5. 0.00004 cm	6. 0.000035 cm
7. 0.000344 m	8. 0.001 m	9. 0.000047 m
10. 0.0000389 m	11. 0.000055359 km	12. 0.00001995 km
13. 0.07096 m	14. 0.351 m	15. 6.03 m
16. 3.837 m	17. 0.0000687 km	18. 0.00007046 km
19. 23.88 m	20. 58.01 m	21. 2989 mm
22. 9125 mm	23. 0.372 cℓ	24. 2 cℓ
25. 4.81 dℓ	26. 0.015 ℓ	27. 300 mℓ
28. 0.0014 ℓ	29. 0.0016 ℓ	30. 0.017 ℓ
31. 300 cc	32. 75 cc	33. 3550 cc
34. 5230 mℓ	35. 27,000 mℓ	36. 41,000 mℓ
37. 18,000 cc	38. 103,000 cc	39. 1 cc
40. 11 cc	41. 500 mℓ	42. 450 mℓ
43. 30,000 mcg	44. 200 mcg	45. 0.000824 g
46. 0.000007 g	47. 110 g	48. 75,000 g
49. 2900 g	50. 0.004008 kg	51. 0.0027 kg
52. 0.00002 kg	53. 0.00004 kg	54. 0.000000016 kg
55. 0.00000002 kg	56. 0.05 kg	57. 0.423 kg
58. 600 g	59. 15 kg	60. 0° C
61. 4.4° C	62. 15.6° C	63. 21.1° C
64. 27.3° C	65. 37.1° C	66. 65.6° C
67. 84.4° C	68. 196.3° F	69. 122° F
70. 42.8° F	71. 50° F	72. 99.5° F
73. 140.2° F	74. 166.1° F	75. 191.7° F

Chapter 7

Section 7.1 Review Problems (p. 167)

1. *i*	3. *c*	5. *l*
7. *x*	9. 5	11. 50
13. 1	15. 500	17. *lii*
19. *i$\overline{ss}$*	21. 8	23. 321
25. 2001		

Section 7.2 Review Problems (p. 172)

1. sixty minims
3. twenty-five fluidrams
5. three fluidounces
7. seven quarts
9. one and one-half gallons
11. ℞ *ccclx*
13. f℥ *xv*
15. pt *ii* or O *ii*
17. qt *xiii*
19. 1 fluidounce or f℥ *i*
21. 128 fluidounces or f℥ *cxxviii*
23. 1 pint or pt *i*
25. 1 quart or qt *i*
27. 2 quarts or qt *ii*
29. ½ pint or pt *$\overline{ss}$*
31. 1 gallon or gal *i*
33. 1920 minims or ℞ *mcmxx*
35. 256 fluidrams or f℥ *cclvi*
37. 128 fluidrams or f℥ *cxxviii*
39. 40 fluidrams or f℥ *xl*

Section 7.3 Review Problems (p. 178)

1. six grains
3. sixteen drams
5. twenty-five pounds
7. gr *i$\overline{ss}$*
9. ℥ *xv*
11. ℔ *ii*
13. 1½ drams or ℥ *i$\overline{ss}$*
15. 1½ ounces or ℥ *i$\overline{ss}$*
17. 2 pounds or ℔ *ii*
19. 1 ounce or ℥ *i*
21. 1 pound or ℔ *i*
23. 1440 grains or gr *mcdxl*
25. 480 drams or ℥ *cdlxxx*
27. 12 ounces or ℥ *xii*
29. 160 grains or gr *clx*
31. 26 drams or ℥ *xxvi*
33. 600 grains or gr *dc*
35. 360 grains or gr *ccclx*
37. 420 grains or gr *cdxx*
39. 64 drams or ℥ *lxiv*

Section 7.4 Review Problems (p. 188)

1. 0.48 mℓ
3. 15 mℓ
5. 22.5 mℓ
7. 1440 mℓ
9. 166.6 gtt or 167 gtt
11. 2 tsp
13. 2 tbsp
15. 1½ glassfuls
17. 11 minims or ℞ *xi*
19. 2⅖ tsp
21. 16 fluidrams or f℥ *xvi*
23. 1 fluidounce or f℥ *i*

25. 2 fluidounces or f℥ ii
27. 32 fluidrams or f ℨ xxxii
29. 12⅘ fluidrams or 12⅘ f ℨ
31. 225 minims or ♏ ccxxv
33. 2250 minims or ♏ mmccl
35. 1 gtt
37. 83.3 gtt or 83 gtt
39. 1 tsp
41. 5 tbsp
43. 12 tsp
45. 3/16 glassful
47. 1½ glassfuls
49. 1⅕ tsp

Chapter 7 Review Problems (p. 191)

1. s̄s̄
2. v
3. xvii
4. xlix
5. ccxxiv
6. cmi
7. mxlvi
8. mm
9. mmmviii
10. mmmdcccxcix
11. 12
12. 37
13. 50½
14. 54
15. 103
16. 438
17. 506
18. 1046
19. 1401
20. 3890
21. 14
22. 19
23. 36
24. 45
25. 12
26. ninety minims
27. nine fluidrams
28. sixteen pints
29. five and one-half pints
30. one-half gallon
31. ♏ cdlxxx
32. f ℨ iis̄s̄
33. f ℨ xxvi
34. qt xi
35. qt vii
36. ¾ fluidounce or ¾ f ℥
37. 2 fluidounces or f ℥ ii
38. 32 fluidounces or f ℥ xxxii
39. 1 fluidounce or f ℥ i
40. 1½ pint or pt is̄s̄
41. 3 pints or O iii
42. 2400 minims or ♏ mmcd
43. 120 minims or ♏ cxx
44. 180 minims or ♏ clxxx
45. 512 fluidrams or f ℨ dxii
46. 768 fluidrams or f ℨ dcclxviii
47. 2 fluidrams or f ℨ ii
48. 120 fluidrams or f ℨ cxx
49. twelve drams
50. fourteen drams
51. twenty-two pounds
52. twenty-eight pounds
53. gr ī
54. gr iiis̄s̄
55. ℨ xi
56. ℨ xxvi
57. 1 dram or ℨ ī
58. 2 drams or ℨ ii
59. 1½ ounces or ℥ is̄s̄
60. 4 ounces or ℥ iv
61. 1920 grains or gr mcmxx
62. 960 grains or gr cmlx
63. 30 ounces or ℥ xxx
64. 4 ounces or ℥ iv
65. 0.78 mℓ
66. 0.6 mℓ
67. 10 mℓ

68. 7.5 mℓ
69. 82.5 mℓ
70. 480 mℓ
71. 120 mℓ
72. 33.3 gtt or 33 gtt
73. 150 gtt
74. 2½ tsp
75. ½ tbsp
76. 3 tbsp
77. 8 tbsp
78. 17 minims or ♏ xvii
79. 8 minims or ♏ viii
80. 8 fluidrams or f ʒ viii
81. 16 fluidrams or f ʒ xvi
82. ½ fluidounce or f ʒ ss
83. ¼ fluidounce or ¼ f ʒ
84. 1 fluidounce or f ʒ ī
85. 1½ fluidounces or f ʒ iss
86. 4 fluidounces or 4 f ʒ iv
87. 24 fluidounces or f ʒ xxiv
88. 3 gtt
89. 5 gtt
90. 112.5 gtt or 113 gtt
91. 300 gtt
92. 3 tsp
93. 5 ¾ tsp
94. 2 tbsp
95. 5 tbsp
96. 3 tsp
97. 18 tsp
98. ⅜ glassful
99. 1 glassful
100. 2 2/15 tbsp

Chapter 8

Section 8.1 Review Problems (p. 199)

1. 45 gr
3. 154 lb
5. 44 lb
7. 130 mg
9. 42.5 g
11. 3189.4 gr
13. 0.8 g
15. 4 g
17. 45.5 kg
19. 7.7 gr
21. 3.5 oz
23. 1.8 oz
25. 255 gr

Section 8.2 Review Problems (p. 203)

1. 157.5 in.
3. 128 in.
5. 3.8 cm
7. 15.2 cm
9. 25.4 cm
11. 51.2 ♏
13. 136 ♏
15. 15 mℓ
17. 270 mℓ
19. 480 mℓ
21. 3840 mℓ
23. 360 mℓ
25. 0.6 mℓ
27. 393.7 cm
29. 5 in.
31. 30 in.
33. 1558.1 ♏
35. 3375.8 ♏
37. 3.9 f ʒ
39. 14.8 f ʒ
41. 0.2 pt
43. 3 pt
45. 128 ♏

Section 8.3 Review Problems (p. 206)

  1. suppository   **3.** Sig.   **5.** left eye

  7. a.l.   **9.** immediately   **11.** by mouth

  13. do not repeat   **15.** ℞

  17. Propoxyphene® HCl 32 mg: Take 2 capsules every 6–8 hours as needed for pain.

  19. Amoxil® 250 mg caps.: Give 1 capsule by mouth 3 times daily before meals until gone.

Section 8.4 Review Problems (p. 210)

  1. SGPT   **3.** bathroom privileges   **5.** PT

  7. subcutaneous   **9.** IPPB   **11.** PPD

  13. CSF   **15.** diabetes mellitus   **17.** intramuscular

  19. electrocardiogram   **21.** electrocardiogram

  23. Laboratory tests: complete blood count.

  25. Start ampicillin 500 mg orally four times daily for 7 days.

Chapter 8 Review Problems (p. 213)

1. 69 gr	**2.** 112.5 gr	**3.** 165 lb
4. 35.2 lb	**5.** 325 mg	**6.** 487.5 mg
7. 113.4 g	**8.** 49.6 g	**9.** 744.2 gr
10. 2551.5 gr	**11.** 0.7 g	**12.** 1.1 g
13. 68.2 kg	**14.** 56.8 kg	**15.** 11.5 gr
16. 1.75 gr	**17.** 2.6 oz	**18.** 3.5 oz
19. 2 oz	**20.** 196.9 in.	**21.** 472.4 in.
22. 3.0 cm	**23.** 7.9 cm	**24.** 48 ♏
25. 115.2 ♏	**26.** 129.6 ♏	**27.** 320 ♏
28. 15 mℓ	**29.** 300 mℓ	**30.** 450 mℓ
31. 120 mℓ	**32.** 5760 mℓ	**33.** 720 mℓ
34. 2.5 m	**35.** 40.8 m	**36.** 3 in.
37. 15 in.	**38.** 1298.4 ♏	**39.** 2596.8 ♏
40. 3635.5 ♏	**41.** 3.0 f℥	**42.** 7.9 f℥
43. 11.8 f℥	**44.** 15.8 f℥	**45.** 1.1 pt
46. 1.5 pt	**47.** 6.9 pt	**48.** 10.8 pt
49. ♏	**50.** rectal suppository	**51.** four times daily
52. as needed	**53.** capsule	**54.** a.d.
55. ad lib.	**56.** aq.	**57.** every other hour
58. p.c.	**59.** each eye	**60.** drop
61. one half	**62.** O.S.	**63.** every
64. Rept.		

65. Lorelco® 250-mg tablets: Take 2 tablets twice a day with the morning and evening meal.
66. Aldomet® 250-mg tablets: Take 1 tablet twice daily for 48 hours, then increase to 2 tablets twice daily.
67. Norpace® 150-mg capsules: Take 2 capsules immediately, then 1 capsule every 6 hours.
68. Sorbitrate® 2.5 mg sublingual tablets. Place 1 or 2 tablets under tongue as needed for angina pain.
69. Dyazide® capsules: Take 1 capsule twice daily after meals.
70. Prednisone 5-mg tablets: Take 4 tablets twice daily for 2 days, 3 tablets twice daily for 2 days, 2 tablets twice daily for 2 days, 1 tablet twice daily for 2 days, 1 tablet daily until gone.
71. K-Lyte® tablets: Dissolve 1 tablet in 6 ounces of water twice daily.
72. SOB
73. T and A
74. Rhesus factor
75. to keep open
76. NG
77. electrolytes
78. milliequivalents
79. Dx
80. fasting blood sugar
81. headache
82. gastrointestinal
83. EEG
84. urinary tract infection
85. WBC
86. H_2O
87. chronic obstructive pulmonary disease
88. P
89. per rectum
90. BC
91. Diagnosis: Nausea and vomiting
92. Laboratory tests: Complete blood count, electrolytes
93. Nothing by mouth
94. Intravenous: Normal saline 500 mℓ to keep open with potassium chloride 30 milliequivalents now.
95. Place nasogastric tube, low suction.
96. Arrange for barium enema tomorrow morning.
97. 5 gr
98. 2 gr
99. 5.1 cm
100. 2.8 oz

Chapter 9

Section 9.1 Review Problems (p. 223)

1. 2 tablets. Give 2 Furadantin® 50-mg tablets four times daily.
3. 2 capsules. Give 2 chloral hydrate gr $vii\overline{ss}$ capsules by mouth at bedtime.
5. 2 capsules. Give 2 Cleocin® 75-mg capsules by mouth every 6 hours.
7. 2 capsules. Give 2 Vibramycin® 50-mg capsules by mouth every day.
9. ½ tablet. Give ½ Periactin® 4-mg tablet by mouth four times daily as needed for itching.

11. 2 capsules. Give 2 Benadryl® 25-mg capsules orally at bedtime.

13. 3 tablets. Give 3 Isoniazid® tablets once daily in the morning.

15. 2 tablets. Give 2 ferrous sulfate 300-mg tablets by mouth three times daily after meals.

Section 9.2 Review Problems (p. 228)

1. Take 4 mℓ Lasix® oral solution 10 mg/mℓ once daily.

3. Take 0.25 mℓ Haldol® 2 mg/mℓ three times daily.

5. Give 2 tsp Benadryl® 12.5 mg/5 mℓ at bedtime as needed for sleep.

7. Give 4 tsp Phenergan® syrup 0.25 mg/5 mℓ immediately for nausea.

9. Take 4 tsp ampicillin oral suspension 125 mg/5 mℓ every 6 hours.

11. Give 2 mℓ Thorazine® concentrate 30 mg/mℓ three times daily and as needed.

13. Take 4 mℓ Dilantin-125 Pediatric® elixir 125 mg/5 mℓ three times daily.

Chapter 9 Review Problems (p. 230)

1. 4 tablets. Give 4 Aldomet® 125-mg tablets orally three times daily for blood pressure.

2. ½ tablet. Take ½ Reserpine® 0.25-mg tablet orally once daily.

3. 5 tablets. Take 5 Apresoline® 10-mg tablets by mouth four times daily.

4. ½ tablet. Take ½ Sudafed® 60-mg tablet orally four times daily for inner ear pressure.

5. 3 capsules. Give 3 Dilantin® 100-mg capsules by mouth each day.

6. ½ tablet. Give ½ Inderal® 20-mg tablet by mouth four times daily.

7. 2 tablets. Give 2 quinidine sulfate 100-mg tablets four times daily.

8. 2 tablets. Give 2 Sudafed® 30-mg tablets orally three times daily for congestion.

9. 2 tablets. Take 2 penicillin G 125-mg tablets every 6 hours.

10. 4 tablets. Take 4 V Cillin K® 200,000-unit tablets four times daily, before meals and at bedtime.

11. 2 tablets. Take 2 Coumadin® 2.5-mg tablets orally each day.

12. 2 tablets. Take 2 Lanoxin® 0.125-mg tablets by mouth once daily.

13. 2 capsules. Give 2 Keflex® 500-mg capsules orally four times daily.

14. 2 tablets. Give 2 Inderal® 10-mg tablets by mouth twice daily.

15. Give ½ tsp Depakene® syrup 250 mg/5 mℓ twice daily.

16. Give 2 tsp Symmetrel® syrup 50 mg/5 mℓ two times daily.

17. Take 1.2 mℓ Mylicon® drops 40 mg/0.6 mℓ after meals and at bedtime.

18. Take 2 tsp Velosef® oral suspension 125 mg/5 mℓ four times daily.

19. Take 2 tsp Ilosone® liquid 125 mg/5 mℓ four times daily.

20. Give 4 tsp NegGram® 250 mg/5 mℓ four times daily.

21. Give 2 tsp Sudafed® liquid 30 mg/5 mℓ every 4 hours as needed.

22. Give 37.5 mℓ Elixophyllin® elixir 80 mg/15 mℓ every 6 hours with food.

380 Answers

23. Take 18.75 mℓ Kaon-Cl® 20% liquid 40 mEq/15 mℓ diluted in 10 fluidounces of water daily.
24. Take 7.5 mℓ Feosol® elixir 220 mg/5 mℓ three times daily after meals.
25. Take 1 tbsp Decadron® elixir 0.5 mg/5 mℓ twice daily after meals.

Chapter 10

Section 10.1 Review Problems (p. 243)

1. Give 2 mℓ Lanoxin® 0.25 mg/mℓ intravenously every 6 hours for 2 doses.

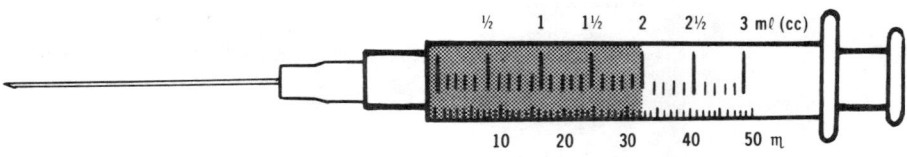

3. Give 1.5 mℓ Gynergen® 0.5 mg/mℓ subcutaneously at onset of attack.
5. Give 1.2 mℓ Kantrex® 1 g/3 mℓ intramuscularly every 8 hours. 1.2 mℓ of air.

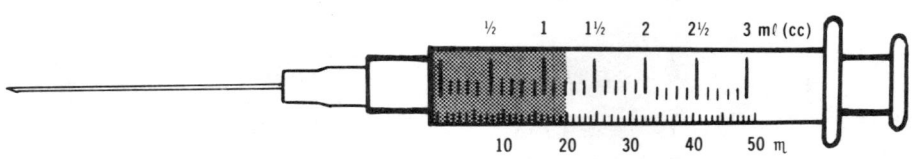

7. Give 0.25 mℓ ACTH 80 U/mℓ subcutaneously four times daily. 0.25 mℓ of air.

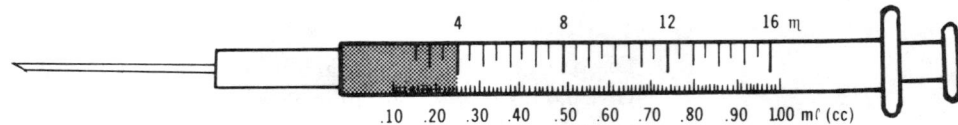

9. Give 1.5 mℓ Demerol® 50 mg/mℓ intramuscularly every 3 to 4 hours as needed for pain.

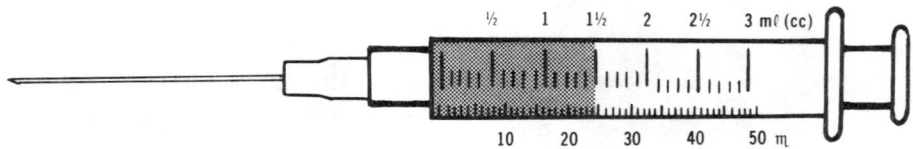

11. Give 3 mℓ Robaxin® 100 mg/mℓ intramuscularly every 8 hours as needed (inject in gluteal region).

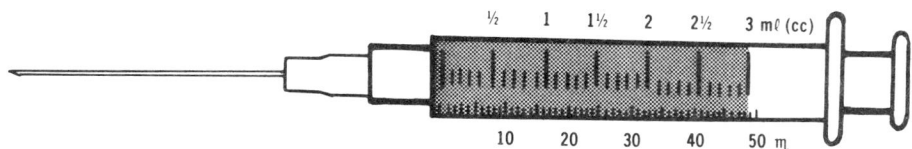

13. Give 1 mℓ Methergine® 0.2 mg/mℓ intramuscularly after delivery of placenta.

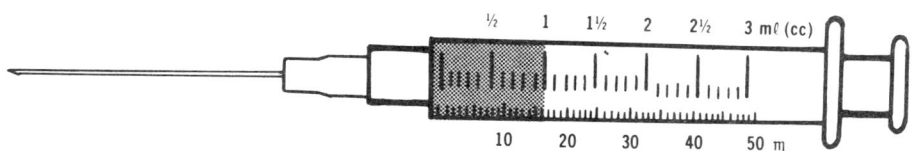

15. Give 3 mℓ Terramycin® 50 mg/mℓ intramuscularly every 24 hours. 3 mℓ of air.

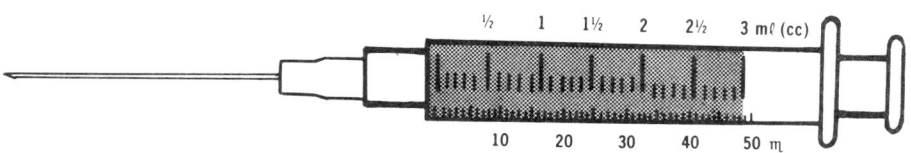

17. Give 2 mℓ Tigan® 100 mg/mℓ intramuscularly three times daily (deep injection into upper outer quadrant of the gluteal region).

19. Give 0.7 mℓ ephedrine sulfate 50 mg/mℓ subcutaneously now.

Section 10.2 Review Problems (p. 251)

1. Dissolve one 60-mg codeine sulfate tablet in 1 mℓ of sterile water for injection. Give 0.75 mℓ of this solution, which contains 45 mg of drug, subcutaneously three times daily as needed.
   1 tablet
   0.75 mℓ

3. Dissolve one 0.4-mg scopolamine HBr tablet in 0.5 mℓ of sterile water for injection. Give 0.4 mℓ of this solution, which contains 0.32 mg of drug, subcutaneously every 8 hours.
   1 tablet
   0.4 mℓ

5. Dissolve one gr-*i* codeine phosphate tablet in ℥ *xvi* of sterile water for injection. Give ℥ *ii* of this solution, which contains ⅛ gr of drug, subcutaneously every 10 hours as needed.
   1 tablet
   ℥ *ii*

7. Dissolve one 0.6-mg atropine sulfate tablet in 1 mℓ of sterile water for injection. Give 0.33 mℓ of this solution, which contains 0.2 mg of drug, intramuscularly pre-op.
   1 tablet
   0.33 mℓ

382   Answers

9. Dissolve two ¹/₂₀₀-gr atropine sulfate tablets in ♏ v of sterile water for injection. Give ♏ v of this solution, which contains ¹/₁₀₀ gr of drug, intramuscularly every 11 hours.
2 tablets
♏ v

Section 10.3 Review Problems (p. 258)

1. Give 2 mℓ penicillin G potassium 500,000 U/mℓ intramuscularly every 4 hours.

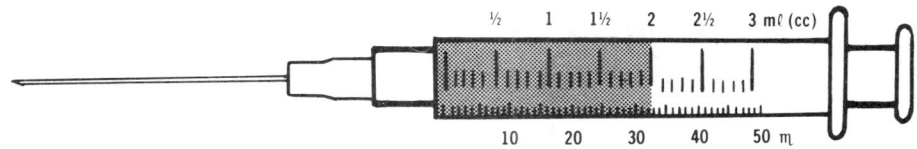

3. Give 1 mℓ Unipen® 250 mg/mℓ intramuscularly every 4 hours.

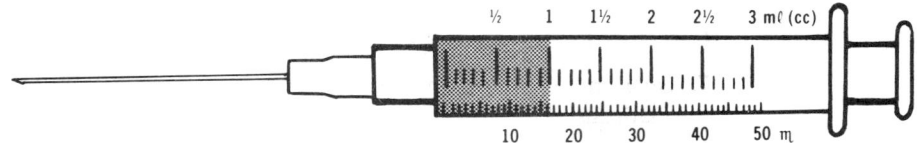

5. Give 2 mℓ ampicillin sodium 125 mg/mℓ intramuscularly every 6 hours.

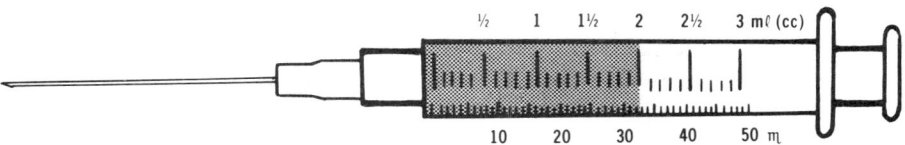

7. Give 1.5 mℓ penicillin G potassium 1,000,000 U/mℓ intramuscularly every 2 hours.

9. Give 3 mℓ carbenicillin 1 g/2.5 mℓ intramuscularly every 6 hours.

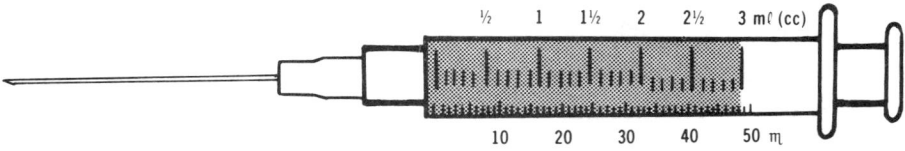

Chapter 10 Review Problems (p. 263)

1. Give 2 mℓ Vasodilan® 5 mg/mℓ intramuscularly twice daily.

2. Give 0.5 mℓ Phenergan® 50 mg/mℓ intramuscularly every 4 to 6 hours as needed for nausea and vomiting.

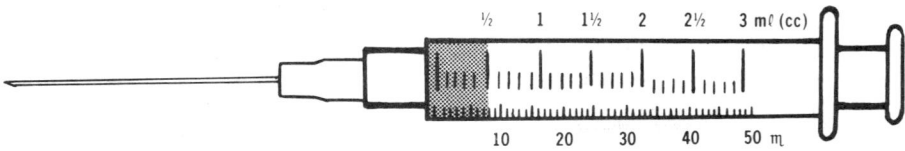

3. Give 0.6 mℓ Haldol® 5 mg/mℓ intramuscularly every 4 to 8 hours as needed for acute agitation.

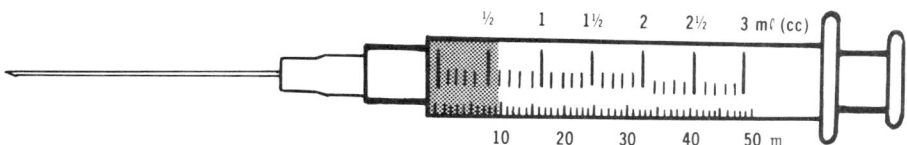

4. Give 3 mℓ Nembutol® 50 mg/mℓ intramuscularly at bedtime.
5. Give 0.2 mℓ Isuprel® 1:5000 subcutaneously now.

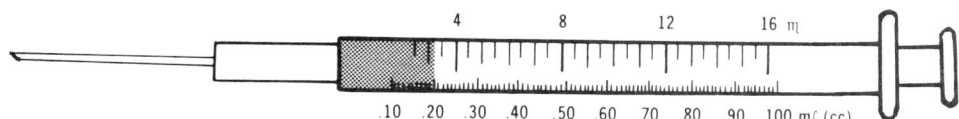

6. Give 0.3 mℓ Aramine® 10 mg/mℓ subcutaneously now.
7. Give 2 mℓ Compazine® 5 mg/mℓ intramuscularly now (deeply into the upper outer quadrant of the buttock).
8. Give 2 mℓ Vistaril® 25 mg/mℓ intramuscularly now (injected into the gluteus maximus region).

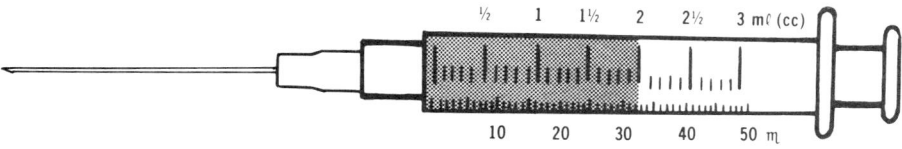

9. Give 0.75 mℓ Apresoline® 20 mg/mℓ intramuscularly now, repeat as needed.

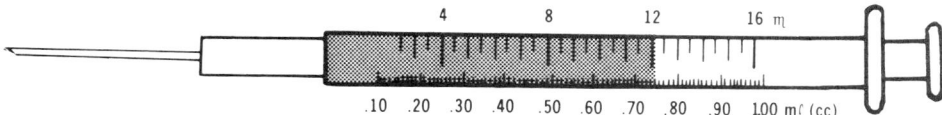

10. Give 0.25 mℓ Brethine® 1 mg/mℓ subcutaneously into lateral deltoid area.

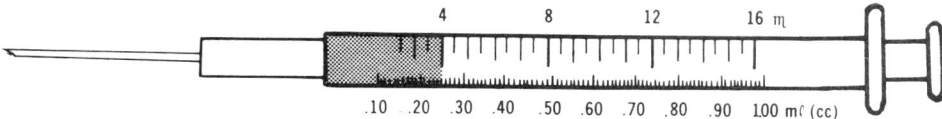

11. Give 0.25 mℓ Theelin aqueous® 2 mg/mℓ intramuscularly today, repeat in 3 days.
12. Give 1.2 mℓ Pronestyl® 500 mg/mℓ intramuscularly every 6 hours.

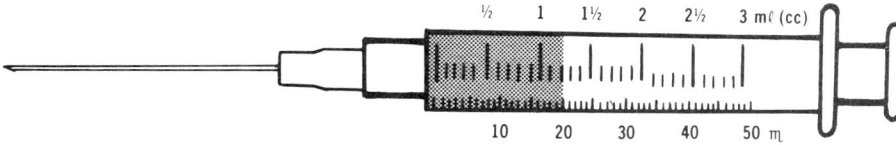

13. Give 0.5 mℓ Dilaudid® sulfate 2 mg/mℓ subcutaneously every 4 to 6 hours as needed for severe pain.

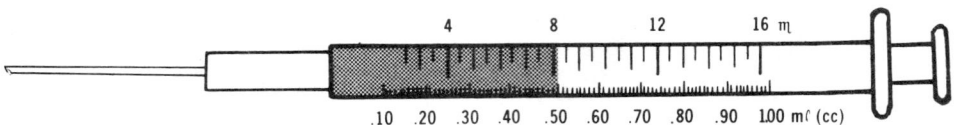

14. Give 0.67 mℓ Talwin® lactate 30 mg/mℓ subcutaneously every 3 to 4 hours as needed.

15. Give 2 mℓ Wyamine® 15 mg/mℓ intramuscularly immediately.

16. Give 2 mℓ Benadryl® 10 mg/mℓ intramuscularly (deeply) now.

17. Give 0.25 mℓ Pitressin® 20 U/mℓ intramuscularly three times daily as needed.

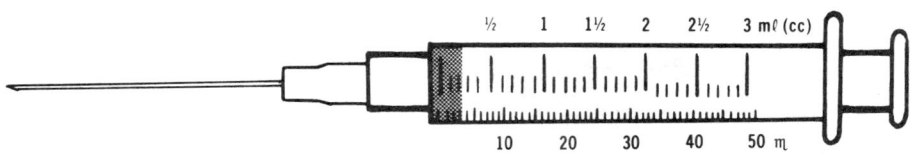

18. Dissolve one 0.4-mg atropine sulfate tablet in 4 mℓ of sterile water for injection. Give 1 mℓ of this solution, which contains 0.1 mg of drug, intramuscularly every 8 hours as needed.
    1 tablet
    1 mℓ

19. Dissolve four ¼-gr codeine phosphate tablets in ♏ viii of sterile water for injection. Give ♏ viii of this solution, which contains gr i of drug, subcutaneously now.
    4 tablets
    ♏ viii

20. Dissolve one 1/100-gr atropine sulfate tablet in ♏ x of sterile water for injection. Give 6.67 ♏ of this solution, which contains 1/150 gr of drug, subcutaneously every 3 to 4 hours.
    1 tablet
    6.67 ♏

21. Dissolve two 0.4-mg scopolamine HBr tablets in 0.8 mℓ of sterile water for injection. Give 0.5 mℓ of this solution, which contains 0.5 mg of drug, subcutaneously every 8 to 12 hours as needed.
    2 tablets
    0.5 mℓ

22. Give 1.8 mℓ ampicillin 250 mg/mℓ intramuscularly every 6 hours.

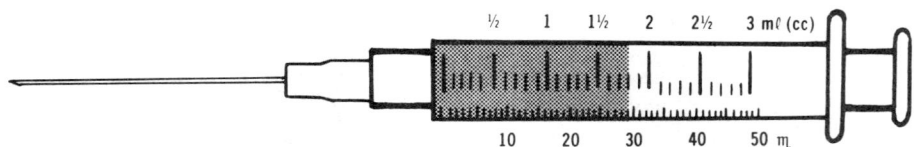

Chapter 12 385

23. Give 2.4 mℓ Geopen® 1 g/3 mℓ intramuscularly every 6 hours.

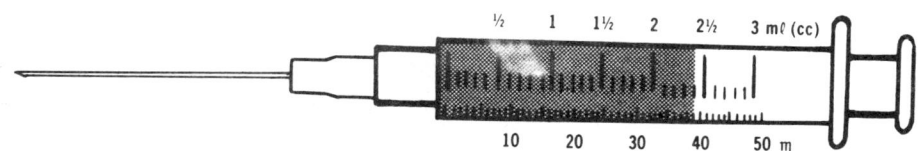

24. Give 1.1 mℓ Kefzol® 225 mg/mℓ intramuscularly every 6 hours.

25. Give 1.67 mℓ Loridine® 300 mg/mℓ intramuscularly every 12 hours.

Chapter 11

Section 11.1 Review Problems (p. 273)

    1. 2 gtt/min   **3.** 18 gtt/min   **5.** 28 gtt/min

    7. 83 gtt/min   **9.** 21 gtt/min

Section 11.2 Review Problems (p. 277)

    1. 4.8 hr   **3.** 7.3 hr   **5.** 8.3 hr

    7. 12.5 hr   **9.** 10 hr

Section 11.3 Review Problems (p. 280)

    1a. 15 mℓ

    1b. 22 gtt/min

    1c. Give aminophylline 375 mg daily diluted in 50 mℓ dextrose 5% in water over 45 minutes.

Chapter 11 Review Problems (p. 282)

    1. 42 gtt/min   **2.** 63 gtt/min   **3.** 56 gtt/min

    4. 42 gtt/min   **5.** 13 gtt/min   **6.** 4.6 hr

    7. 3.9 hr   **8.** 2 hr   **9.** 16.7 hr

    10a. 80 gtt/min

    10b. Give Mandol® 1 g diluted in 100 mℓ dextrose 5% in water "intravenous piggyback" every 8 hours over 20 minutes.

    10c. 40 minutes

Chapter 12

Section 12.1 Review Problems (p. 292)

    1.

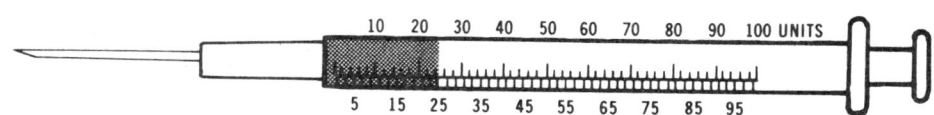

3. 48 units to be given with the U-100 syringe

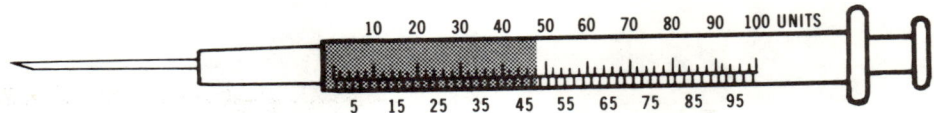

5. 0.8 ml

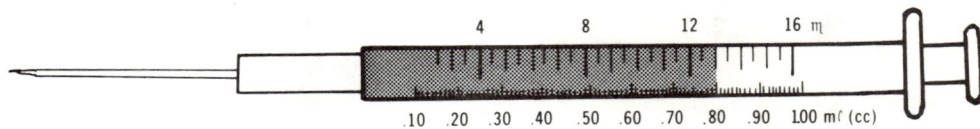

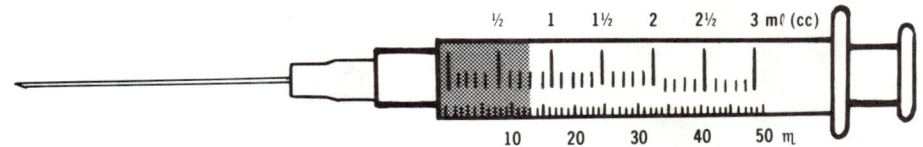

7. 12 ㎖

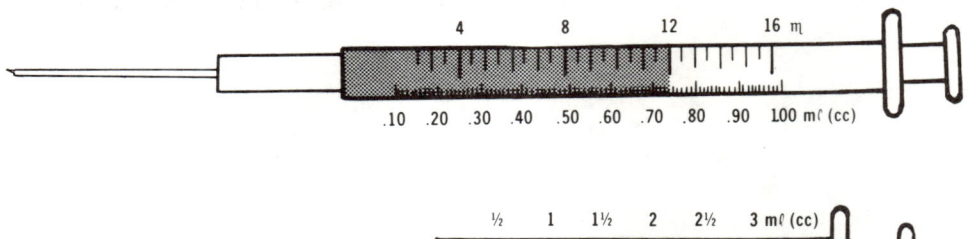

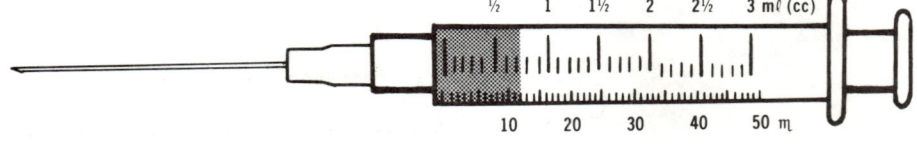

Section 12.2 Review Problems (p. 298)

1. 0.5 ml

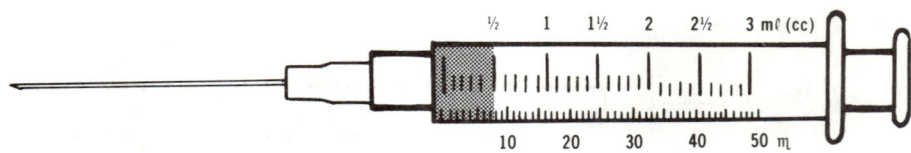

3. 7500 U

5. 4000 U

Chapter 12   387

Chapter 12 Review Problems (p. 301)

1.

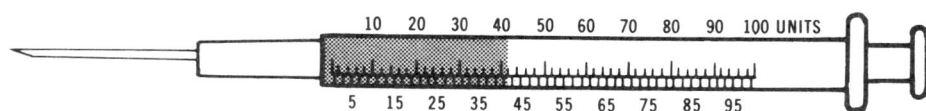

2.

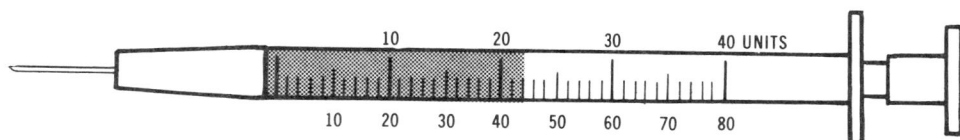

3.

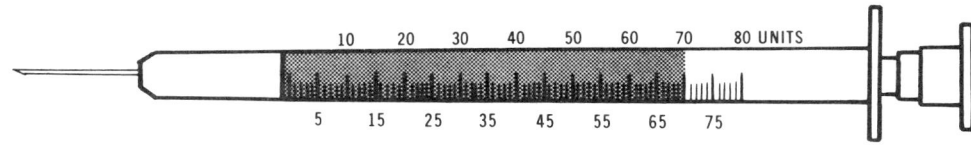

4. 5 units to be given with U-40 syringe

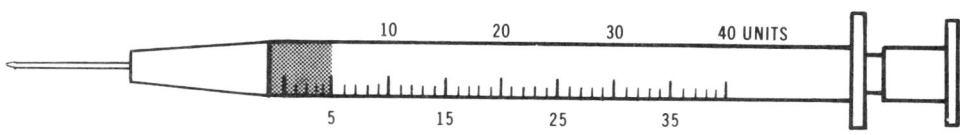

5. 1.6 units using U-40 syringe scale
3.2 units using U-80 syringe scale

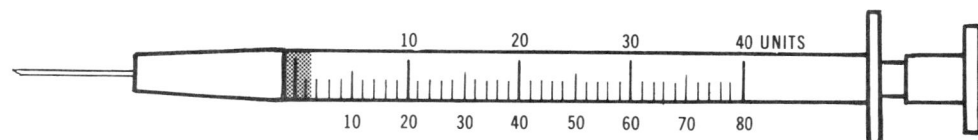

388 Answers

6. 9 ɱ

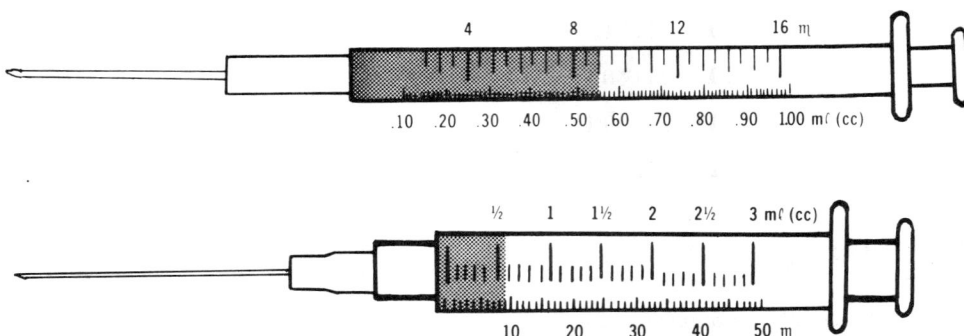

7. 0.4 mℓ

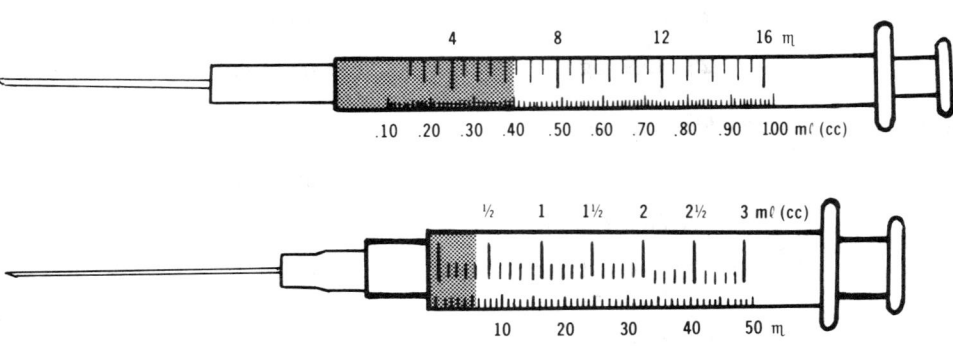

8. 0.9 mℓ

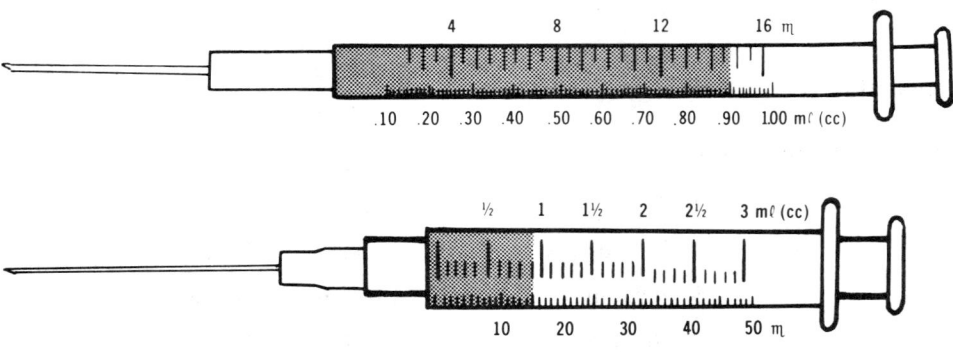

9. 0.375 mℓ

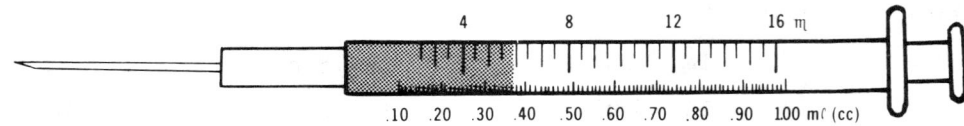

10. 2400 U
11. 4000 U
12. 20,000 U
13. 16,000 U
14. 3750 U

Application I

Section A Review Problems (p. 318)

1. 13.3 mg
3. 10 mg
5. 62.5 mg
7. 6.7 mg
9. 33.3 mg
11. 3.1 mℓ
13. 0.44 m²
15. 0.61 m²
17. 0.53 m²
19. 3.5 mg
21. 32.5 mg
23. 100 mg
25. 12.9 mg
27. 17 microdrops/min
29. 63 microdrops/min

Section B Review Problems (p. 330)

1. 20 mℓ/min
3. 58 mℓ/min
5. 31 mℓ/min
7. 58 mℓ/min
9. Ccr = 110 mℓ/min
   Dose = 1–2 g q. 4 hr
11. Ccr = 65 mℓ/min
    Dose = 250 to 500 mg q. 12 hr
13. Ccr = 48 mℓ/min
    Dose = 125 to 250 mg q. 12 hr
15. Ccr = 30 mℓ/min
    Dose = 1–2 g every 8–12 hours

Application II

Review Problems (p. 341)

1. 25 mEq/ℓ of K and 25 mEq/ℓ of acetate ions
3. 77.6 mEq/ℓ of Na and 77.6 mEq/ℓ of Cl ions
5. 169.6 mEq/ℓ of Na and 169.6 mEq/ℓ of lactate ions
7. 10.1 mEq/ℓ of K and 10.1 mEq/ℓ of Cl ions

Application III

Review Problems (p. 356)

1a. 15 mℓ
1b. 3.8 mℓ
1c. 0.5 mℓ
1d. 4.6 mℓ
3a. 15 mℓ
3b. 2.2 mℓ

Index

Abbreviations:
   common medical, 207–208
   Latin, 204–205
   used in drug administration, table of, 359
Addends, 3
Addition:
   of decimals, 72
   of fractions, 28
   of mixed numbers, 44
   of whole numbers, 3
Amount of prescribed, 132
Amount of pure drug, 128
Amount of solution, 128
Amount of stock, 132
Ampule, 236
Anticoagulant, 295
Apothecaries' measurement system, 163–194
   fluid measure in, 167–173
      changing from one fluid measure to another in, 170
   gallon, 168
   graduates, 168
   ounce, 173
   pint, 168
   pound, 173
   Roman numerals in, 164–166, 167
   quart, 168
   weights, 174
Approximate equivalents, table of, 360
Aseptic, 235
Aseptic technique, 235
Atomic weights, table of selected, 340, 360

Base, 123
Body-surface area, 311
Boiling point, 157

Cancellation, 50
Capsule, 219
Carrying, 4
Celsius, 157
Centigrade, 158
Centiliter, 147

Centimeter, 141
Changing (*see* Conversion)
Clark's Rule, 310–311
Clot, 295
Colon, 98
Comparison of fractions, 41
Complex fraction, 23
Composite number, 31
Conversion, 108
   of Celsius to Fahrenheit, 158
   of decimals to fractions, 68
   of decimals to percents, 118
   of Fahrenheit to Celsius, 158
   of fractions to decimals, 67
   of fractions to percents, 120
   of improper fractions to mixed numbers, 7
   of mixed numbers to improper fractions, 25
   of percents to decimals, 119
   of percents to fractions, 121
   table for units of measure, 111
   table for units of time, 110
   of units from one measurement system to another, 196–204
   of units of length, 143, 200
   of units of volume, 148, 170, 200
   of units of weight, 153, 175, 197
Creatinine, 321
   clearance, 322
   Clearance Nomogram, 322
Cross product, 39
Cubic centimeter, 147
   in external solutions, 128
Cubic meter, 146

Decigram, 152
Deciliter, 147
Decimal part, 14
Decimal point, 63
Decimals, 62–96
   addition, 72–74
   changing to fractions, 68–70
   changing to percents, 118
   division of, 86–89

multiplication of, 83–86
reading and writing of, 63–65
rounding off of, 79–82
subtraction of, 74–76
Decimeter, 141
Dekagram, 152
Dekaliter, 147
Dekameter, 141
Denominator, 22
Density, 180
Diabetes mellitus, 285
Difference, 6
Diluent, 219
Diluting stock solutions, 133
Dimensional analysis, 108–109
Dimensional quantities, 108
Dividend, 12
Division:
  of decimals, 86
  of fractions, 52
  of mixed numbers, 52
  of whole numbers, 11
Divisor, 12
Dram, 173
Drop, 180
Drops per minute, 271

Electrolytes, 335–336
Electrolyte solutions, 335–342
Embolus, 295
English system of measurement, 174, 196
  inch, 196
  ounce, 196
  pound, 196
Enteric, 219
Equal fractions, 39
Equivalents, approximate, table of, 360
Equivalent weight, 339
External solution, 127
Extremes, 101

Fahrenheit, 157
Fluidounce, 168
Fluidram, 168
Fractional part, 13
Fractions, 21–61
  addition of, 28
  comparison of, 41
  complex, 23
  conversion to decimals, 67
  conversion to percents, 120
  division of, 52
  equal, 39
  improper, 22
  least common denominator of, 32
  like, 28
  multiplication of, 49
  proper, 22
  raising to higher terms, 30
  reducing, 26
  subtraction of, 37
  unlike, 30

Freezing point, 157
Fried's Rule, 309–310

Gallons, 168
Geriatrics, 321
Glassful, 179
Grain, 173
Gram, 152
  in external solution, 127

Hectogram, 152
Hectoliter, 147
Hectometer, 141
Heparin, 295–299
Higher terms, raising fractions to, 30
Household system of measurement, 163–194
  fluid measure in, 179
  table of equivalents, 180
Hyperalimentation, 344
Hypodermic tablets, 247

Improper fraction, 22
  changing to mixed numbers, 25
Incompatibility, 270
Injectable drugs, packaging of, 236
Injection, 235
Insulin, 285–295
  action speed, table of, 285
  Lente, 285
  NPH, 285
  Protamine zinc, 285
  Regular, 285
  review of concentrations, table of, 289
  Semilente, 285
  Ultralente, 285
International System of Units (SI), 141
Intramuscular injection, 235
Intravenous admixture, 277
Intravenous fluids, 270
Intravenous injections, 235, 270 (*see also* I.V.)
Ion, 335
I.V.:
  administration set, 271
  infusions, 277
  piggybacks, 277
  running time, 274

Kelvin, 157
Kilogram, 151
Kiloliter, 147
Kilometer, 141

Large-volume parenteral, 270
Latin abbreviations, 207–208
Least common denominator, 31
Length, changing units of, 143, 200
Like fractions, 28
Liter, 146
Lowest terms, reducing fractions to, 26

Means, 101
Medical abbreviations, 207–208

Index

Meter, 142
Metric prefixes, 141
Metric system of measurement, 140-162
   changing units of length in, 143-145
   changing units of volume in, 148-151
   changing units of weight in, 153-156
   graduates, 147
   length, 141
   volume, 147
   weight, 151-153
Microdrop, 316
Microgram, 152
Micron, 142
Milliequivalent, 339-340
Milligram, 152
Milliliter, 147
   in external solutions, 128
Millimeter, 141
Minim, 168
Minuend, 6
Mixed numbers, 23
   addition of, 44
   changing to improper fractions, 24
   division of, 52
   multiplication of, 50
   subtraction of, 52
Multiple-dose vial, 236
Multiplicand, 8
Multiplication:
   of decimals, 83
   of fractions, 49
   of mixed numbers, 50
   by powers of ten, 85
   of whole numbers, 8
Multiplier, 8

Nonelectrolyte, 335
Numbers:
   composite, 31
   fractional, 22
   mixed, 23
   prime, 31
   whole, 3
Numerator, 22

Oral dosages, 219
Oral solutions, 223-229
Ounce, 173, 196

Pancreas, 285
Parenteral administration of drugs, 235
Parenteral containers, 236
Part, 124
Partial product, 8
Pediatrics, 309
Percent, 118
   changing to decimals, 119-120
   changing to fractions, 121-122
Percentage, 124
Percentage-strength solution, 127
Percent-proportion formula, 124

Piggyback, 277-278
Pint, 168
Place value, 63
Pound, 173
Powers of ten, 85, 89
Precipitates, 270
Prefilled syringes, 236
Primary L.V.P., 278
Prime numbers, 31
Product, 8
Proper fraction, 22
Proportion, 101
Pulmonary embolism, 295

Quart, 168
Quotient, 12

Raising fractions to higher terms, 30
Rate, 123
Ratio, 98
Ratio-strength solution, 131
Reconstitution, 224, 252
Reducing fractions to lowest terms, 26
Reducing ratios to lowest terms, 99
Remainder, 12
Roman numerals, 164-167
Rounding off decimals, 80

Scored tablets, 219
Serum creatinine level, 321
Single-dose vial, 236
SI units, 141
Solute, 127
Solutions, oral, 223-229
Solvent, 127
Solving proportions, 102
Solving word problems, 15
Stock solution, 132
Strength, 128
   of prescribed, 132
   of solution, 128
   of stock, 132
Subcutaneous injection, 235
Subtraction:
   of decimals, 74-77
   of fractions, 37-39
   of mixed numbers, 45-48
   of whole numbers, 5-7
Subtrahend, 6
Sum, 3
Superior vena cava, 344
Surface-area formula, 315

Tablespoonful, 179
Tablet, 219
   hypodermic, 247
Teaspoonful, 179
Temperature, 157
Thrombophlebitis, 295
Thrombus, 295
Total parenteral nutrition, 344
Triple point, 157

Unit abbreviations, table of, 359
Units method (dimensional analysis), 108–109
Unlike fractions, 30
Unscored tablets, 219

Valence, 336–339
Vial, 236

Viscosity, 180
Volume, changing units of, 148, 170, 200

Weight, changing units of, 153, 175, 197
West Nomogram, 311–317
Whole numbers, 3
Word problems, solving, 15

Young's Rule, 310